Micro- and Nanobubbles

Micro- and Nanobubbles

FUNDAMENTALS AND APPLICATIONS

edited by
Hideki Tsuge

CRC Press
Taylor & Francis Group
Boca Raton London New York

CRC Press is an imprint of the
Taylor & Francis Group, an **informa** business

Micro- and Nanobubbles: Fundamentals and Applications

First published 2014 by Pan Stanford Publishing Pte. Ltd.

Published 2023 by CRC Press
Taylor & Francis Group
6000 Broken Sound Parkway NW, Suite 300
Boca Raton, FL 33487-2742

ISBN 13: 978-981-4463-10-2 (hbk)

Visit the Taylor & Francis Web site at
http://www.taylorandfrancis.com

and the CRC Press Web site at
http://www.crcpress.com

British Library Cataloguing-in-Publication Data
A catalogue record for this book is available from the British Library.

Contents

Preface

Prof. H. Ohnari (2007) developed microbubble generators to produce micrometer-size bubbles and used them for aquacultures of oysters in Hiroshima, scallops in Hokkaido, and pearls in Mie Prefecture. Microbubbles show growth promotion effects, which have been highlighted in the literature.

Following are the characteristics and features of microbubbles compared with millimeter- and centimeter-size bubbles:

1. Small bubble size: Microbubbles are generally of the micrometer order. Depending on the working field, their size is 10–40 μm for bioactivity and less than 100 μm for fluid physics.
2. Slow rising velocity: For instance, a 10 μm microbubble rises only 20 cm in 1 h.
3. Decreasing friction drag: Decreasing friction drag of microbubbles has been experimentally verified by blowing microbubbles from the bottom of a big ship.
4. High pressure inside microbubbles and large gas dissolution: The inner pressure values of bubbles whose diameters are 1 μm and 100 nm are 3.87 and 29.7 atm, respectively, under 1 atm surrounding pressure of the bubble. With decreasing bubble size, the pressure inside the bubble increases so that the partial pressure of the dissolved gas component, which is the driving force of dissolution, increases and the gas dissolves easily.
5. Negatively charged surface of microbubbles: Microbubbles are negatively charged with an average ζ potential of –30 to –40 mV independent of the bubble diameter. The surface charge of the gas–water interface is strongly affected by the pH of the water. There is no possibility of coalescence of microbubbles even in the case of dense microbubble water. By charging the surface of suspended materials positively and adsorbing on

the negatively charged surface of microbubbles, we can apply it to the separation operation.

6. Bioactivity effect: Ohnari found growth promotion, observed opening phenomena, measured the increase of blood flow rate of oyster in Hiroshima, scallop in Hokkaido, and pearls in Mie Prefecture. He ascertained the effect of bioactivity.

Okajima and Harada (2007) found medically the following: By transcutaneous electrical nerve stimulation of microbubble, insulin-like growth factor-1 (IGF-1) was produced in large volumes in living substances. By the stimulations of microbubbles through light and heat, capsaicin-sensitive sensory nerves, which detect temperature and pain, are stimulated and calcitonin gene–related peptide (CGRP) is delivered from the ends of sensory nerves. As CGRP promotes the production of NO and prostaglandin from endothelial cells and IGF-1 is produced, finally the increase in blood flow is promoted. The effect of microbubbles resembles the capsaicin effect made by bathing in an acidic hot spring and taking ginger extract.

The first edition of *The Latest Technology on Microbubbles and Nanobubbles* was published in Japanese in September 2007, and its second revised edition was published in Japanese in May 2010. This time, we decided to publish the English version of this book. This book is organized into three parts:

Part 1: Fundamentals
Part 2: Application of Microbubbles
Part 3: Application of Nanobubbles

Part 1 covers the characteristics, formation, analysis, and sonochemistry of microbubbles. Part 2 introduces readers to the application of microbubbles in the environment, engineering, medicine, stock raising, agriculture, and marine industry. Part 3 focuses on nanobubbles and their application in the medical field.

In this direction, in November 2013, the Fine Bubble Industries Association was established in Japan for the promotion of international standardization mainly for measurement methods of nanobubbles in the first stage (http://www.fbia.or.jp/).

We hope these new technologies will be utilized globally.

Following are a few resources on microbubbles and nanobubbles:

NTS (2006) *Latest Technology on Minute Bubbles*. NTS Book Co., Japan (in Japanese).

Ohnari H. (2007) Today's subjects of microbubble technology, *Cem. Eng. Jpn.* **71**, 154–159 (in Japanese).

Okajima K., Harada N. (2007) Bioactivity of microbubbles, in *The Latest Technology of Microbubbles and Nanobubbles*, CMC Book Co., Japan, pp. 90-108 (in Japanese)

Special issue (2006) Application to environmental technology of microbubbles and nanobubbles, *Eco Ind.* **11**, 5–63 (in Japanese).

Special issue (2007) Microbubbles, *Chem. Eng. Jpn.* **71**, 154–188 (in Japanese).

Special issue (2007) Microbubbles and clean technology, *Clean Technol.* **17,** 1–31 (in Japanese).

Special issue (2009) Various usages of micro- and nanobubbles, *Mater. Integr.* **22**, 2–48 (in Japanese).

Special issue (2010) Characteristics and application of microbubbles and nanobubbles, *Bull Soc. Sea Water Sci. Jpn.* **64**, 3–38 (in Japanese).

Techno-system (2005) *Concepts in Basic Bubble and Foam Engineering*. Techno-system Co., Japan (in Japanese).

Tsuge, H. (ed.) (2007) *The Latest Technology on Microbubbles and Nanobubbles*, 1st ed. CMC Books Co., Japan (in Japanese).

Tsuge, H. (ed.) (2010) *The Latest Technology on Microbubbles and Nanobubbles II*, 2nd ed. CMC Books Co., Japan (in Japanese).

Hideki Tsuge

June 2014

Part I

Fundamentals

Chapter 1

Characteristics of Microbubbles

Hideki Tsuge
Keio University, Tokyo 166-0004, Japan
htsuge@jcom.home.ne.jp

By comparing microbubbles (MBs) with millimeter- or centimeter-sized bubbles, the following characteristics of MBs are listed:

1. **Small bubble size**
   The size of MBs is generally μm order; it depends on the working field, that is, 10–40 μm for bioactivity field and less than 100 μm for fluid physics field.
2. **Slow rising velocity**
   The rising velocity of an MB depends on the physical properties of liquids. For MBs of 100 μm diameter, Re number ($=dU\rho/\mu$) is nearly 1, and its shape is spherical. Such MBs behave as fluid (clean) spheres or solid (unclean) spheres. By the experimental results of the MBs' rising velocity, both fluid and solid spherical behaviours were observed, while the rather many experimental results following the Stokes equation [Eq. (1.1)] applicable to solid spheres are reported[1].

$$U = \rho g d^2/18\,\mu, \tag{1.1}$$

Micro- and Nanobubbles: Fundamentals and Applications
Edited by Hideki Tsuge

ISBN 978-981-4463-10-2 (Hardcover), 978-981-4463-11-9 (eBook)
www.panstanford.com

where U is the rising velocity, ρ is the liquid density, g is the gravitational acceleration, d is the bubble diameter and μ is the liquid viscosity. For example, 10 μm MBs rise only 20 cm in 1 h.

3. **Decreasing friction drag**

   Friction drag of gas–liquid two-phase flow decreases with increasing MB volume fraction in vertical pipe flow [2]. Such experiments of decreasing friction drag are now verified by blowing MBs from the bottom of big ships [3].

4. **High pressure inside MBs (self-compression effect)**

   Using Young–Laplace equation (Eq. (1.2)), the pressure in a bubble (ΔP) whose diameter is d increases to that larger than the surrounding pressure due to the surface tension σ.

$$\Delta P = 4\sigma/d \tag{1.2}$$

For example, the inner pressures of bubbles whose diameters are 1 μm and 100 nm are 3.87 and 29.7 atm, respectively, using Eq. (1.2) (surface tension of water at 20°C: 72.8 mN/m, the surrounding pressure of bubble is 1 atm). With decreasing bubble size, the pressure in the bubble increases. With decreasing bubble size, the partial pressure of dissolved gas component, that is, the driving force of dissolution, increases and the gas dissolves easily.

5. **Large interfacial area**

   Interfacial area of bubbles divided by volume A/V is obtained by Eq. (1.3). With decreasing bubble diameter d, A/V increases and contributes to gas dissolution fraction.

$$A/V = 6/d \tag{1.3}$$

6. **Large gas dissolution**

   Mass transfer rate from gas to liquid, or dissolving rate N (mol/s), is written by Eq. (1.4) when the gas phase mass transfer resistance is neglected.

$$N = k_L A(p - p^*)/H, \tag{1.4}$$

where k_L is the liquid phase mass transfer coefficient (m/s), A is the bubble surface area (m^2), p is the partial pressure of dissolved component in bubble (Pa), p^* is the partial pressure of gas phase equilibrium with dissolved component in liquid (Pa) and H is the Henry constant ($p = HC$).

The liquid phase mass transfer coefficient k_L is written by Eq. (1.5) when Re < 1, and the rising velocity of spherical bubble follows Stokes law, where D_L is the gas diffusion coefficient in liquid phase, d is the bubble diameter and U is the bubble rising velocity [4].

$$k_L = \frac{D_L}{d}\left[1+\left(1+\frac{dU}{D_L}\right)^{1/3}\right] \tag{1.5}$$

When bubble diameter changes from 1 μm to 1 mm for oxygen–water system (oxygen diffusion coefficient $D_L = 2.60 \times 10^{-9}$ m^2/s in water), the liquid phase mass transfer coefficient k_L becomes 10^{-4}–10^{-3} m/s by Eq. (1.5). Mass transfer rate N is obtained by Eq. (1.4) using the data of k_L and bubble inner pressure. When a 1 mm bubble segregates into bubbles whose diameters are 10 μm and 100 nm, bubble numbers increase by 10^6 and 10^{12} times, respectively. When the surface area of the 1 mm bubble is defined as 1, the surface areas of 10 μm and 100 nm bubbles increase to 6×10^4 and 6×10^{10}, respectively. When the mass transfer rate of the 1 mm bubble during 1 mm rise is considered as 1, mass transfer rates of 10 μm and 100 nm bubbles become 10^8 and 10^{18}, respectively, due to the decrease in rising velocity. By using MBs, the dissolution rate increases rapidly.

7. **Dissolution and contraction**

   Ohnari describes the contraction of MB as follows [5]:

   For the swirling-type MB generator, the pressure at gas suction becomes 0.06 MPa less than the atmospheric pressure. MBs are formed under the combined pressure of atmospheric pressure and static pressure; they are pressurised by liquid and begin to contract. Due to high pressure and high temperature, contraction and dissolution of MBs take place. As shown in Fig. 1.1, a 20 μm MB contracts and under 10 μm, it contacts rapidly and disappears. On the other hand, MBs larger than 65 μm will expand further [6].

8. **Negatively charged surface of MB**

   Takahashi measured the surface potentials of MBs using the experimental set-up of electrophoresis cells shown in Fig. 1.2 [1]. MBs formed by the generator in the water reservoir were introduced to the electrophoresis cell (1.0 mm thick, 23.0 mm high and 75.0 mm wide). A transfer switch was used to change the direction of the electric potential in the electrophoresis cell every 1 s. As MBs are charged negatively, MBs rise in a zigzag manner as shown in Fig. 1.3.

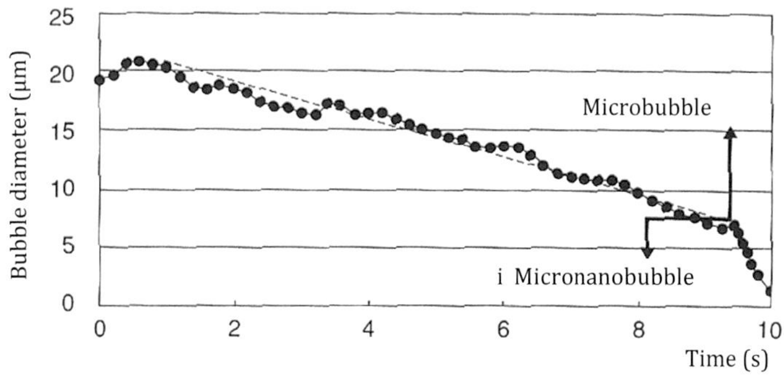

Figure 1.1 Time course of MB diameter.

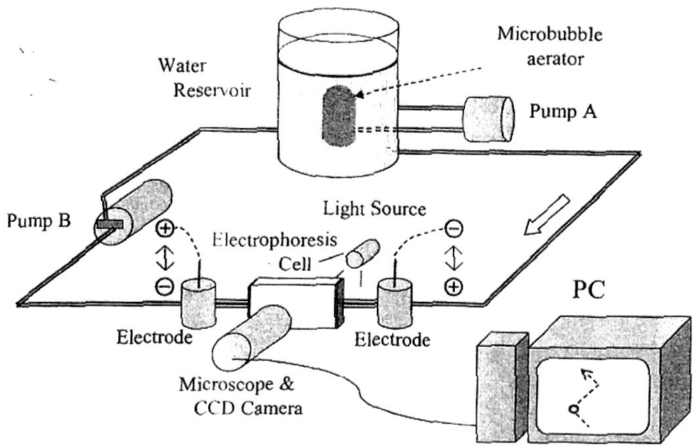

Figure 1.2 Experimental set-up of electrophoresis cell.

By measuring the vertical rising velocity of MBs, the bubble size is obtained using Stokes equation, and from the horizontal velocity, the ζ potential of MBs is obtained using Smoluchowski equation (Eq. (1.6)).

$$\varsigma = \mu u/\varepsilon, \tag{1.6}$$

where ζ is the zeta potential (V), μ is the viscosity of water (kg/ms), u is the mobility (m^2/sV) and ε is the permittivity of water (s^2C^2/kgm^3). Figure 1.4 shows the effect of bubble size on the ζ potential of MB in distilled water. MBs are negatively charged with an average ζ potential between −30 and −40 mV independent of bubble diameter.

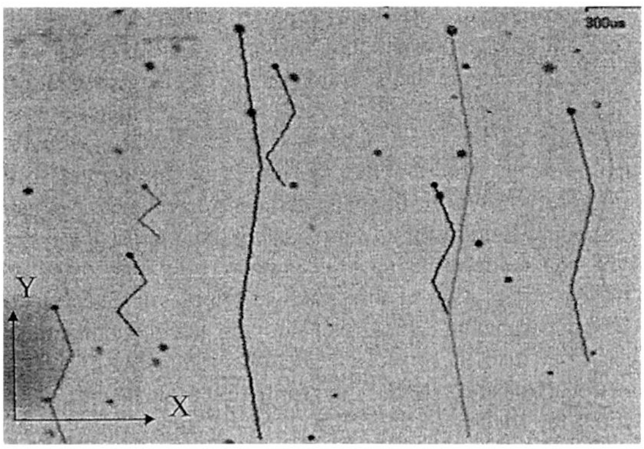

Figure 1.3 Rising loci of MBs.

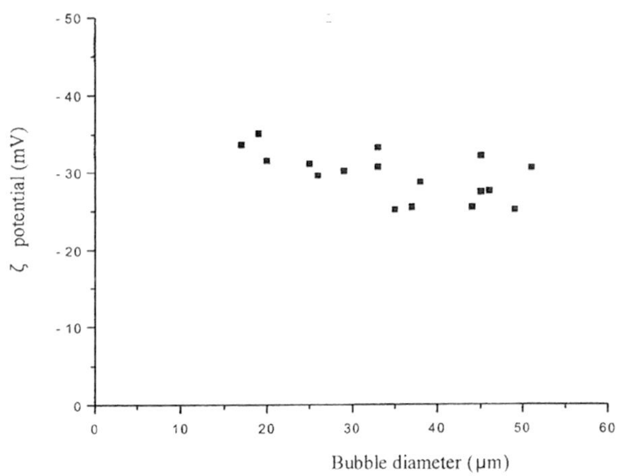

Figure 1.4 Effect of bubble diameter on ζ potential.

The gas–water interface is negatively charged; so, OH^- must play an important role in electrical charge. Takahashi explained the charging mechanism by using more

OH^- compared to H^+ ions at the gas-water interface [1]. Most researchers have explained the adsorption of OH^- onto the interface by the difference in hydrogen energy between H^+ and OH^- or by the orientation of water dipoles at the interface. Hydrogen atoms pointing towards the water phase and oxygen atoms towards the gas phase causes attraction of anions to the interface.

As shown in Fig. 1.5, the ζ potential of MB depends on the pH of the water adjusted by HCl and NaOH. The surface charge of the gas–water interface was strongly affected by the pH of the water. According to Takahashi, at strong acidic pH, ζ potential is 0 mV, while at strong base pH, it is −100 mV [1].

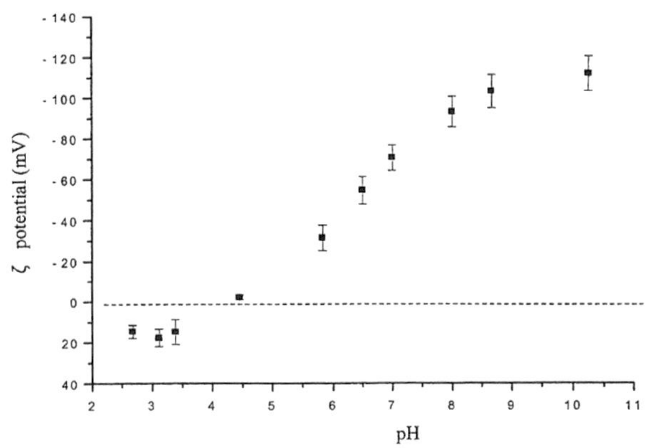

Figure 1.5 Effect of pH on ζ potential.

As the surface of MB is charged negatively, it shows that there is no possibility of coalescence of MB in the case of dense MB water. By charging the surface of suspended materials positively and adsorbing on the negatively charged surface of MB, we can apply it to the separation operation.

9. **Crushing and formation of free radicals**

   Takahashi et al. introduced electrodes in water, crushing MBs by the shock waves formed and free radicals were created [7].

   By comparing electron spin resonance, the number of formed free radicals by crushing shock waves on MBs is two or three orders larger compared with the case of ultrasound radiation in water. Under 50 μm MB, free radicals are produced in several minutes during compression at high concentrations of MBs. As the ζ potential increases during compression, ion concentration around the MB increases so that excess ions are formed, and free radicals are created [7].

10. **Bioactivity effect**

    Ohnari observed growth promotion [5], shell opening and increase in rate of blood flow of oysters in Hiroshima, scallops in Hokkaido and pearls in Mie Prefecture, thus ascertaining the effect of bioactivity.

    Okajima and Harada observed the following medical findings [8]: by transcutaneous electrical nerve stimulation of MB, insulin-like growth factor-1 (IGF-1) was produced in living substances. By the stimulations of light and heat of MB, capsaicin-sensitive sensory nerves which detect temperature and pain are stimulated and calcitonin gene related peptide (CGRP) is released from the sensory nerves ends. As CGRP promoted the productions of NO and prostaglandin from vascular endothelial cells, IGF-1 is produced, and finally, increase in blood flow is promoted. The effect of MBs resembles the effect of capsaicin made using acidic hot springs and ginger extracts.

11. **Change of liquid properties by bubbling of MBs**

    Himuro found that by bubbling MBs in the city water at 25°C, liquid viscosity and surface tension decrease and electrical conductivity increases [9]. It is due to the change in cluster by cleavage of the hydrogen bond among water molecules and ionisation of chemicals in the city water.

References

1. Takahashi M (2005) ζ potential of microbubbles in aqueous solution: electrical properties of the gas-water interface, *J. Phys. Chem. B*, **109**, 21858–21864.

2. Serizawa A, Inui T, Eguchi T (2005) Flow characteristics and pseudo-laminarization of vertically upward air-water milky bubbly flow with micro bubbles in a pipe, *JJMF*, **19**, 335–343 (in Japanese).

3. Kodama Y (2007) Drag reduction of ship by microbubbles, *Chem. Eng. Jpn.,* **71**, 186–188 (in Japanese).
4. Clift R, Grace JR, Weber ME (1978) *Bubbles, Drops, and Particles*, Academic Press, New York.
5. Ohnari H (2007) Todays subjects of microbubble technology, *Chem. Eng. Jpn.,* **71**, 154–159 (in Japanese).
6. Ohnari H (2006) *All on Microbubbles*, Nippon Jitsugyo Shuppan Co., Japan (in Japanese).
7. Takahashi M, Chiba K, Li P (2007) Free-radical generation from collapsing microbubbles in the absence of a dynamic stimulus, *J. Phys. Chem. B,* **111**, 1343–1347.
8. Okajima K, Harada N (2007) Bioactivity of Microbubbles, in *The Latest Technology of Microbubbles and Nanobubbles*, CMC Book Co., Japan, 90–108 (in Japanese).
9. Himuro S (2007) Characteristics of physical chemistry of microbubbles, *Chem. Eng. Jpn.,* **71**, 165–169 (in Japanese).

Chapter 2

Formation and Analysis of Microbubbles

2.1 General

Hideki Tsuge
Keio University, Tokyo 166-0004, Japan
htsuge@jcom.home.ne.jp

2.1.1 Introduction

The formation methods of microbubbles (MBs) are divided into two cases: one is the gas is accompanied with liquid introduced and another is blowing the gas in quiescent liquid. In this chapter, MB formation methods and measuring methods of MB diameter are discussed. In the following sections, the formation methods by several makers and measuring methods of MB diameter and ζ potential are shown.

2.1.1.1 With liquid flow

2.1.1.1.1 Rotational liquid flow method

Ohnari developed MB generator (M2-LM type, produced by Nanoplanet Research Institute Corp.) as shown in Fig. 2.1, which is cylindrical type (diameter 50 mm, height 100 mm) pressurized water is fed from the bottom water pipe by pump and rotational

Micro- and Nanobubbles: Fundamentals and Applications
Edited by Hideki Tsuge

ISBN 978-981-4463-10-2 (Hardcover), 978-981-4463-11-9 (eBook)
www.panstanford.com

liquid flow is produced in the cylinder [1]. By Bernoulli's theorem, the central part of the cylinder becomes low pressure produced by rotational liquid flow, and air is sucked automatically. Air swirling flow is formed, and MBs are produced near the outlet of cylinder. Liquid flow rate is about 12 L/min, and the swirling speed of two-phase flow in the cylinder is 300–600 round/s. The ratio of gas flow rate/liquid flow rate is 1/7–1/15. M2-LM type generator was used for the aquaculture of oysters in Hiroshima, and good growth promotion was reported by TV news.

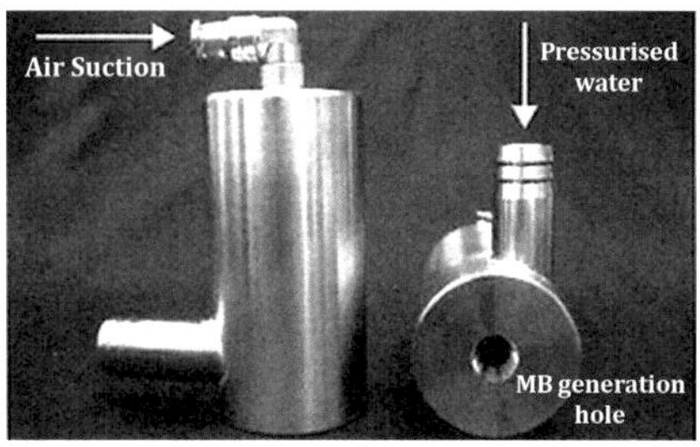

Figure 2.1 Rotational liquid flow MB generator (M2-LM).

MB size distribution by M2-LM type generator is 10–50 μm, and the mode size is 15 μm. The concentration of MB by M2-LM is thinner than that by dissolved air flotation method. Li and Tsuge examined the following MB generating method [2]. The combination of the centrifugal pump (Nikuni Co. Ltd.) and the rotating-flow MB generator (M2-M, Nanoplanet Research Institute Corp.) was used. Water and air are sucked simultaneously by the pump. Some air dissolves into water before transporting to the generator. The pressurised water is then decompressed through the MB generator with a high rotating velocity.

Recently the inside of generator is divided by the partition, and the high-quality MB generator is produced by Aqua Air Co. [3].

2.1.1.1.2 Static mixer type

As shown in Fig. 2.2, static mixer (OHR Fluid Engineering Institute Corp.) induces helically two-phase flow by guide vane and mushroom-type projections (current cutter) and destructs to MB water (MB size 5–50 μm) whose production is maximum 1500 L/min [4].

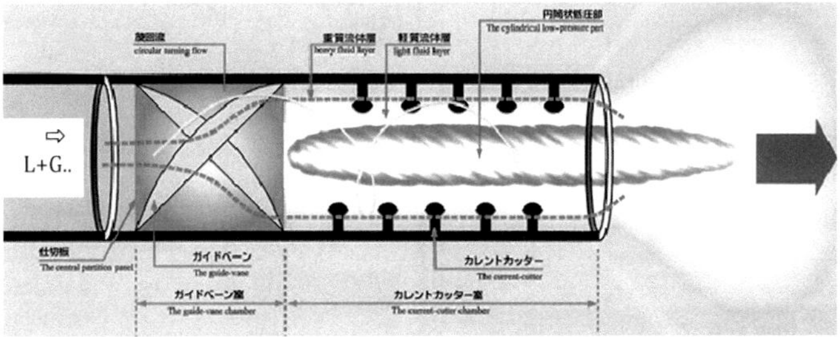

Figure 2.2 Schematic diagram of OHR static mixer.

2.1.1.1.3 Ejector type

Ejector-type nozzle ejects liquid in high speed as aspirator, and gas is absorbed by the back pressure at the outlet, finally produces MBs by turbulent mixing and shearing mechanism. The average MB size produced is 424 μm [5].

2.1.1.1.4 Venturi type

When gas and liquid are simultaneously flowed in the squeezing part of Venturi tube, the sudden change of liquid flow rate produces shock wave and large bubbles are destroyed to MBs. To avoid bubble coalescence, 50 ppm 3-pentanol is added, in which case 100 μm MBs are mainly produced [6].

2.1.1.1.5 Dissolved air floatation type

Dissolved air floatation MB generator is shown in Fig. 2.3. About 3–4 atm pressurized air is dissolved into water and is flushed through the nozzle; the supersaturated air produces MBs in the water. The concentration of MBs is high, and the liquid is like milk. MB size distribution has two peaks as shown in Fig. 2.4.

2.1.1.2 Without liquid flow

2.1.1.2.1 Capillary type

Kukizaki sintered porous glass made of loamy soil and the nanobubbles (NBs) whose bubble diameter is 720 nm are formed from the sintered membrane of average pore size 84 nm [7]. To avoid the bubble coalescence, surface active agent is often added.

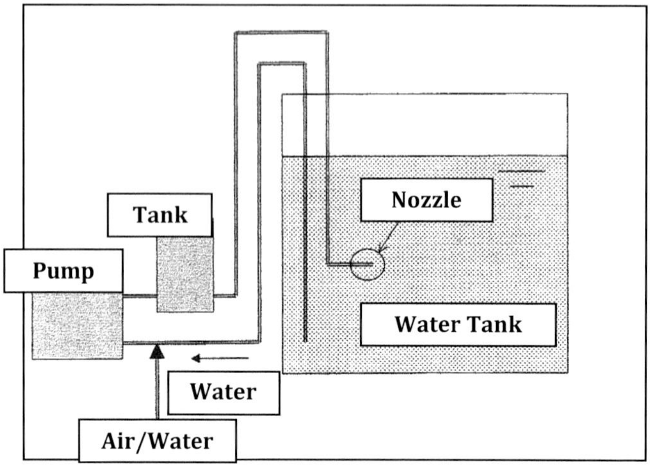

Figure 2.3 Schematic diagram of dissolved air floatation MB generator.

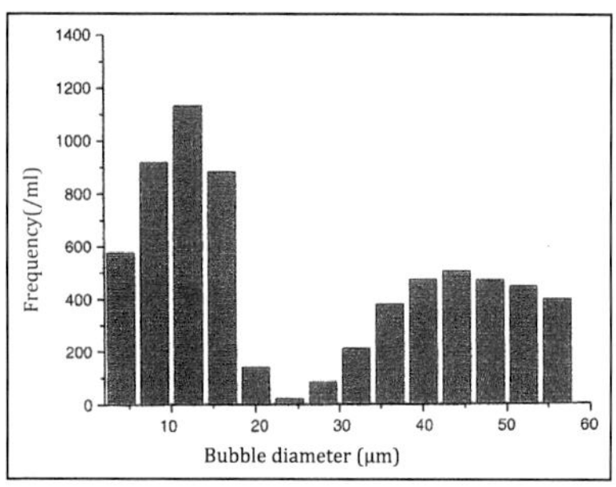

Figure 2.4 Bubble size distribution of dissolved air floatation MB generator.

2.1.1.2.2 Rotating type

The rotor made of sintered material is rotated at high speed in the surrounding stator, and gas is automatically absorbed. Average 50 μm MBs are formed, and the reaction crystallization was conducted by Onoe et al. [8]. The merit of the generator is no need of liquid pump. The rotating-type generators are produced by Fuki Works Co., Nomura Electronics Co. and Royal Electric Co. and have been used for the purification of lakes and marshes.

2.1.1.2.3 *Ultrasound type*

Makuta et al. radiated ultrasound to thc bubbles formed from the needle and produced same size MB (about 10 μm) in high viscosity liquid (100 cP) such as silicon oil [9]. This method is applicable for the production of homogeneous MBs and is proposed for the application of medical materials.

2.1.1.2.4 *Vapour condensation type*

When the mixed vapours of nitrogen and steam are blowing from nozzle into water, the steam condensates and non-condensable nitrogen MBs obtained. The mode diameter of MBs is 20–40 μm. MB size is affected by gas composition, nozzle diameter and vapour flow rate [10].

2.1.1.2.5 *Electrolysis type*

Tsuge et al. made electrolysis of water and found the effect of vibration of low frequency vibrating plates and electrolyte [11]. Average MB size is 35–55 μm, and its size distribution is 15–100 μm.

2.1.1.3 Measuring method of MB size

The measurement methods of MB are outlined in this section, and the details are shown in the following chapters.

2.1.1.3.1 *Visualization method*

Digital microscope and CCD (Charged Coupled Device) camera are used for the photograph of MBs, and processes graphically and MB sizes are measured. MBs less than 40 μm need microscope. Measurable MB size is larger than several μm.

2.1.1.3.2 *Laser diffraction/scattering method*

By radiation of laser beam, the diffraction scattering light of small bubbles scatters forward, backward and cross direction [12]. Such scattering pattern was measured, and by comparing the scattering pattern obtained by Mi scattering theory, the MB size distribution is obtained. MB size distribution between 10 nm and several mm is measurable.

2.1.1.3.3 Coulter counter method

When the electrodes are inserted in electrolyte solution surrounded by the partition wall having small hole and voltage is applied, the electrical resistance is controlled by the resistance in the small hole [13]. When the MBs pass through the hole, the electrical resistance increases so that the number and size of MB can be measured. The measurable range is 0.2–600 μm, whereas for more wide size ranges hole diameter should be changed.

2.1.1.3.4 Dynamic light scattering method

It is used for the measurement of NBs [3]. By Brownian motion, the observed NBs are diffusing and the laser light strength fluctuates. As the diffusivity of bubbles depends on bubble size, the bubble size is expressed by the fluctuation factor of laser light. By autocorrelation of fluctuation of scattering light, size distribution is obtained. As the exclusion of dust is necessary in measurement of NBs, membrane filter is used. To distinguish dust and NB, further check is necessary.

References

1. Ohnari H (2006) *All on Microbubbles* (Nippon Jitsugyo Shuppan Co., Japan) (in Japanese).
2. Li P, Tsuge H (2006) Water treatment by induced air floatation using microbubbles, *J Chem Eng Japan,* **39**, 896–903.
3. Takahashi M (2009) Fundamentals and engineering applications of microbubbles and nanobubbles, *Material Integration,* **22**, 2–19 (in Japanese).
4. Uematsu H (2006) Advantage and Possibility of OHR Microbubble, *J. Resources and Environment,* **42**, 100–103 (in Japanese).
5. Nakatake Y, Watanabe K, Eguchi T (2007) Combustion improvement for diesel engines with ejector-type micro-bubble mixed fuel, *Trans Japan Soc Mechanical Eng,* **73**, 2368–2374 (in Japanese).
6. Fujiwara A (2006) Generating method of microbubbles by Venturi tube, *Eco Industry,* **22**, 27–30 (in Japanese).
7. Kukizaki M, Nakajima T, So G, Obama Y (2004) Formation and control of size of uniformly dispersed nanobubbles from porous glass membrane, *Kagaku Kogaku Ronbunshu*, **30**, 654–659 (in Japanese).

8. Onoe K, Matsumoto M, Akiya T (2002) Reactive crystallization of calcium carbonate using CO_2 micro-bubbles, *Bull Soc Sea Water Sci Japan,* **56**, 357–361 (in Japanese).
9. Makuta T, Takemura F, Hirahara E, Matsumoto Y, Shoji M (2004) Generation of micro gas bubbles of uniform diameter in an ultrasonic field, *Trans Japan Soc Mechanical Eng.,* **70**, 2758–2767 (in Japanese).
10. Terasaka K, Saito J, Toda Y, Kobayashi D (2009) Microbubble generation using direct-contact condensation of mixed vapor, *Progress in Multiphase Flow Research,* **4**, 103–110 (in Japanese).
11. Tsuge H, Ogawa T, Ohmasa R (2008) Microbubble formation by electrolysis using a new mixing equipment with low frequency vibratory fins, *J Chem Eng Japan,* **41**, 557–561.
12. Maruyama M (2007) *Laser Diffraction · Scattering Method,* The Latest Technology of Microbubbles and Nanobubbles CMC Book Co. Japan, pp. 31–41 (in Japanese).
13. Fujiwara A (2007) *Electrical Resistance Method,* The Latest Technology of Microbubbles and Nanobubbles CMC Book Co. Japan, pp. 42–45 (in Japanese).

2.2 New Type of Microbubble Generator by Two-Phase Flow Method

Osamu Matsumoto
Aqua Air corporation, Tokyo Japan
officematsumoto@apricot.ocn.ne.jp

2.2.1 Development Concept

The fundamental concept of this development was to make a useful microbubble generator in a wide variety of practical fields. Many kinds of generators were developed at the time, but most of them were too small for the practical usages. Considering the usages, the generator must be stable and reliable for generating microbubbles in the variety of practical sites for the operations of such as waste water treatment and environmental remediation.

The technical properties of the new developed device are as follows: easy on-site installation due to the simple structure, short setting time and effective operation with little consumption electricity. Also, it is a very important point to predict the practical effect at the operation sites from the results of laboratory tests that the devices must demonstrate the same performance regardless of their size.

Photo 2.1 Generation of millimetre bubbles.

The microbubble generator can produce any size of bubbles, from nanobubble to microbubble, according to the purposes, and we can offer the devices of various sizes according to your offer.

2.2.2 Method of Bubble Generation

Our microbubble generator is classified to the two-phase (gas and water) turbulent type, and the generator uses ultra-high-speed turning flow method by multiple cylinder structure.

2.2.3 Structural Feature

The ultra-high-speed swirl flow along the centre line of the device is the most important operation for the stable generation of a large amount of tiny bubbles, and this is a patented work (Japan Patent No. 3890076).

As shown in Photo. 2.1, the water is introduced into the apparatus by a pump, and the circulating rate of the water flow is accelerated as the water is going through the multiple cylinders from the outer to the centre one. A gas was automatically introduced from the gas inlet owing to the pressure drop caused by circulating flow, and a tornado-like twirl of gas formed along the centre axis was forced out from the outlet with circulating water. The mixture of gas and water was dispersed by the force of circulation, and the shearing force generated at the outlet separated the mixture into fine bubbles.

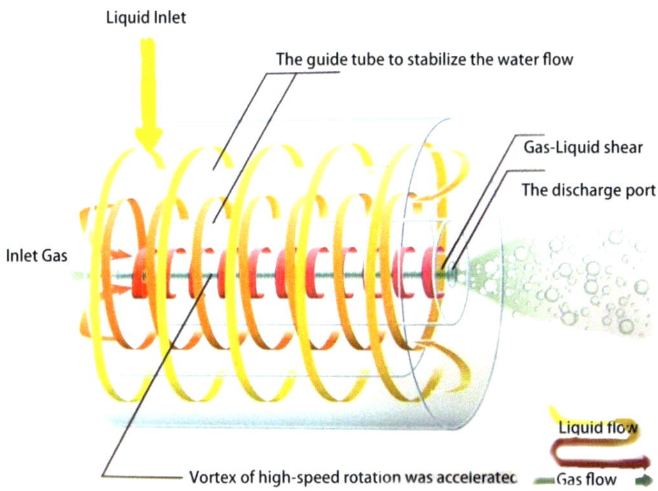

Figure 2.5 Schematic diagram of the generator.

It is a superior point of this apparatus to generate a large amount of fine bubbles through a hydrodynamic mechanism characterised by the control of ultra-fast twirling speed.

The schematic diagram of the microbubble generator is shown in Fig. 2.5. The apparatus have been designed to accelerate the circulating flow in the cylindrical body, and the interior region is separated by two built-in cylinders with different radii. The circulating speed is increasing and, in the case of our ϕ45 mm bubble generator, the speed finally reaches about 1000 rpm at the outlet according to the evaluation by an ultra-high speed camera.

2.2.4 Characteristics of the Microbubble Generator

- The typical multiple cylinder structure of the generator enables the apparatus to generate a fair amount of microbubbles, and the stable operation of the internal swirling vortex flow controlled by the built-in cylinders permits the usages at a wide variety of sites from a small laboratory scale to a large industrial and environmental scale.

- Since there are no moving parts with very simple structure, there is no mechanical trouble; users can choose the generators of various materials, such as plastics, ceramic and stainless steel, depending on the applications.
- Excellent operating efficiency of the apparatus, which works only for the injection pump.
- By adjusting the amount of injecting gas volume, the bubbles changes their diameter from millimetre scale to nanometre scale dimension.
- Not only the liquid–gas combination but also the liquid–liquid combination is also available. So the apparatus can be used to generate fine emulsification, dispersion, fine mist and fine bubbles.
- Gas or liquid as an inhalation medium is automatically taken into the apparatus through the gas inlet caused by the pressure drop due to the high-speed circulation of water flow.

2.2.5 Appearances of the Generator: Shape and Dimensions

The apparatus is available in various sizes from ϕ10 mm to bigger than ϕ600 mm.

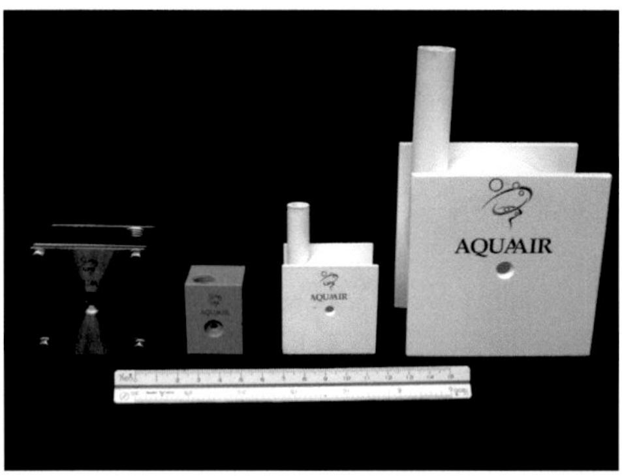

Photo 2.2 Microbubble generator ϕ20 mm–ϕ90 mm.

Photo 2.3 Microbubble generator ϕ300 mm for shrimp ponds.

2.2.6 Operating Status of the Bubble Generator

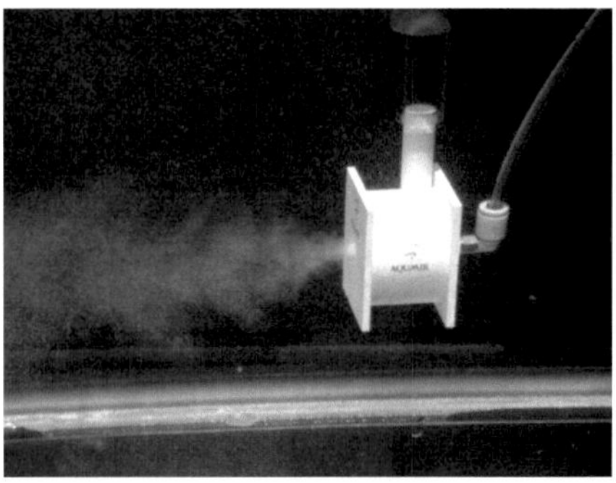

Photo 2.4 Microbubble generator ϕ45 (standard type) in operation.

Photo 2.5 Function test of ϕ300 at laboratory.

Photo 2.6 Operation of Microbubble generator ϕ300 at a shrimp ponds.

2.2.7 Practical Applications of the Microbubble Generators

2.2.7.1 Big-sized bubble generator

Photo 2.7 Photograph of the pond before the installation of microbubble generator.

1. Installation of a microbubble generator at a shrimp pond (2.5 m water depth 100 m × 100 m): Microbubble generator ϕ300 was installed in the pump and operated from the beginning of October to the end of November. It was a little bit late to examine the direct effect on microbubble on the

growth of shrimp, but we recognised the effect on the colour of water suggesting a good condition for plankton in the pond owing to the enough oxygen supply.

2. It is expected to use the generators for the improvement of water quality at lakes, source of drinking water, by the environmental remediation due to oxygen supply to bottom.

2.2.7.2 Medium-sized bubble generator

1. As shown in the pictures below, medium-type generators were used to activate shrimps for marketing in a water reservoir (1 m depth 15 m × 1 m). Fairly good results were obtained for maintaining the freshness of shrimps.
2. It is expected to be used as a new type of cleaning in position. Tests were conducted at plants for food and cosmetic and succeeded to reduce the amount of chemicals and the operating temperature.
3. Ozone microbubble is expected for wastewater treatments.
4. You can use the generators to produce emulsion very effectively by introducing liquid instead of gas through the gas inlet. The emulsion produced by the generator is stable because of the surface electricity of suspended emulsions.

Photo 2.8 Medium-sized bubble generation device status (ϕ75) set.

Photo 2.9 Medium-sized bubble generation device operating conditions (ϕ75).

2.2.7.3 Small-sized bubble generation equipment

1. This type of generator is very useful for laboratory tests because of low electricity consumption.
2. The generator is small enough to be embedded into consumer products and operates efficiently as eco-products in future.
3. We are planning to manufacture a much smaller generator to be used at a variety of technical fields.

We are now conducting several kinds of tests for the future applications such as surface cleaning and wastewater treatment by using variety of gases including ozone and carbon dioxide. Since the apparatus can be used to generate emulsion, we are also conducting experiments regarding the mixing efficiency and the stabilisation of fine emulsions. Most of these tests are carried out according to the agreements of confidentiality with many private companies.

2.2.8 Measurement of Size Distribution of Microbubbles

The size distributions of microbubbles generated by the generators were evaluated by a liquid particle counter at National Institute of Advanced Industrial Science and Technology.

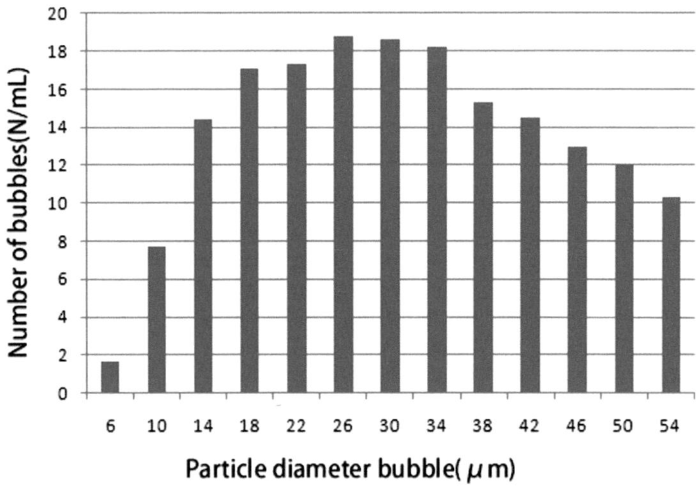

Figure 2.6 Size distribution of microbubbles generated by the generator of ϕ10 mm (Shower-head type) in tap water. This generator is very small but can produce a lot of microbubbles constantly.

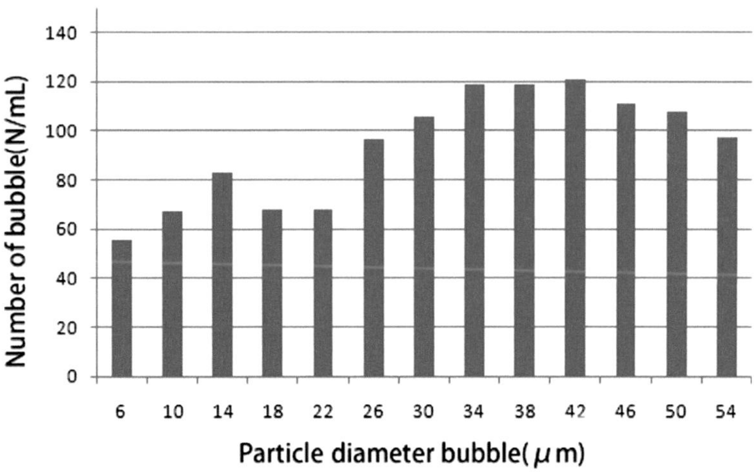

Figure 2.7 Size distribution of microbubbles generated by the generator of ϕ40 mm in seawater. The generators can produce a fairly high concentration of microbubbles because of the effects of electrolytes in seawater. It has been recognised that there are two peaks in the size distribution in seawater.

2.2.9 Experimental Study for Shrimp Farming Technology

We have been conducting the experimental study for shrimp farming using the microbubble technology for the future applications. Laboratory cultivation was carried out, and it has been demonstrated that the survival rate would be doubled compared with the conventional methods. We hope to contribute to make the sustainable society all over the world by using our microbubble technology.

2.3 Measurement and Analysis

2.3.1 Simulation of Microbubble Coalescence and Repulsion

Tomoaki Kunugi[a] and Yukihiro Yonemoto[b]

[a]*Department of Nuclear Engineering, Kyoto University, Yoshida, Sakyo, Kyoto 606-8501, Japan*

[b]*Priority Organization for Innovation and Excellence, Kumamoto University, Kumamoto-shi, Kumamoto, Japan*

kunugi@nucleng.kyoto-u.ac.jp, yonemoto@mech.kumamoto-u.ac.jp

The mechanism of bubble coalescence/repulsion is still unclear, and this problem of bubble coalescence/repulsion is a key issue when it comes to numerical research on multiphase flows. Experimental studies have revealed that not only the concentration of the electrolyte dissolved in water but also the existence of ion pairs plays an important role in bubble coalescence. However, no appropriate numerical models have been developed until now. The bubble coalescence mechanism may be related to thermodynamics, electromagnetics, hydrodynamics, heat and mass transfer and so on. In this chapter, microbubble interaction on the basis of a new gas–liquid interfacial model with the coalescence model is discussed, and the mass transfer is also considered in the numerical simulation. The results indicated that the contamination at the interface had an important effect on microbubble coalescence and repulsion.

2.3.1.1 Introduction

On a macroscopic scale corresponding to bulk fluid flows, a gas–liquid interface has been characterised by discontinuous physical properties, such as density, viscosity and so on. On the other hand, on a microscopic scale corresponding to molecular motion, the physical properties at the interface continuously vary from the gas phase to the liquid phase. This implies that interfacial phenomena can be characterised by various time and space scales. Therefore, the modelling of the gas–liquid interface, whose physical properties vary during transition from micro- to macroscopic scale, has been known as a big problem in multiphase flow science and engineering. Especially, the mechanism of bubble coalescence/repulsion has not yet been clarified. At the gas–liquid interface, many complicated phenomena, such as evaporation, condensation, electrokinetics, and heat and mass transfer, occur. Interfacial interactions such as bubble coalescence/repulsion may be affected by these interfacial phenomena.

Experimental research on bubble coalescence/repulsion mainly focuses on the effects of the concentration of the electrolyte on the behaviour and coalescence threshold of the bubbles in the liquid [1–3]. These researches revealed that the ions adsorbed at the gas–liquid interface, the combination of these ions and electrolyte concentration were important factors in bubble coalescence/repulsion. Contamination at the interface could greatly influence bubble coalescence/repulsion because these contaminations might be related to the heat and mass transfer and electrochemistry around the interface [4,5]. A coalescence model for a combination of simple ions has been theoretically proposed [6]. It focused on determining the impact of repulsive interaction based on the results from the formation of an electric double layer around the interface on bubble coalescence. However, a detailed mechanism has not yet been elucidated.

On the other hand, in theoretical and numerical studies, interfacial interactions were considered on the basis of hydrodynamic lubrication theories [7,8]. In such lubrication theories, bubble coalescence was modelled based on a drainage mechanism, where the liquid existing between two bubbles drained from the liquid film. This theory was based on the premise of the existence of a liquid film. Recently, the contamination at the interface has been considered in the lubrication theory, and the

Marangoni effect on the drainage mechanism has been discussed [9]. Contaminants such as adsorbed materials or ions at the interface were strongly related to heat and mass transfer of the liquid, and they might affect the timing of bubble coalescence [9,10]. Moreover, the interfacial interaction of bubbles in fluid flow would be also affected by the fluid vortical motion around the bubbles. However, in the research on the interfacial interactions based on the lubrication theories, the formation mechanism of the liquid thin film has not been discussed yet. Thus, there are still problems remaining with respect to the clarification of the mechanism of bubble coalescence/repulsion.

Because of the background mentioned above, focusing on the effect of an electrostatic potential due to contamination at the interface on the microbubble interaction, we have been conducted the numerical simulation of the microbubble interaction. In this chapter, some numerical results of microbubble coalescence/ repulsion process are shown, and the phenomenon is discussed.

2.3.1.2 Electrostatic effect on microbubble interaction and mass transfer

In our previous study [11], the numerical simulation of bubble interaction was conducted by considering an interaction force as a kind of a Coulomb force. With respect to the bubble interaction in the numerical simulation, a distance between the centres of two bubbles was calculated using a labelling method, which was a popular manner in the image analyses. It was modelled that the interaction force acts on the bubble surface cells when the distance between two bubble centres became smaller than a certain distance. Then, the bubble velocity after the interaction was calculated by satisfying the momentum and kinetic energy conservations. The calculated velocity was set on the numerical cells of bubble surface. Figure 2.8 depicts a numerical simulation result of the microbubble interaction: the diameter of the microbubble was 100 m, and the comparison of numerical results to the experimental observation. In the numerical result, as shown in Fig. 2.8, one bubble is sliding downwards along the surface of the other bubble. It is found that the numerical result retrieves the bubble behaviour well compared with the experimental observation. This result indicated the importance of an interfacial electrostatic potential in the microbubble interaction. To model the microbubble coalescence and repulsion physically in the numerical simulation, the electric effect of ions and contamination adsorbed at the surface on the bubble interaction must be considered from the thermodynamic point of view. Moreover, the contamination surrounding the bubbles in fluid such as mass transfer cannot be ignored.

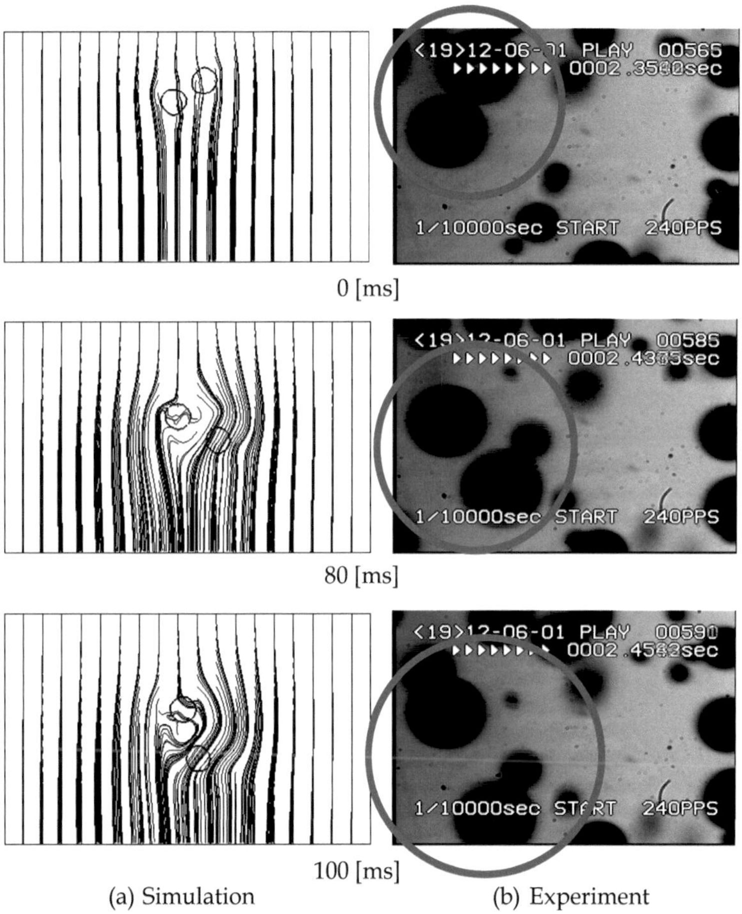

Figure 2.8 Interaction of microbubbles.

Thus, in order to elucidate the mechanism of bubble coalescence and repulsion, we have developed a new thermodynamic and mathematical interfacial model on the basis of the phase field theory [12] associated with the multiscale concept ranging from micro- to macroscopic scales in order to consider the multiscale physics at the interface. In this modelling, the interface was assumed to be a finite thickness like a thin liquid membrane, and the free energy including the electric charge due to a contamination was derived at the microscopic scale, that is the interface thickness. The free energy was incorporated into the Navier–Stokes equation as source terms based on the Chapman–Enskog expansion [13]. Finally, the multiscale multiphase flow equation was derived [14]. On the other hand, at the macroscopic point of view, the interfacial jump condition based on thermodynamics was also derived using the multiscale multiphase flow equation [15]. In that work, the jump condition at the interface was mainly characterised by the curvature related to the shape of the interface, which means that the interface is a mathematical surface: it means zero thickness. Therefore, the jump condition is considered as a macroscopic interfacial equation. Especially, a theoretical consideration of the interfacial interaction on the basis of the thermodynamic interfacial jump condition [16] indicated that the electrostatic force caused by the contamination adsorbed at the interface distorted the shape of bubble surface and formed a liquid thin film between interfaces.

2.3.1.3 Modelling and simulation of microbubble coalescence and repulsion

Many pieces of experimental evidence revealed the importance of the electrolytes and the adsorbed ions at the interface with respect to bubble coalescence [1–3]. This indicates that the mass transfer cannot be ignored to consider interfacial interactions. According to the Henry's law, the amount of a gas dissolved in a liquid is proportional to the gas pressure at the bubble surface in equilibrium when the system is at a constant temperature. It is very difficult to make the perfectly purified water in which there is no contamination. Thus, a relationship between the dissolved gas and the contamination in the liquid would be very important in the interfacial interaction. In our study, the coalescence model of microbubbles has been developed by applying a conventional nucleation theory [17] combined with the concept of the contamination in the liquid regarding to the mass transfer considered between the interfaces. The numerical results of microbubble interactions on the basis of the multiscale multiphase flow equation, and the coalescence model [18,19] are discussed in the following sections.

Figure 2.9 depicts the results of microbubble interactions obtained by two-dimensional numerical simulation. In the simulation, the coalescence model and the electrostatic potential at the interface were considered. In case 1, both the electrostatic potential at the interface and the coalescence model were not considered. In case 2, both models were considered. In case 1, the microbubbles at 1.50 ms were just contacted before the coalescence. Then, two microbubbles coalesced with each other at 2.00 ms. On the other hand, in case 2, the microbubbles maintained a constant distance at 1.50 ms. After that, two microbubbles coalesced with each other at 2.25 ms. Obviously, the timing of coalescence in case 2 delayed from case 1. This behaviour was also observed in our experiment [10].

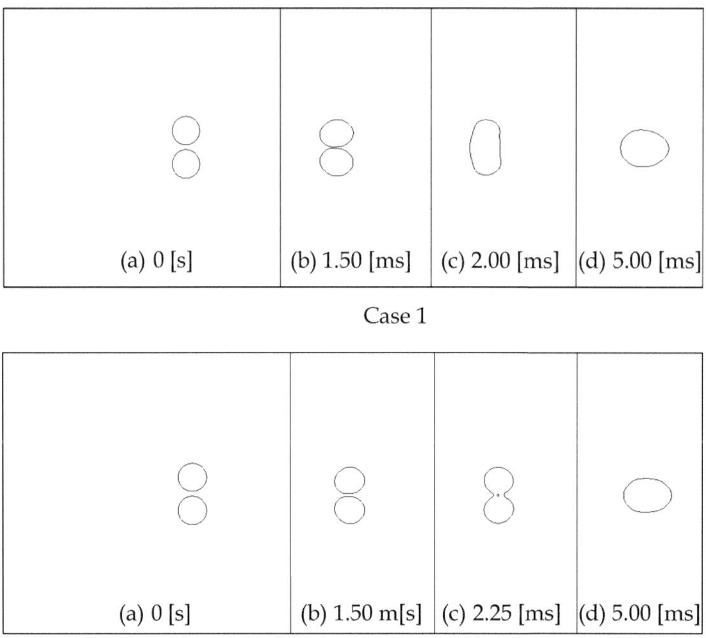

Case 2:

Figure 2.9 Numerical results of microbubble coalescence (cases 1 and 2: electrostatic potential 0 (V) and -1.0×10^{-5} (V), electric conductivity 103.146 (S/cm)).

In the three-dimensional numerical simulation, similar results were obtained in a simulation where the model of the interfacial electrostatic potential was only considered. Especially, the thin liquid film was observed between two microbubble interfaces as shown in Fig. 2.10a. This result reveals that the pressure in the liquid film is lower than the bulk pressure around the microbubbles as shown in Fig. 2.10b. This means that the bulk liquid flows into the thin liquid film to satisfy the continuity and maintain a certain thickness of the liquid film. Eventually, the general drainage explanation based on the lubrication theory [7], which is based only on hydrodynamics, cannot explain the premise of the existence of a liquid film between two microbubbles. Therefore, the present numerical results could prove that the contamination at the interface (electrostatic potential) is very important with respect to the bubble interaction.

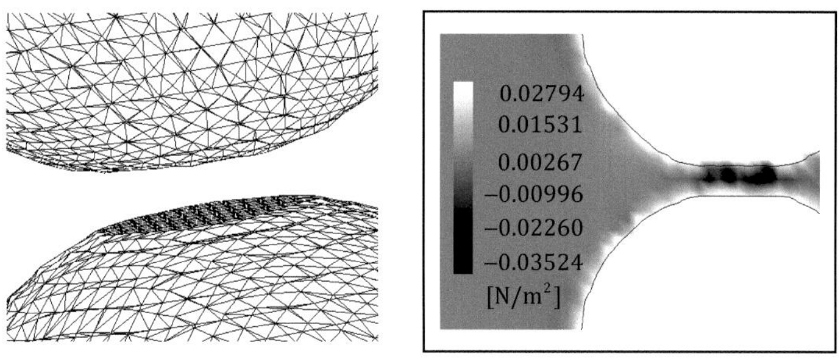

Figure 2.10 Microbubble interaction in three-dimensional simulation. (a) Liquid film in three dimensions. (b) Pressure distribution in liquid film.

2.3.1.4 Summary

The present numerical results indicate the possibility that bubble interactions on the mesoscopic scale can be simulated by using the multiscale multiphase flow equation. At the present, our interfacial model does not perfectly consider the various physical phenomena occurring at the interface. Thus, there may be some processes with respect to bubble coalescence in addition to those considered in the present coalescence model. However, at least, it can be seen that there are two coalescence processes: one process is related to only the electrostatic potential, and the other is related to mass transfer in addition to the electrostatic potential. In order to understand the mechanism of microbubble coalescence, the theoretical and experimental investigations of the phenomena should be continued.

References

1. Craig VSJ (2004) Bubble coalescence and specific-ion effects, *Curr Opin Colloid Interface Sci,* **9**, 178.
2. Henry CL, Craig VSJ (2008) Ion-specific influence of electrolytes on bubble coalescence in nonaqueous solvents, *Langmuir,* **24**, 7979.
3. Marčelja S (2004) Short-range forces in surface and bubble interaction, *Curr Opin Colloid Interface Sci,* **9**, 165.
4. Joshi KS, Braumann A, Jeelani SAK, Blickenstorfer C, Naegeli I, Windhab EJ (2009) Mechanism of bubble coalescence induced by surfactant covered antifoam particles, *J Colloid Interface Sci,* **339**, 446.
5. Auster ND, Gunde R, Mäder R, Windhab EJ (2009) Binary coalescence of gas bubbles in the presence of a non-ionic surfactant, *J Colloid Interface Sci,* **333**, 579.
6. Marčelja S (2006) Selective coalescence of bubbles in simple electrolytes, *J Phys Chem* B, **110**, 13062.
7. Reynolds O (1886) On the theory of lubrication and its application to Mr. Beauchamp Tower's experiments, including an experimental determination of the viscosity of olive oil, *Philos Trans R Soc Lond,* **177**, 157.
8. Kaur S, Leal LG (2009) Three-dimensional stability of a thin film between two approaching drops, *Phys Fluids,* **21**, 072101.
9. Leshansky AM (2001) On the influence of mass transfer on coalescence of bubbles, *Int J Multiphase Flow,* **27**, 189.
10. Yonemoto Y, Yanagisawa H, Kawara Z, Kunugi T (2008) Coalescence of microbubble, *J JSEM,* **8**(1), 38–44, ISSN 13464930.
11. Matsumoto Y, Fukami T, Kunugi T, Serizawa A (2001) Fluid-dynamic structure of air-water bubbly flow with micro-bubbles, in *Proceedings of the 4th International Conference on Multiphase Flow*, New Orleans LA, USA, CDROM_332.
12. Cahn JW, Hilliard JE (1958) Free energy of a nonuniform system. I. Interfacial energy, *J Chem Phys,* **28**(2), 258–267.
13. Chapman S, Cowling T (1970) *The Mathematical Theory of Non-uniform Gases,* 3rd ed, Cambridge University Press, Cambridge.
14. Yonemoto Y, Kunugi T (2010) Multi-scale modeling of the gas-liquid interface based on mathematical and thermodynamic approaches, *Open Transport Phenom J,* **2**, 69–79.

15. Yonemoto Y, Kunugi T (2011) Macroscopic gas-liquid interfacial equation on the basis of thermodynamics and mathematical approaches, in *Mass Transfer—Advanced Aspects* (Hironori N ed), InTech, Rijeka, Croatia.

16. Yonemoto Y, Kunugi T (2008) Analytical consideration of thermodynamic jump condition, *Progress in Multiphase Flow Research*, **3**, 27–34 (in Japanese).

17. Abraham FF (1974) *Homogeneous Nucleation Theory,* Academic Press, NY, USA.

18. Yonemoto Y, Kunugi T (2010) Fundamental numerical simulation of microbubble interaction using multi-scale multiphase flow equation, *Microgravity Sci Technol*, **22**, 397–405.

19. Yonemoto Y, Kunugi T (2009) Fundamental study of microbubble coalescences based on multi-scale multiphase flow equations, *Trans JSME Ser B*, **75**, 1790–1797 (in Japanese).

2.3.2 Physical Property Evaluation by ζ-Potential

Yuka Yoshida, Norihito Kawaguchi, and Nobuhiko Kubota
Advanced Applied Science Department, Research Laboratory,
IHI Corporation, Yokohama 235-8501, Japan
yuka_yoshida@ihi.co.jp

2.3.2.1 Introduction

Application of various technologies using microbubbles is expected. In these days, some medical treatments and industry applications have already started. However, when the microbubbles are applied, many unknown physical properties, for example, the electrification state of the surface of the microbubbles and a collapse phenomenon, are still left.

In this study, it is investigated to elucidate the electrification state of the surface of the microbubbles by measuring the ζ-potential of the microbubbles.

The ζ-potential of the microbubbles is reported by Takahashi [1]. He measured the ζ-potential of air microbubbles and obtained the results that the ζ-potential of air microbubbles does not depend on the size of the microbubbles and depends on the changing of pH of a solvent. Then, in order to investigate the properties of the microbubbles other than of the air, the ζ-potential measurement of the microbubbles of argon gas was tried. Moreover, the ζ-potential measurement of air microbubbles was also tried for comparison with Takahashi's measurement result.

2.3.2.2 Experiment

2.3.2.2.1 Measuring method and conditions of ζ-potential

The ζ-potential was measured with air and argon microbubbles. Microbubble generator (4-MDG-045, Aura Tech) was used for creation of the microbubbles. Ion-exchanged water was used as a solvent, and water temperature was set to 25 ± 5°C. Chloride was used for adjustment of pH, and it was set to pH 3. The equipment is shown in Fig. 2.11 using microscopic electrophoresis as the measuring method of the ζ-potential of the microbubbles [2]. The voltage of 20 V/cm was applied to the observation cell of the microbubbles.

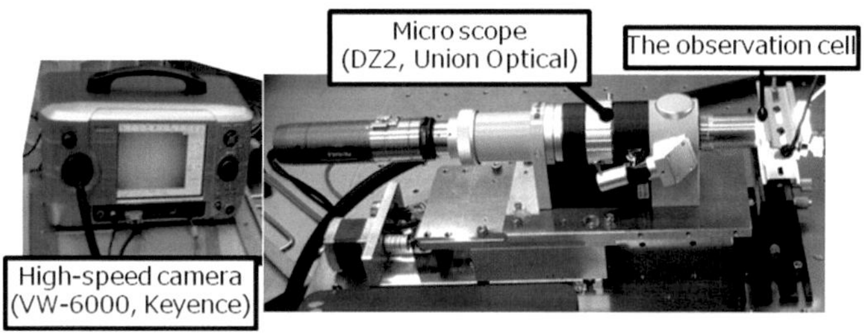

Figure 2.11 Measuring equipment of the microscopic electrophoresis.

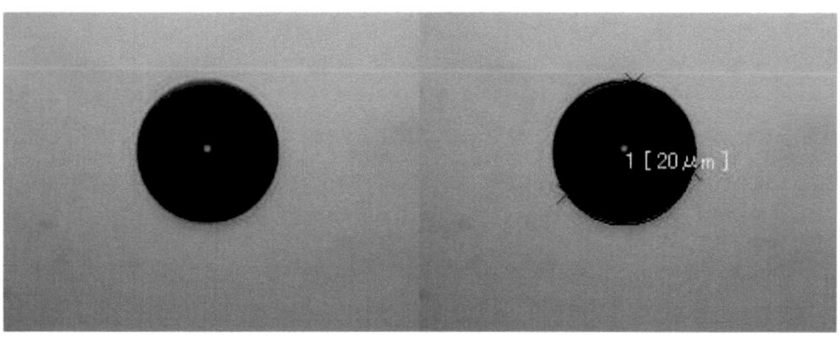

Figure 2.12 Bubble size measurement.

The size of the microbubbles used for calculation of the ζ-potential is 20–30 μm, and it obtained directly from the picture acquired on the ζ-potential measurement as shown in Fig. 2.12. The ζ-potential of the polystyrene particle was measured with this measuring method and the commercial measuring device (ζ-potential measuring device by Otsuka Electronics), the result was compared and the validity in this measuring method was checked.

2.3.2.2.2 Calculation method of ζ-potential

The ζ-potential can be obtained by observing and measuring the moving rate (electrophoresis rate) of the particles (microbubbles) according to the given voltage slope and using the Smoluchowski equation (Eq. 2.1) in the case of using microscopic electrophoresis [3]. In this equation, Z is ζ-potential, n is viscosity of the solvent, u is electrophoresis rate and e is dielectric constant of the solvent.

$$Z = \frac{nu}{e} \tag{2.1}$$

However, when the voltage is applied to the observation cell, not only the particles (microbubbles) of a measuring object but also an internal liquid will flow in response to the influence of the voltage, and the electrophoresis rate of the particles (microbubbles) obtained by actual measurement become an apparent rate [4]. When quartz is used for the observation cell under these pH conditions, the flow of the internal liquid is almost negligible since the isoelectric point of the quartz is between pH 2 and 3.

2.3.2.3 Results and discussion

2.3.2.3.1 The ζ-potential of polystyrene particles

The polystyrene particles were measured in pH 3 using both the microscopic electrophoresis and the ζ-potential measuring device, and the result were compared. The results are shown in Fig. 2.13. In this figure, a horizontal axis indicates the ζ-potential (mV), and the vertical one indicates the distance from the centre of the observation cell in the depth direction. Zero means the centre of the cell.

As a result of measurement, it has checked that the ζ-potential measured by both the techniques of the polystyrene particle were almost the same value, which were between –20 mV and –10 mV.

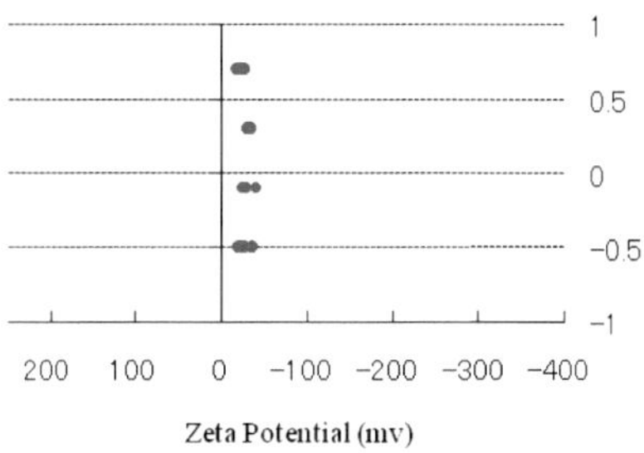

(a)

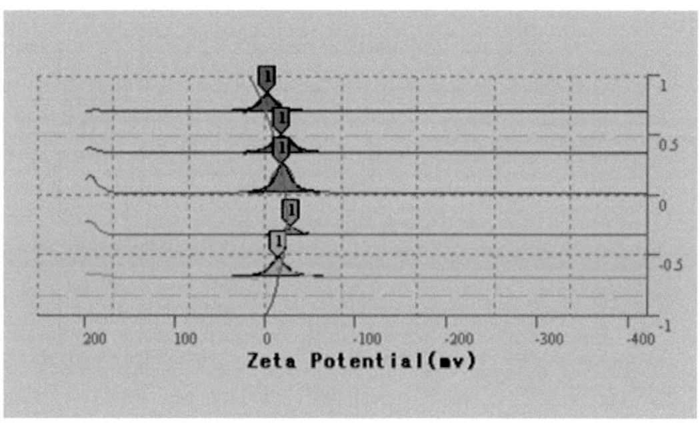

(b)

Figure 2.13 The ζ-potential measurement results of the polystyrene particles. (a) The results of the microscopic electrophoresis. (b) The results of the ζ-potential measuring device.

2.3.2.3.2 The ζ-potential of the air and argon gas microbubbles

The measurement results of the air and argon gas microbubbles in pH 3 are shown in Fig. 2.14.

As the measurement results shown in Fig. 2.14, both the ζ-potentials of the air and argon gas microbubbles were in the range of 10–20 mV. This result is corresponded with the measurement result of the ζ-potential of air microbubbles by Takahashi. Under the conditions of pH 3, any difference of the ζ-potential was not observed in the microbubbles of the air and the argon gas.

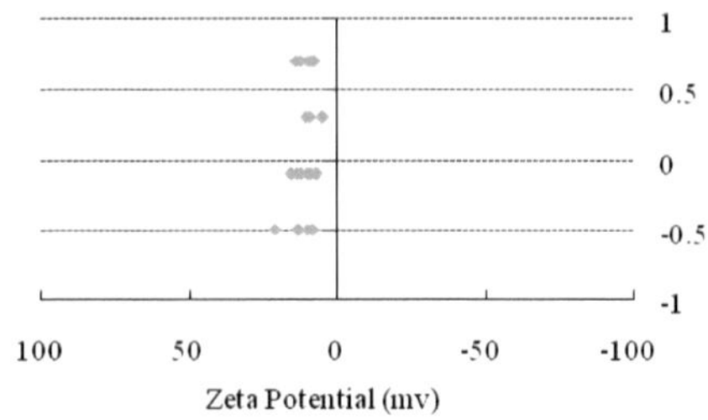

(a)

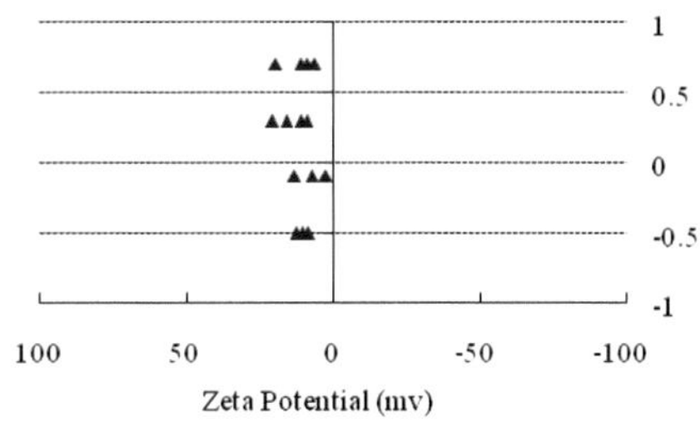

(b)

Figure 2.14 The measurement results of the ζ-potential of the microbubbles in pH 3. (a) The ζ-potential of the air microbubbles. (b) The ζ-potential of the argon gas microbubbles.

2.3.2.4 Summary

The ζ-potential of the microbubbles of the argon gas was measured, and the behaviour was investigated. As a result, the difference was not observed in the ζ-potential of the microbubbles of the air and the argon gas under the conditions of pH 3. Still only one condition of the ζ-potential under the conditions of pH 3 of argon microbubbles was measured. So, another measurement with other pH conditions and other gas will be performed in near future to elucidate the electrification state of the surface of the microbubbles.

References

1. Takahashi M (2005) The ζ Potential of Microbubbles in Aqueous Solutions: Electrical property of the gas-water interface, *J Phys Chem B*, **109** (46), 21858–21864.

2. Kitahara (1995) *Zeta Potential*, Scientist Press Co. Ltd. (in Japanese).
3. Furusawa (2004) Zeta(ζ)-potential measurements, *Bunseki*, **5**, 247–254 (in Japanese).
4. Mori (1980) A unified theory of determining the electrophoretic velocity of mineral particles in the rectangular micro-electrophoresis cell, *Fusen*, **27** (3), 117–126 (in Japanese).

2.3.3 Laser Diffraction/Scattering Method

Mitsuru Maruyama
Global Application Development Center, Shimadzu Corporation, Kanagawa, Japan
maruyama@shimadzu.co.jp

In recent years, microbubbles and nanobubbles have become a focus of attention, with applications such as the following:

i. Pharmaceuticals field
ii. Purification and cleaning
iii. Sterilization

In the pharmaceutical field, in particular, micro- and nanobubbles have been increasingly used for drug delivery systems and ultrasonic contrast media. In addition, while there are many precedents in the context of purification, applications to cleaning (particularly the cleaning of electronic equipment) remain forthcoming.

One of the significant factors determining the cleaning efficacy and other characteristics of these bubbles is their particle size distribution. Although a variety of methods are used for equipment for measuring particle size distributions, analysers using the laser diffraction/scattering method feature a number of benefits. These include a short measurement time (a few seconds to a few dozen seconds), a wide measurement range (a few dozen nanometres to several millimetres), good reproducibility and simple operability. As a result, through the 1990s, this type of particle size analysers became the norm. At first, the focus was on wet measurements, in which particles are suspended in a liquid. However, dry measurements, in which particles are in a gaseous phase, are now also performed. Recently, micro and nanobubble measurements have also become available, thereby making strides towards an epoch-making methodology, completely distinguished from conventional particle size analysers by the breadth of applicable fields.

In terms of basic principles, micro- and nanobubble measurements use a simple method, in which the particles are exposed to a laser beam, and the particle size is specified from the scattered light generated as a result, using the dependence of particle size on the spatial intensity distribution pattern. Applications of this method are subsequently expected in a variety of fields. In addition, although a red laser is conventionally used as the light source, there are now also analysers that use short wavelength (blue) light sources, or even shorter ultraviolet lasers so as to enable the measurement of tinier particles, thereby extending the measurement range.

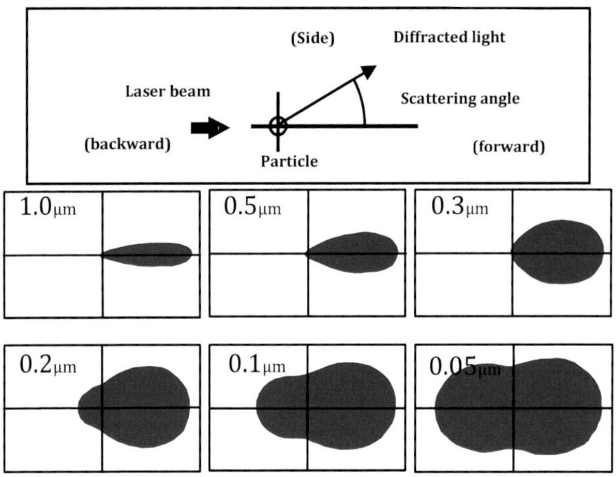

Figure 2.15 Relationship between particle size and light intensity distribution pattern for scattered light.

2.3.3.1 Measurement principles

When particles are exposed to laser light, scattered light is emitted from the particles. This spatial scattered light intensity distribution pattern (the change in scattered light intensity in accordance with the scattering angle) changes in accordance with particle size as shown in Fig. 2.15. For example, scattered light emitted from 5 μm particles is concentrated in the forward direction. As the particle size becomes smaller, the angle of the scattered light emitted becomes larger, and when the particles are 0.3 μm or smaller, the light scattered to the sides and backwards also becomes stronger. Furthermore, when the size drops to 0.1 μm, the intensity of the light scattered in the backward and forward directions becomes essentially the same. In this way, since the scattered light's light intensity distribution pattern changes in accordance with particle size, it conversely becomes possible to specify the particle size from the light intensity distribution pattern.

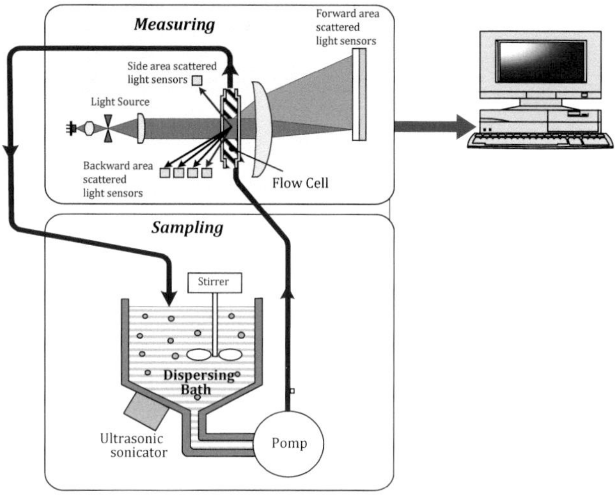

Figure 2.16 Typical system configuration.

2.3.3.2 System configuration

In actual measurements, targets are groups of particles consisting of a number of particles of different sizes, which are dispersed in liquid or gas. If a group of particles is exposed to laser light, the scattered light emitted from the group of particles will be a superposition of the scattered light generated by the individual particles. A typical system configuration is shown in Fig. 2.16. Basically, it consists of an optical system (scattered light detector), a dispersion bath and a processor. The laser beam generated by the laser light source is converted to a slightly wider beam using the collimator lens and is then shone on the group of particles in a cell with a light path length of a few millimetres. The suspension filling the dispersion bath is circulated by a circulation pump. The light scattered in the forward direction is condensed by the lens and is united into a ring-shaped diffracted/scattered image on the detecting plane at the focal distance.

This is detected by ring sensors with detector elements arranged in concentric circles. Light scattered to the sides and backwards is detected by the side scattered light sensor and backscattered light sensors, respectively. In this way, light intensity distribution data is obtained by detecting the light intensity distribution pattern using a variety of detector elements.

2.3.3.3 Calculating the particle size distribution

The diffraction/scattering phenomenon due to the group of particles can be represented by the following (vector and matrix) equation:

$$\mathbf{s} = \mathbf{A}\mathbf{q}, \tag{2.2}$$

where

$$\mathbf{s} = \begin{bmatrix} s_1 \\ s_2 \\ \vdots \\ \vdots \\ s_m \end{bmatrix} \qquad \mathbf{q} = \begin{bmatrix} q_1 \\ q_2 \\ \vdots \\ \vdots \\ q_n \end{bmatrix} \tag{2.3}$$

$$\mathbf{A} = \begin{bmatrix} a_{1,1} & a_{1,2} & \cdots & \cdots & a_{1,n} \\ a_{2,1} & & & & \vdots \\ \vdots & & a_{i,j} & & \\ \vdots & & & & \vdots \\ a_{m,1} & \cdots & & \cdots & a_{m,n} \end{bmatrix} \tag{2.4}$$

Here, "s" is the light intensity distribution data (vector). The elements "s_i" (i = 1, 2, ... m) are the light intensity (incident light quantity) detected by the various detection elements in the forward scattered light sensor and by the side scattered light sensor and backscattered light sensors. "q" is the particle distribution (frequency distribution %) vector.

The particle size range targeted for measurement (maximum particle size: x_1, minimum particle size: x_{n+1}) is divided into "n" segments, and the respective particle size sections become $[x_i, x_{i+1}]$ (j = 1, 2, ... n). The "q" elements "q_j" (j = 1, 2, ... n) are the particle quantities corresponding to the particle size sections $[x_i, x_{i+1}]$. Normally, a volumetric standard is used:

$$\sum_{j=1}^{n} q_j = 100\% \tag{2.5}$$

In other words, the data are normalised so that the total will equal 100 %.

"**A**" is the coefficient matrix for converting the particle distribution "q" (vector) to the light intensity distribution "s" (vector).

In terms of the physical meaning, the elements in "A", "$a_{i,j}$" ($i = 1, 2, \ldots m, j = 1, 2, \ldots n$), are the incident light quantity corresponding to the "ith" element of scattered light from the group of particles with the unit particle quantity belonging to the particle size section $[x_i, x_{i+1}]$. In other words, the numerical matrix "**A**" represents the very scattering phenomenon itself due to the group of particles. The value of "$a_{i,j}$" can be theoretically calculated in advance. This calculation uses Fraunhofer diffraction theory if the particle size is substantially larger than the wavelength of the laser light (10 times or more). However, for regions smaller than this, it is necessary to use Mie scattering theory. Fraunhofer diffraction theory can effectively be considered an excellent approximation to Mie scattering theory for very small-angle scatter in the forward direction when the particle size is significantly larger than the wavelength.

To calculate the elements in the coefficient matrix "**A**" using Mie scattering theory, it is necessary to set the absolute refractive index (complex number) for the particles and the media (liquid media) in which they are dispersed.

Thus, the following equation is obtained to find the least squares solution for the particle distribution (vector) "q" based on Eq. (2.2).

$$\mathbf{q} = (\mathbf{A}^T \mathbf{A})^{-1} \mathbf{A}^T \mathbf{s} \tag{2.6}$$

Note that "$\mathbf{A}^{\mathrm{T}}$" is the transposed matrix for "**A**", and "$(\ )^{-1}$" is the inverse matrix.

On the right side of Eq. (2.6), each of the light intensity distribution (vector) "**s**" elements is a numerical value detected with the forward scattered light sensor and side scattered light sensor. Also, the coefficient matrix "**A**" can be calculated in advance using Fraunhofer diffraction theory and Mie scattering theory. Accordingly, if Eq. (2.6) is calculated using this already-known data, then the particle size distribution (vector) "q" is obtained.

Equation (2.6) is the basic method for calculating the particle size distribution from the light intensity distribution data with the laser diffraction/scattering method. However, just executing this equation as is leads to quite large errors, so in reality, the calculations performed on a computer are quite complicated and take into consideration a variety of conditions. In many cases, restrictions are added to the calculation, for example, to prevent the particle quantity from being a negative number, and to ensure that the particle size distribution is to some extent continuous.

2.3.3.4 System configuration for the measurement of microbubbles

The configuration of the system for measuring particle size distribution via the laser diffraction/scattering method has already been described, but there are some changes when microbubbles are targeted for measurement. This is because microbubbles and nanobubbles are very unstable and prone to collapse due to the particle characteristics. For this reason, as shown in Fig. 2.17, it is preferable that micro- or nanobubble size distribution measurements are made directly, without circulating the material. If the bubbles are still unstable, then to ensure stability, one method is to add a small amount of surfactant or ethanol.

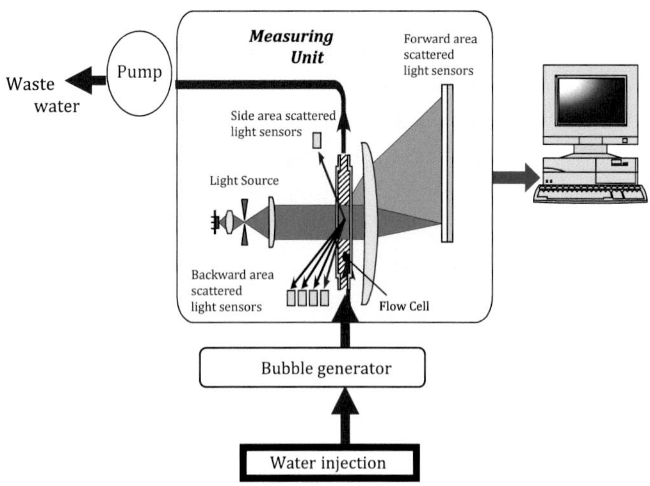

Figure 2.17 System configuration for bubble measurements.

In terms of stabilizing and maintaining bubbles, it is also possible to measure them as is in a batch cell. Even in this case, inadvertent circulation may lead to bubbles being created or destroyed.

2.3.3.5 Measurement example

2.3.3.5.1 Effectiveness of a dispersing agent

Figure 2.18 shows the bubble size distributions obtained with the same bubble generator when they are generated with purified water only and with 1% ethanol added to purified water. It is evident that when the dispersing agent is added, the microbubbles stabilise, and small-particle bubbles are generated.

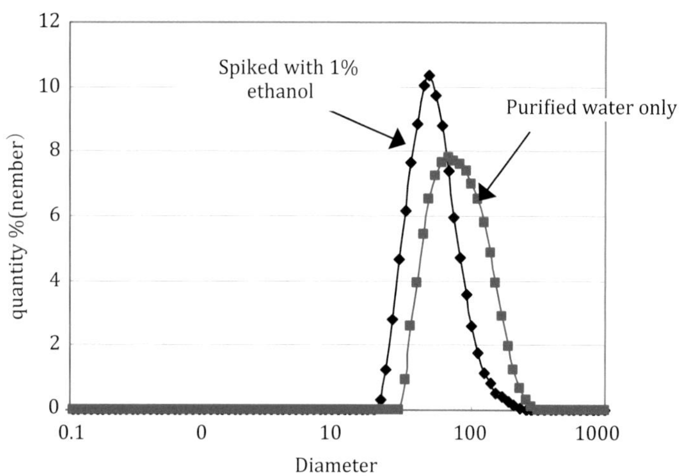

Figure 2.18 Comparison of bubble size with purified water only and purified water spiked with 1% ethanol.

2.3.3.5.2 Bubble generation

The microbubble generation process was measured at 1 s intervals using two different pore bubble generation methods, and the respective 30 s measurement results are shown in Figs. 2.19 and 2.20. It is evident that with generation method A, microbubbles with a stable bubble size distribution are generated from the start, but with generation method B, stable bubble generation is not obtained. As in this case, since bubble generation may become unstable, measurements at short intervals must be considered as well.

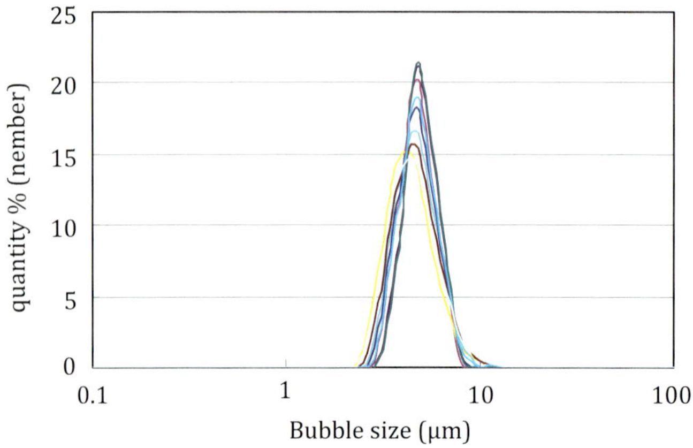

Figure 2.19 Temporal development of bubbles generated with generation method A.

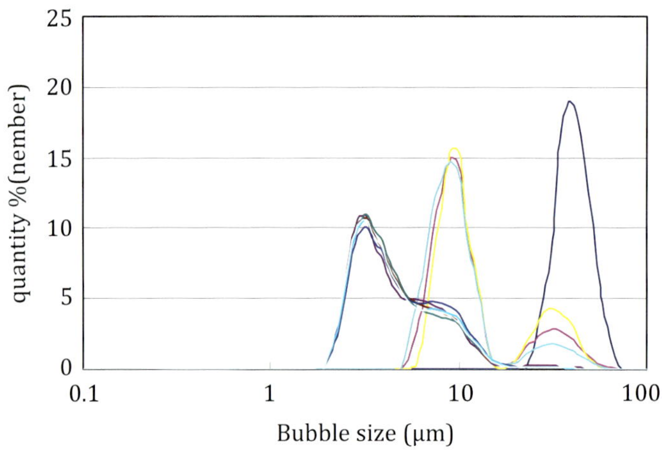

Figure 2.20 Temporal development of bubbles generated with generation method B.

2.3.3.5.3 Possibility of calculating the bubble concentration (Void ratio)

Normally, with particle size distribution measurements using the laser diffraction method, as indicated in Section 2.3.3.3, not the absolute number of particles but the relative distribution is measured. However, it is possible to obtain the bubble concentration (void ratio) by measuring the scattered light intensity.

The scattered light intensity distribution, shown in Fig. 2.21 is the scattered light intensity detected with scattered light sensors arranged at a variety of angles. In normal measurements, this is used to check the optimal measurement concentration. However, if the particle quantity doubles, so too does the intensity, so it is possible to estimate the bubble concentration by comparison to a sample with a known concentration. Note that if the bubble size distribution differs significantly, then the scattered light intensity distribution diagram and intensity will change. Accordingly, calibration using a sample with similar particle sizes is necessary.

However, a general rule regarding this bubble concentration calculation is that the scattered light has not saturated the sensitivity of the light receiving sensor, so the limit concentration is about 0.1%.

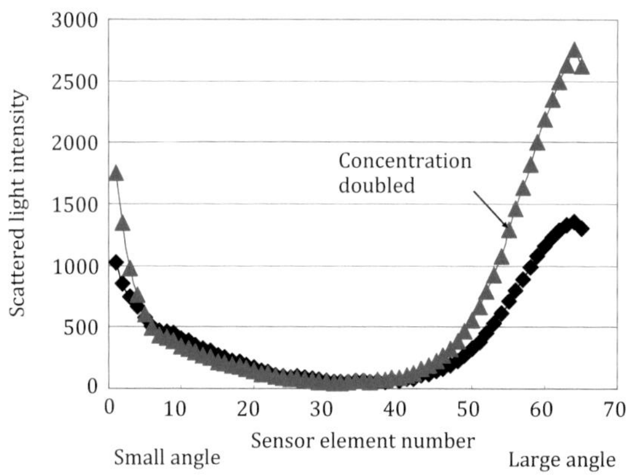

Figure 2.21 Light intensity distribution data for scattered light changing in accordance with particle (bubble) concentration.

2.3.3.5.4 Definition of particle quantities for bubble size distributions

Standards of expression for investigating distribution quantities in bubble size distributions include particle number standards for investigating the number of particles and volumetric standards for investigating weights or volumes. Accordingly, even for the same aggregate of micro particles, bubble size distributions will differ considerably depending on the definitions of representative particle sizes (representing the size of the individual particles) and particle quantity (number of particles, length, surface area and volumetric standard). JIS Z 8819-1 specifies how to represent these particle size analysis results. With the laser diffraction method, measurements are generally performed using a volumetric standard. However, when it is assumed that the particles are spherical, the conversion can be done with other standards. These conversions can be performed with the following equations:

Number of particle size divisions: m
Particle size: X_j ($j = 1, 2, \ldots\ldots m+1$)
Average particle size for each section: Z_j ($j = 1, 2, \ldots\ldots m$)
Difference % (volumetric standard): q_j ($j = 1, 2, \ldots\ldots m$)
Difference % (particle number standard): r_j ($j = 1, 2, \ldots\ldots m$)

$$y_j = \frac{\log_{10} x_j + \log_{10} x_{j+1}}{2}$$

$$z_j = 10^{y_j}$$

$$p_j = \frac{q_j}{(z_j)^3}$$

$$s = \sum_{j=1}^{m} p_j$$

$$r_j = \frac{p_j}{s} \times 100$$

For example, if 1 μm particles and 10 μm particles are considered, then even though the ratio will be 1:1 with a particle number standard, it will be 1:1000 with a volumetric standard. Accordingly, caution is needed, as a very different impression will result even for the same particle distribution measurement target depending on whether a volumetric standard or particle number standard is used.

In general, bubble size distributions are often expressed using either a particle number standard or a volumetric standard. Figure 2.22 shows the bubble size distributions using both the volumetric standard and particle number standard. Looking at the volumetric standard results, it is evident that approximately 20 μm microbubbles are generated. From the particle number standard results, however, it is evident that a very large number of comparatively small 5 μm microbubbles are generated.

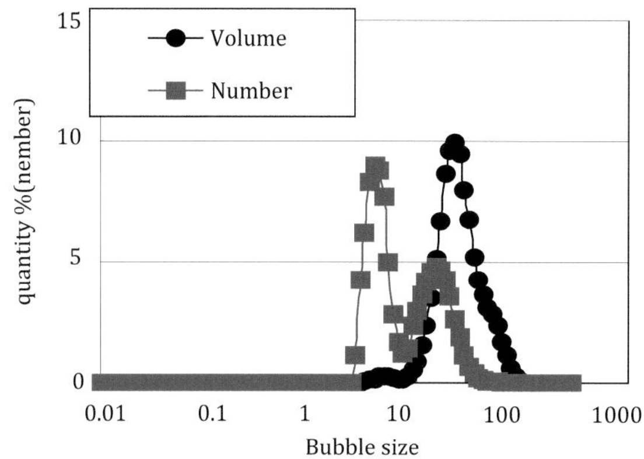

Figure 2.22 Difference in the expression of bubble size distributions due to distribution standards (point of reference).

2.3.3.6 Considerations regarding measurement

2.3.3.6.1 Considerations regarding the measurement path

Bubbles are very unstable. Accordingly, stable measurements cannot be performed if pumps that produce pulses along the measurement path are used.

Also, bubbles may tend to adhere to oily components due to their characteristics, so stable measurements cannot be performed if hydrophobic materials are used in the measurement path.

2.3.3.6.2 Considerations regarding the measurement concentration

If the particle concentration is large, as shown in Fig. 2.23 light scattered from one particle will tend to shine onto other particles and be further scattered. This phenomenon is referred to as multiple scattering. In this case, the particles will tend to appear smaller than at appropriate concentrations. For this reason, while results will differ depending on the light path length (cell length) and particle size distribution, normally, a concentration less than a few 100 ppm is used. Recently, however, analysers have been developed that is capable of measuring even high-concentration samples on the order

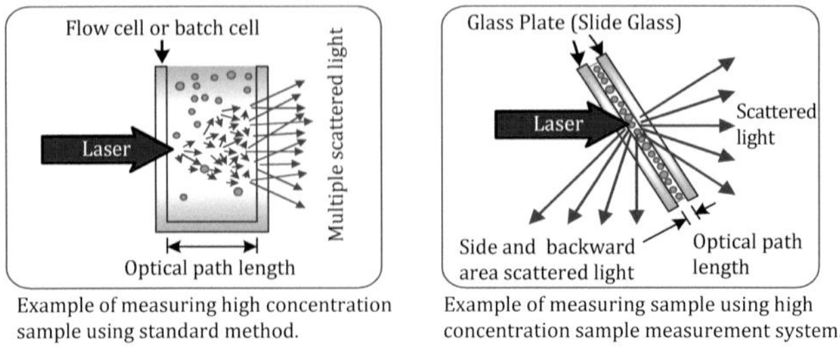

Figure 2.23 Approaches to high-concentration measurements.

of a few percent to 20% by volume, by making the light path length as short as possible.

On the contrary, the scattered light intensity becomes extremely small when the particle concentration is low, so the S/N ratio will drop, and accurate results will not be obtained. In this case, suggestions include increasing the laser light output or lengthening the light path length.

2.3.3.6.3 Contamination of the optical system

If the optical system becomes contaminated (particularly by the adhesion of contaminants to the cell), coherency will worsen due to light scattered by the adhering particles. Accordingly, there will be an impact on the particle size distribution even if blank calibration is performed. Also, bubbles may tend to adhere to contaminants due to their characteristics, leading particle sizes to appear large, so caution is needed.

References

1. The Society of Powder Technology, JAPAN (ed) (1998) *Powder Technology Handbook*, 2nd ed, Nikkan Kogyo Shimbun Ltd.
2. The Society of Powder Technology, JAPAN (ed) (2000) *Terminology Dictionary of Powder Technology*, 2nd ed, Nikkan Kogyo Shimbun Ltd.
3. The Society of Powder Technology, JAPAN (ed) (1994) *Measurements, Techniques of Particle Diameter*, Nikkan Kogyo Shimbun Ltd., pp 145–167.

2.3.4 Electrical Sensing Zone Method (Coulter Counter)

Yiming Yang
Micromeritecs Japan G.K., Kashiwanoha, Kashiwa, Chiba 277-0882, Japan
yiming.yang@micromeritics.com

2.3.4.1 Introduction

Micro- or nanobubbles are expected to be applied in the field of various environmental, industrial, food, and the medical field. Generally, microbubbles and nanobubbles have been classified according to the bubble size, it is said that microbubbles are 10–100 μm, and nanobubbles are less than 1 μm. It is most important to measure the bubble size and the concentration for knowing the function of bubbles, electrical sensing zone method can be used on the measuring the bubble size and the concentration as a standard measurement method. It is the best way as the highest accuracy measurements of bubble size distribution and the concentration, since this method is measuring the volume and number of individual bubble.

Electrical sensing zone method was developed as a technology that measures simultaneously the number and the size of the particles dispersed in the electrolysis solution by Wallace Coulter late in the 1940s, and it was also called the Coulter principle. Initially, the instrument based on this technology has been used to facilitate the analysis of blood cells; it was also a very useful tool in industrial applications. Electrical sensing zone method are described in international standards and national standards in many measuring instruments, from the height of the measurement accuracy, as a criterion in the evaluation of particle size measurement technology, and are frequently used.

This chapter introduces the pretreatment, the display of the important matter, the measurement result, the statistics analysis and the measurement procedure, etc. for measuring the size distribution of the microbubbles and nanobubbles more correctly using this method; at the same time, it explains the newest technology of the electrical sensing zone method.

2.3.4.2 Principle

2.3.4.2.1 Electrical sensing zone method

By the particle size distribution measuring method, called an electrical sensing zone method, the wall which has one penetrated aperture is established into the electrolysis solution, which particles are distributing, an electrode is put on both sides and constant current is sent through the aperture. When particles pass aperture with an electrolysis solution by attracting the electrolysis solution, which dispersed particles by a certain fixed power of absorption from the inner side of fine pores, only the quantity in which the electrolysis solution in fine pores is equivalent to the volume of particles decreases, and the electrical resistance of aperture becomes larger in proportion to this eliminated amount of electrolysis solutions. The current flowing into aperture is not concerned with existence of particles, but the amount of voltage change is proportional to the amount of change of the electrical resistance of aperture. The volume of particles can be measured from the amount of voltage change, and it can ask for the diameter of a ball equivalent of particles from this volume; it can display the diameter distribution of a bubble. Moreover, the bubble concentration in suspension can be quantitatively measured by calculating a voltage change (electric pulse) number by the number with which the bubble passed fine pores counting and attracting fixed capacity (Fig. 2.24).

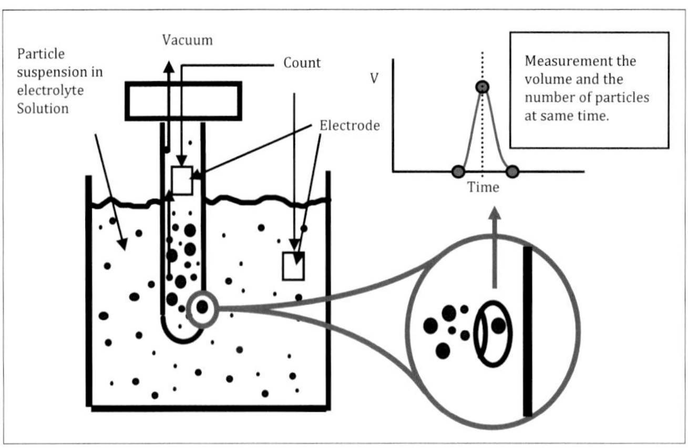

Figure 2.24 The principle of Electrical Sensing Zone method.

In international standard ISO 13319 "Determination of particle size distributions—Electrical sensing zone method", the guidance about particle size distribution measurement of the particles using Coulter principle is provided.

2.3.4.2.2 *Equivalent sphere diameter*

Although it has measured the true volume of particle (bubble) each, since it is common to denote particles (bubble) by a diameter as for an electric detection belt method when applying to the industrial field, it changes particle (bubble) volume into the diameter (ESD) of a ball equivalent (Fig. 2.25). In this case, the diameter of a particle (bubble) turns into a diameter of the ball of the same volume as that particle. In the case of the method of computing from the two-dimensional data (area) of particles, such as an image analysis method, two or more values will be made from one particle (bubble) with a detection position. Temporarily, when the particles (bubble) of the same triangular pyramid are measured, the value of an image analysis method varies, but the method of measuring volume is not subject to the influence of the measurement direction.

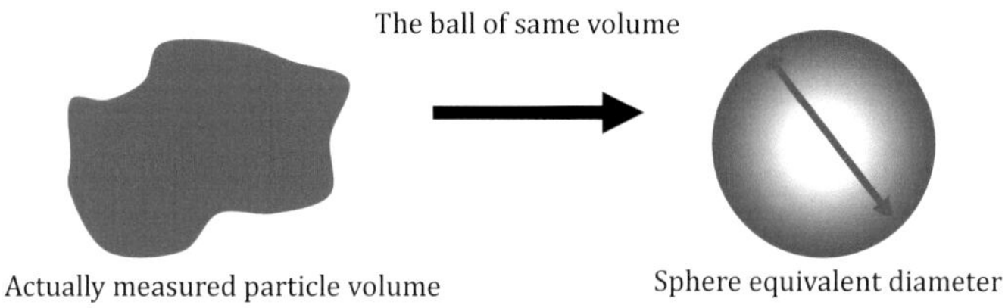

Figure 2.25 Equivalent sphere diameter.

2.3.4.3 Calibration

Particle volume is measured using known standard particles, and it asks for this proportionality factor. This the operation of a series of is calibrating of equipment. Like a formula (2.7), particle volume (V_p) is proportional to the electric pulse (ΔV) obtained when particles pass an aperture.

$$\Delta V = I\Delta R = I\frac{\rho^0}{S^2}V_p \tag{2.7}$$

where ΔV is the electric pulse, I is the current value, S is the cross-sectional area of aperture, ρ^0 is the resistance of electrolysis solution and V_p is the particle volume, (I) of formula (2.7) is a constant current value actually passed, and the value of (ρ^0/S^2) changes in an electrolysis solution, the diameter of an aperture, the form of an aperture, etc. Generally, (K_d) is denoted by formula (2.8) and connected with resistance (ρ^0) of current value (I), the cross-sectional area (S) of an aperture and an electrolysis solution.

$$K_d \propto \frac{S^2}{I \cdot \rho^0} \tag{2.8}$$

In actual calibrating, it measures using the standard particles (monodisperse particle) which the diameter of a particle (or volume) understands correctly, and determines (K_d). Each diameter of a particle (d) is calculated by formula (2.9) by determined (K_d).

$$d = K_d \sqrt[3]{V_p} \tag{2.9}$$

2.3.4.3.1 *Selection of standard particles*

It is necessary to choose calibrating particles according to the diameter of an aperture to be used, and they usually use the standard particles from 10% to 20% of an aperture diameter. The optimal calibrating refers to the calibrating particles which Table 2.1 recommends.

Table 2.1 The optimal calibrating particles is recommended

Aperture diameter (μm)	Range of standard particle diameter (μm)	Diameter of value of standard particles to recommend (μm)
20	2–4	2
30	3–6	3
50	5–10	5
70	7–14	10
100	10–20	10
140	14–28	20
200	20–40	30
280	28–56	30
400	40–80	43
560	56–128	65
1000	60–150	90
2000	90–260	90

2.3.4.4 Sample measurement

Unlike measurement of particles, it is not necessary to pretreat the bubble ultrasonically. Creation of measurement liquid is completion as bubble water is mixed with the electrolysis solution used for an electric detection belt method about measurement of nanobubbles. The rest is set in a measuring device and becomes pushing the start button of equipment. When measuring quantitatively, it is necessary to make regularity the ratio (1:9), which an electrolysis solution mixes with nanobubbles water and measures a fixed quantity in measurement mode to the measuring device side.

Moreover, when measuring microbubbles, since they are unstable, the microbubbles must measure the prepared measurement liquid promptly; and fixed-quantity measurement is the same method as nanobubbles, sets to 1:9, the ratio which an electrolysis solution mixes with bubble water, and measures a fixed quantity in measurement mode to the measuring device side. An example of the optimal diameter of a bubble, an aperture tube and dilution magnification, and a time-base range is shown in Table 2.2.

Table 2.2 Sample and use aperture, dilution factor, measurement range

Sample	Aperture diameter (μm)	Dilution factor	Measurement range (μm)
Nanobubble	20	10 times	0.4–12
Microbubble (below 10 μm)	50	10 times	1–30
Microbubble (above 10 μm)	200	10 times	4–120

2.3.4.4.1 Measurement procedure of nanobubbles

Since a measurement procedure has some differences depending on the model, a concrete procedure is carried out to my having each manual referred to and describes only a fundamental procedure here. The aperture of 20 μm is usually used for nanobubbles measurement.

1. A measuring instrument and the computer for control are turned on, dedicated software is started and it checks that operation of each part is normal.
2. An aperture suitable for a sample is attached, and an electrolyte solution is filled to instrument.

3. Each conditions (counting for bubble, the current value to an aperture, an amplification rate, a calibration factor, calculation method etc.) for measuring are set up.
4. A 90 mL electrolyte solution is put into a measurement container (round-bottom beaker), a container is installed in the predetermined position of a measuring device and it agitates at a speed suitable for a sample.
5. The bubble of 10 mL is extracted and is supplied to the prepared beaker.
6. Measurement is started in mode (100 μL) of volume.
7. The diameter distribution of a bubble and a statistics value are calculated. Various data are printed and saved.

2.3.4.4.2 Measurement procedure of microbubbles

The aperture of 50–200 μm is usually used for measurement of microbubbles.

1. A measuring instrument and the computer for control are turned on, dedicated software is started and it checks that operation of each part is normal.
2. An aperture suitable for a sample is attached, and an electrolyte solution is filled to instrument.
3. Each conditions (counting for bubble, the current value to an aperture, an amplification rate, a calibration factor, calculation method etc.) for measuring are set up.
4. A 90 mL electrolyte solution is put into a measurement container (round-bottom beaker), a container is installed in the predetermined position of a measuring device, and it agitates at a speed suitable for a sample.
5. Using a syringe as much as possible with a large exit, the bubble of 10 mL is extracted and is supplied to the prepared beaker.
6. Measurement is started in mode (100 μL) of volume.
7. The diameter distribution of a bubble and a statistics value are calculated. Various data are printed and saved.

2.3.4.5 Display and statistics analysis of a measurement result

In instrument in recent years, in order to measure the volume of a bubble by a multi-channel, the diameter distribution of a bubble is obtained in real time. Since the bubble diameter and number of a ball are calculated, number standard distribution is correctly convertible for volume standard distribution. Moreover, since the number of channels is changeable, the optimal result can express as the equipment that carries the digital pulse processing capability at any time if needed. In order to measure the diameter distribution of a bubble by high resolution, it is desirable to narrow a time base range or to increase the number of channels.

2.3.4.5.1 Display of the size distribution of a bubble

Size distribution of a bubble recommends what is usually displayed by the number distribution. However, when conducting an experiment which a bubble destroys by giving an ultrasonic wave to a bubble, in order to display bubble distribution, it recommends using number distribution and volume distribution. As shown in Fig. 2.26, while there was a peak of 0.5–0.8 μm before ultrasonic irradiation in 5.2×10^6 number/mL, after ultrasonic irradiation was be to 1.7×10^6 number/mL, and the nanobubbles of the 0.5–0.8 μm decreased, but the total volume of a bubble changed from 7.6×10^5 μm^3/mL to 6.0×10^5 μm^3/mL in the same range of volume distribution, and it has the almost same volume. It was proved that the nanobubbles were united according to an ultrasonic wave. Therefore, it is necessary to use the number distribution and volume distribution for display of the bubble size distribution when the nanobubbles are processed by ultrasonic irradiation, etc.

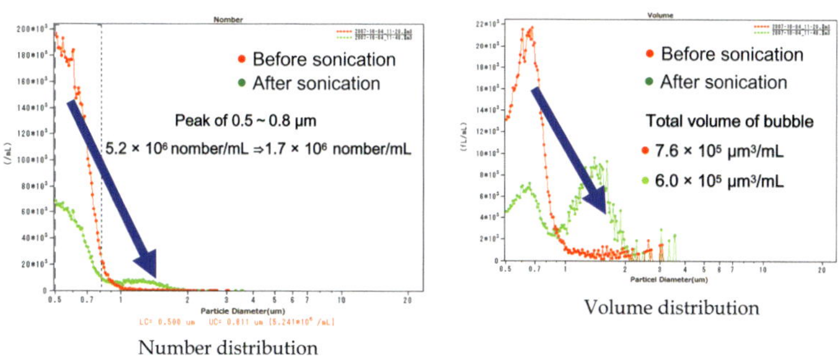

Figure 2.26 Distribution before and after supersonic. (Data are offered from Kyowakisetsu Inc.)

2.3.4.5.2 Statistics of the size distribution for bubble

Statistics and a size distribution display recommends taking the statistics of number distribution and volume distribution, respectively. Usually, the concentration of a bubble is computed by number distribution, and the sum total volume of a bubble is calculated by volume distribution. Moreover, it is desirable to ask also for the central value (the median diameter and the modal diameter) of each distribution and the diameter of an average bubble. With commercial equipment, various statistics calculations are included in software, and a statistics value can be obtained only by menu selection.

2.3.4.5.3 Report and record for result

As for the measurement result of the bubble by electrical sensing zone method, it is desirable to write the matters following at least in a report. Moreover, if it is possible, it is desirable to make digital pulse shape data record with the diameter data of a bubble.

1. Sample information
2. Procedure in dispersion of sample, method, etc.
3. Starting time, measuring time
4. Used aperture diameter
5. Setting conditions of instrument (current of aperture, amplification rate, calibration factor K_d, etc.)
6. Frequency and cumulative size distribution of bubble (number standard, volume standard or area standard)
7. Frequency list of the bubble size distribution, the statistics result of the bubble size distribution

2.3.4.6 Notes on measurement

2.3.4.6.1 Removal of electric noise

It is desirable to install the instrument in environment clean in electromagnetism and to attach a noise filter to the power supply by a case. It is necessary to ensure ground installation of main part. Since a 20 μm aperture is used for measurement of a nanobubble, cautions are required for especially an electric noise.

2.3.4.6.2 Contamination influence form solution

By electrical sensing zone method, although it is the feature to survey the number absolutely of bubble, compared with other measuring methods, the influence of contamination is also high, and cautions are required for mixing of a foreign substance enough in the process of measurement. Washing of tools (beaker, pipette, aperture, etc.) and management of electrolysis solution require cautions enough. The washed tool is an electrolysis solution to be used and is prewashing. Moreover, since even the air bubbles in electrolysis solution may be measured as contamination, when pouring an electrolysis solution into a measurement beaker, it puts in slowly so that a bubble may not form. Since the dissolved gas becomes the air bubble while measuring, it becomes a factor with error when the electrolysis solutions of private manufacturer are used, it is necessary to carry out decompression degassing of the solution beforehand. Decompression degassing is unnecessary when using the exclusive electrolysis solution on the market.

2.3.4.6.3 Washing aperture

At the comparatively small aperture (less than 100 μm), fine pores have easily got blocked with foreign substance particles. In order to prevent this, it is necessary to soak the aperture, which is not used in detergent, it is desirable to also soak the aperture currently used in the electrolysis solution for measurement. If the latest instrument carries the camera that observes the blocked aperture, taking out automatic warning when the aperture will be blocked, and there is a function to remove a jam automatically with contrary pressure. In addition, the block of aperture is removable by one of the methods below.

1. Adverse pressure: Pour liquid in the pressure from the opposite direction to the aperture until it cancels the block.
2. Heating: Liquid is boiled at heating by current of aperture, then the block can be removed.
3. Brushing: It is possible to clear blockages from the aperture tube using the short brush of hair. It is necessary to warn against doing damage to the aperture tube at this time.

4. Clearing blockages by ultrasonic: Where an electrolyte solution is filled to the aperture tube, the aperture portion is dipped in ultrasonic bath of a low power for about 1 s. This operation is repeated if needed. Although this method is effective, ultrasonic damage may be done to the aperture tube, it is necessary to carry out carefully.

Caution: It is better not to use ultrasonic cleaning for the aperture tube of less than 50 μm size.

2.3.4.7 Conclusion

The size distribution measuring method of the bubble is measuring the bubble size using various principles from a dynamic light scattering method to a high-speed image analysis method or the laser diffraction scattering method. However, these are measuring the bubble by two dimensions, and only the electrical sensing zone method is carrying out three-dimensional measurement, which surveys volume of the bubble. Thereby, there is no influence of the detection direction, and it can measure a slight volume change of the bubble by precise and high resolution. Since a bubble can be quantitatively measured while measuring the volume of the bubble, in order to measure the number and the size distribution of micro- and nanobubbles, the electrical sensing zone method is the most outstanding measuring method.

Moreover, since international standards particles are united with the prototype meter which is a standard of length, if a measuring device is proofread using this, the traceability, which used the diameter of a ball equivalent as the base, is also securable.

References

1. Measurement technology for particle size (Nikkan Kogyo Shimbun Ltd, 1994).
2. The light scattering diffraction particle-size-distribution measuring method of laser + polarization (Optical alliance 2000.8).
3. Application of the Electrical Sensing Zone method particle-size-distribution measurement machine in quality control of CMP slurry (Tribology planarization technology for CMP in semiconductor 2007.11).

4. The Association of Powder Process industry and Engineering, Japan (1994) Particulate engineering-the foundation of distribution and application.
5. ISO 13319 Determination of particle size distributions—Electrical sensing zone method.

2.3.5 Measurement of Size and Zeta Potential of Nanobubbles

Kazuhiro Nakashima
Sysmex Corporation, Kobe 651-2271, Japan
Nakashima.Kazuhiro@sysmex.co.jp

2.3.5.1 Introduction

There are various types of nanobubble generation systems that include ultrasonic and gyration, pressure dissolution, detailed hole by enclosing the gas in a solvent. Size and zeta potential of nanobubbles can be controlled by changing electrolyte concentration and pH in the solvent. The Zetasizer Nano of Malvern Instruments (Fig. 2.27) can measure the size and the zeta-potential of nanobubbles by using the dynamic light scattering (DLS) method and the laser Doppler method. This paper addresses the measurement principle and the function, the example of nanobubbles measurement with the Zetasizer Nano.

Figure 2.27 Zetasizer Nano of Malvern instruments.

2.3.5.2 Size measurement

In DLS, the diameter of particles is calculated by measuring Brownian motion of particles in a solvent. The Brownian motion is the random motion of particles, which happened by the collision of the solvent molecules and the particles; it is known that small particles move quickly, and large particles move slowly in a solvent.

The size of the particle is given by the Stokes–Einstein equation Eq. (2.10) with the diffusion coefficient. In DLS, the diffusion velocity of the particles by Brownian motion is measured. This is performed by measuring fluctuation of scattering light intensity using the suitable optical device. Figure 2.28 shows correlation curve of the typical size of particles. Since the Brownian motion of large particles is slow and the fluctuation of scattering light intensity changes slowly, the correlation will persist for a long period of time. Moreover, since the Brownian motion of small particles is fast and the fluctuation of scattering light intensity changes quickly, the correlation will reduce for a short period of time. The diffusion coefficient of particles is calculated by the inclination of the correlation curve, and the diameter of particles is computed using the Stokes–Einstein equation (Eq. (2.10)). Moreover, it is shown that the monodispersity of particles is the rapid inclination of the correlation curve. It is shown that the polydispersity of particle is the slow inclination of the correlation curve. The diameter of particles is calculated by the correlation function using various algorithms.

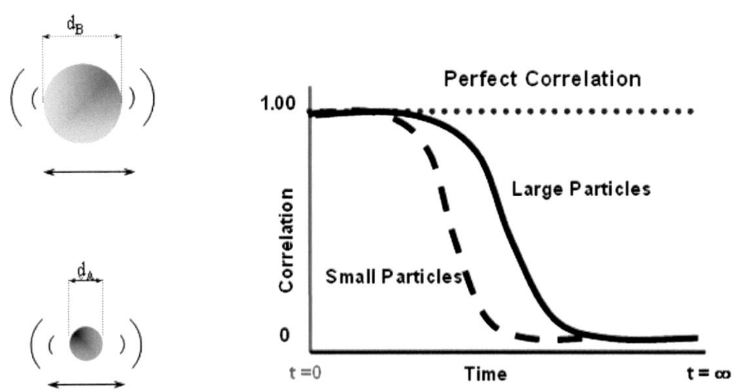

Figure 2.28 Correlation curve of the typical size of particles.

1. Cumulant analysis: The method of estimate of Z-Average (mean diameter) and polydispersity index (PDI) by applying the single exponential function to the correlation function (ISO13321 Part 8).
2. Non-negative least squares (NNLS) analysis: The method of acquire of the diameter distribution by applying the plural exponential functions to the correlation function.

The diameter distribution of particles is the plot of the relative intensity of the scattering light of the various size particles.

$$d(H) = \frac{kT}{3\pi \eta D} \tag{2.10}$$

$d(H)$ = hydrodynamic diameter
D = translational diffusion coefficient
k = Boltzmann's constant
T = absolute temperature
η = viscosity

2.3.5.3 Zeta potential measurement

Increasing the charge of the particle surface, the distribution of ion surrounding the interface domain is affected, and, as a result, the concentration of opposite ion (the electric charge of particles and opposing electric charge ion) will rise near the particle surface. For this reason, the electrical double layer is formed in the surroundings of each particles. Moving the particles by an electric field, the ion in an electric double layer move with the particles, and the potential of outside of the electric double layer is the zeta potential. Figure 2.29 shows schematic representation of zeta potential. The zeta potential is an index of the disperse stabilisation in colloid system. The zeta potential of all suspended particles—positive/negative—when one of big values are shown, the restitution between particles is large and the diameter distribution of particles is stabilised over a long period of time.

Moreover, if the zeta potential of particles is low, the restitution between particles will become small and aggregation of particles, adhesion on a container will occur. If the electric field is impressed, particles can be drawn near to the electrode of an opposite mark, the viscous force influence to particles resist this motion easily. If

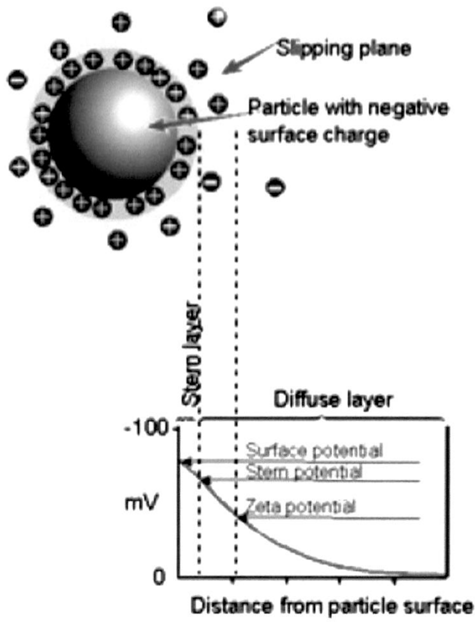

Figure 2.29 Schematic representation of zeta potential.

two power of being opposed to each other reaches equilibrium, the motion of particles become uniform velocity. The factors that influence uniform velocity are the electric field intensity and the voltage slope, the dielectric constant of a solvent, the viscosity of a solvent and the zeta potential.

The relation between the zeta potential and the electrophoretic mobility of particles in an electric field is shown by the following Henry equation (Eq. (2.11)).

$$U_{\mathrm{E}} = \frac{2\varepsilon Z f(\kappa a)}{3\eta} \tag{2.11}$$

U_{E} = electrophoretic mobility
Z = zeta potential
ε = dielectric constant
η = viscosity
$f(\kappa a)$ = Henry's function
κ^{-1} = termed the Debye length
a = radius of particle

The electrophoresis measurement of zeta potential is the general method used to the middle electrolytic concentration of aquarius solvent. In this case, $f(\kappa a)$ set to 1.5, and the formula is called Smoluchowski approximation. The Zetasizer Nano measures by combining M3-PALS technology to the laser Doppler method.

Mixed-mode measurement is the method of changing the reversal cycle of the electric field which is fast-field-reverse (FFR) and slow-field-reverse (SFR) (Fig. 2.30). FFR can measure the electrophoretic mobility of particles without the influence of electric osmosis.

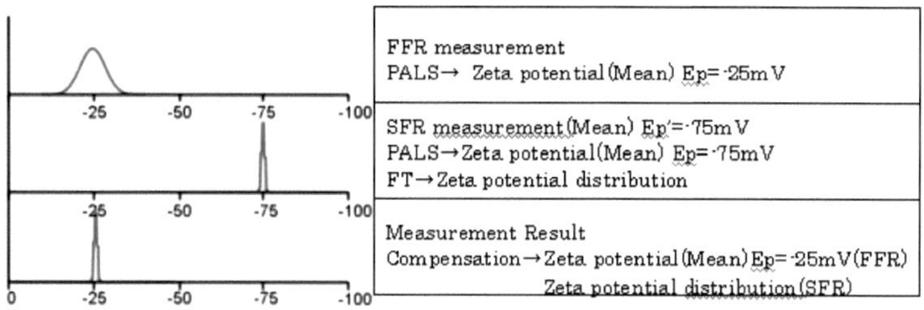

Figure 2.30 Measurement process: series of FFR and SFR.

The Zetasizer Nano calculates the average of electrophoretic mobility by analysing the phase shift of scattering light pitch with the phase analysis light scattering (PALS) method for FFR. SFR can improve the resolution of the electrophoretic mobility, because there are also many amounts of samplings in the electric field and the electric osmosis is stabilised. Then, by conducting frequency analysis using Fourier conversion, while analysing the exact distribution of the electrophoretic mobility, the average of electrophoretic mobility is computed by the PALS method like FFR. Then, Zetasizer Nano calculates the exact distribution of the electrophoretic mobility by analysing the frequency with the Fourier conversion and computes the average of electrophoresis mobility with the PALS method of FFR.

The difference of the average of electrophoretic mobility computed by SFR and FFR is the speed of the electrical osmosis, and it can measure the zeta potential without the influence of an electrical osmosis by rectifying the exact distribution of the electrophoresis mobility obtained by Fourier conversion. PALS analyses the abnormal conditions of the pitch (phase shift) by the interference of reference light and scattering light (Fig. 2.31). In analysing the phase shift of scattering light pitch, the electrophoretic mobility is correctly computed except for the factors (Brownian motion, particle sedimentation, the convection of suspension, etc.). The Zetasizer Nano detects in high sensitivity the electrophoretic mobility of the particles, which are the low-charge particle surface in a high-concentration electrolyte or non-polar solvent.

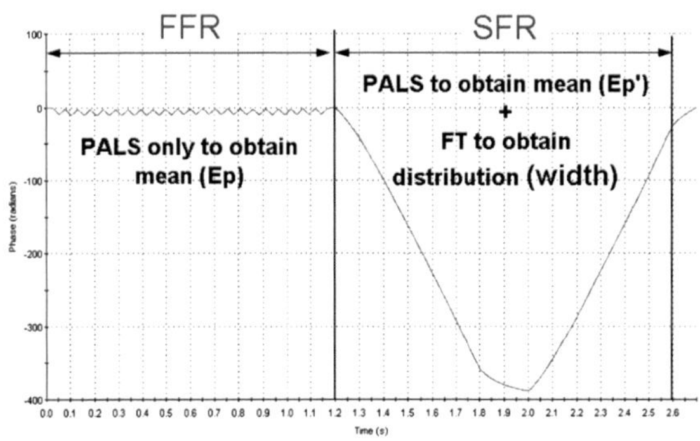

Figure 2.31 Phase plot.

2.3.5.4 Nanobubble measurement

Figures 2.32 and 2.33 show the diameter and the zeta potential of the nanobubbles that are generated from the nanobubble generation equipment BUVITAS made from Kyowa corporation by using Zetasizer Nano ZS (green laser: the wavelength of 532 nm, the dip cell used). According to Fig. 2.32, it was thought that the correlation curve was the reliability data from the intercept of correlation coefficient showed high value near 1.0 and the high S/N ratio without the influence of the noise.

Moreover, since the correlation curve of attenuation intensity changed slowly, the correlation will persist for a long period of time. Z-Average (mean diameter) computed from the cumulant analysis was 168 nm, and PDI was 0.398 as the large size and polydispersity result.

The mean diameter and the distribution computed from the NNLS analysis showed that the main peak was 168 nm, and the half of width was 105 nm as comparatively broad distribution.

From the result of the cumulant analysis and the NNLS method, comparatively broad distribution ware obtained.

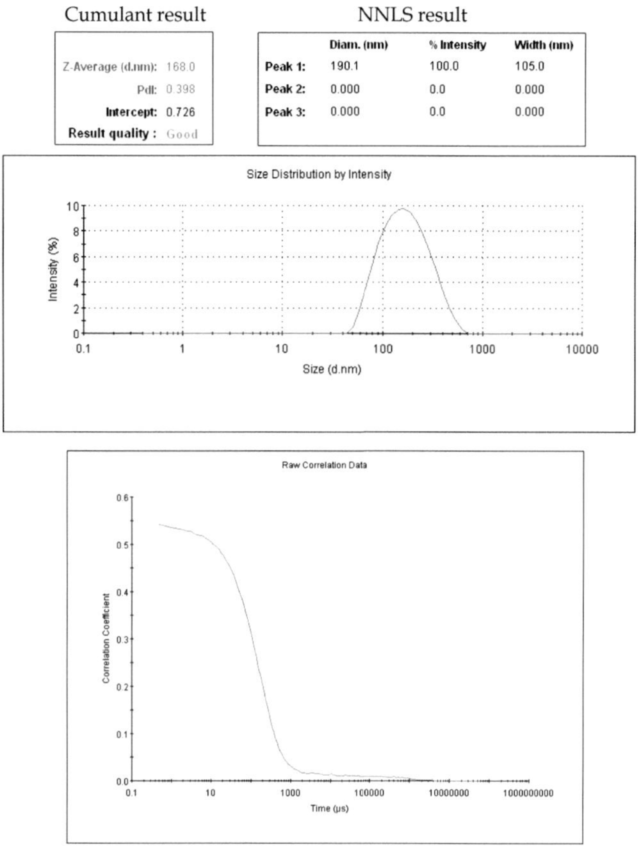

Figure 2.32 Result of diameter distribution of nanobubbles (upper: distribution, lower: correlation curve).

According to Fig. 2.33, the reliability data were obtained because nanobubble followed the change of an electric field FFR and SFR and the phase was clear. It was obtained that the distribution of zeta potential was comparatively as sharp as the average of zeta potential was −13.7 mV and the standard deviation of zeta potential distribution was 2.93 mV.

Since there was no diameter parameter of particles in the Henry (Eq. (2.11)), the zeta potential distribution was comparatively sharp as the mobility of the particles in the electric field were gathered in spite of the diameter distribution was broad.

			Mean (mV)	Area (%)	Width (mV)
Zeta Potential (mV):	-13.7	Peak 1:	-13.7	100.0	2.93
Zeta Deviation (mV):	2.93	Peak 2:	0.00	0.0	0.00
Conductivity (mS/cm):	0.175	Peak 3:	0.00	0.0	0.00
Result quality :	Good				

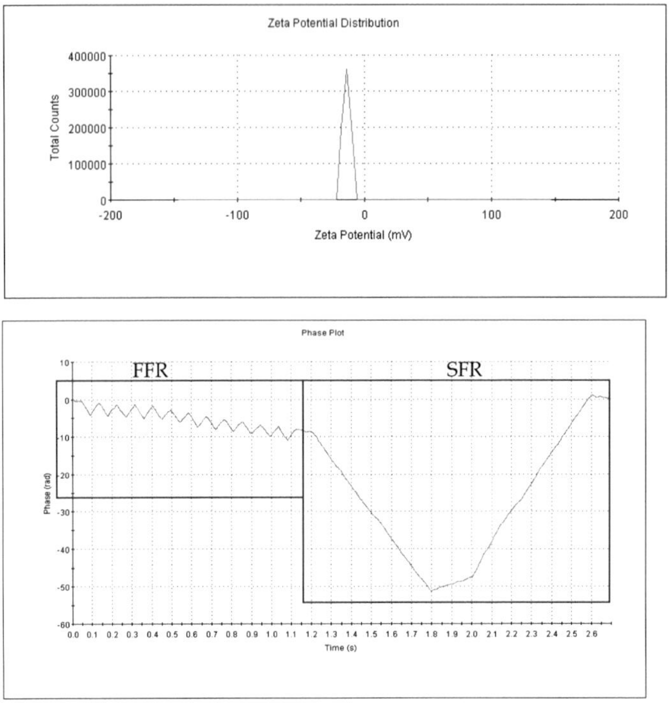

Figure 2.33 Result of zeta potential of nanobubbles (upper: zeta potential distribution, lower: phase plot).

2.3.6 Particle Size Distribution of Nanobubbles by NanoSight System

Ayako Irie

Advanced Technology Div. Quantum Design Japan, Tokyo, 171-0014, Japan

irie@qdj.co.jp

The information of nanobubble (NB) size and concentration become important to verify the effect of NB. In existing systems, transmission electron microscopy, dynamic light scattering and laser diffraction system, it is difficult to measure the NB size and concentration at the same time. Assumed size and concentration measurement of NB was performed using the "NanoSight system" which is able to track individual nanoparticles moving under Brownian motion and relates the movement to a particle size according to the following formula derived from the Stokes–Einstein equation. As a result, it checked that the model size of particle size distribution was 72 nm, and the total concentration was 3.28E8 particles/mL. From this result, it was suggested that it is possible to measure the size and concentration of NB using "NanoSight system". Therefore, initial studies using the NanoSight system would suggest that it is useful for NB generator's optimisation, NB's physical studies and so on.

2.3.6.1 Introduction

Microbubble/nanobubble (NB) is of interest due to a specific ability which a bubble has, that is cleaning effect. Various application examinations are done in many kinds of business fields. However, scholarly knowledge is low, for this reason, this research is still developing. It is clear that the information of bubble size and concentration is very important to study the effect that bubble have, but the measurement method is not established.

There are already well-defined techniques for the analysis of nanoparticles. Transmission electron microscopy is used for observation of the particle shape and for measurement the size of the nanoparticles. This is not suitable to measure NB, as samples must be analysed in a reduced pressure environment. Therefore, particle size distribution system, this is the technology that considers that bubble is solid particles and measures them, is made available for measurement the NB size.

Particle size distribution system is developed to measure the particle size distribution of particles. That means the result from existing particle size distribution system is displayed "frequency". So, it cannot be counted the particle number directly, and this system needs to input a refractive index for every measurement particle. A refractive index has a close relation to density. There is a question: "May I use the refractive-index value of the gas in atmospheric pressure to a NB with unknown internal pressure?" When measuring known solid particles, it is satisfactory, but to measure a strange substance like an NB, it is necessary to pay attention to the input of physical properties information.

Dynamic light scattering (DLS): it is known the photon correlation method. Measure fluctuations within the scattered light intensity from a sample result from the Brownian motion of the particles within a sample. Application of the Stokes–Einstein equation relates the speed of particle movement to particle size in a sample of known viscosity and temperature. DLS performs well in a mono-modal populations but is highly influenced by a small number of larger particles that scatter a disproportionately large amount of light compared with a smaller particle (light scatter is proportional to the ^6 of the radius in the Rayleigh regime).

It is thought that an NB, size is under 1000 nm, has distribution width, and exists in liquid. Moreover, in NB liquid, particulates other than the bubble, that is dirt, which arose at the time of bubble generation may be included. Thus, it is difficult to measure the actual size distribution and concentration of NB. We introduce measurement of an NB's size, concentration and zeta potential (ZP) using a "NanoSight system", which uses a unique technology called "Nanoparticle Tracking Analysis (NTA)".

2.3.6.2 About NanoSight

2.3.6.2.1 Principles of operation

NTA uses the properties of both light scattering and Brownian motion in order to obtain particle size distributions of samples in liquid suspension. A laser beam (of arbitrary wavelength but typically those available from laser diodes operating at 635, 532, 488 or 405 nm) is passed through a prism edged glass flat within the sample chamber. The angle of incidence and refractive index of the glass flat is designed to be such that when the laser reaches the interface between the glass and the liquid sample layer above it, the beam refracts to an intense low profile resulting in a compressed beam with a reduced profile and a high power density. The particles in suspension in the path of this laser beam scatter light in such a manner that they can be easily visualised via a long working distance, ×20 magnification microscope objective fitted to an otherwise conventional optical microscope objective onto which is mounted a charge-coupled device (CCD), Electron Multiplying Charge Coupled Device (EMCCD) or high-sensitivity Complementary Metal-Oxide Semiconductor (CMOS) camera, operating at 30 frames per second. This captures a video file of particles moving under Brownian motion within a field of view of approximately 100 μm × 80 μm × 10 μm (Fig. 2.34). NTA can find out the difference between a particle structure and a material from the difference in the dispersion intensity on the surface of a particle [1].

2.3.6.2.2 Tracking in two dimensions

Within the field of view, particles are seen moving under Brownian motion, either directly by eye using the microscope oculars or via the camera. The proprietary NTA software records a video file of the particles viewed and then simultaneously identifies and tracks the centre of each particle on a frame-by-frame basis. The image analysis software then determines the average distance moved by each particle in the *x* and *y* directions. This value allows the particle diffusion coefficient (Dt) to be determined from which, if the sample temperature T and solvent viscosity η are known, then the sphere-equivalent hydrodynamic diameter d of the particles can be identified using the Stokes–Einstein equation (Eq. (2.12)) [1].

$$Dt = \frac{TK_{\mathrm{B}}}{3\pi\eta d} \tag{2.12}$$

where K_{B} is Boltzmann constant.

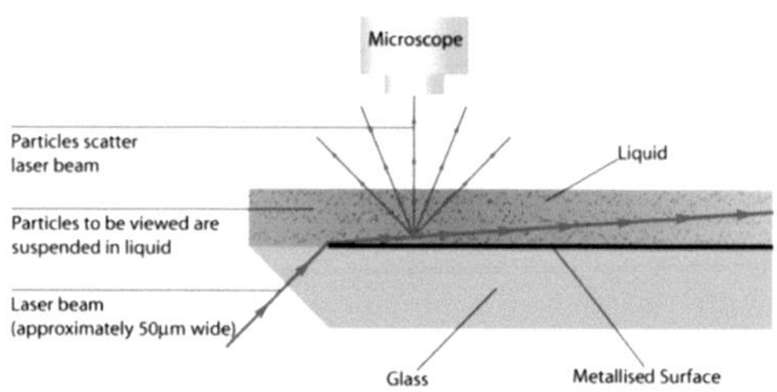

Figure 2.34 Schematic of the optical configuration used in NTA.

2.3.6.2.3 Motive forces

It is possible to apply an electric field through the sample by use of immersed electrodes. This induces electrophoresis of charged particles from which their ZP, ζ, can be determined. This parameter is central to colloidal performance and stability in a wide range of nanoparticulate materials and products, a high surface charge-minimising aggregation of the dispersion because of charge repulsion. Thus, a measure of ZP determines tendency for sample aggregation over time but varies with pH, temperature, concentration and particle size.

NTA is capable of measuring directly, in real time and on a particle-by-particle basis, the electrophoretic velocity and polarity of charged nanoparticles under an applied field. However, unlike conventional high-frequency ZP analysis measuring instruments, applying a constant electric field through an aqueous liquid causes a parabolic flow profile in the sample chamber due to electro-osmosis, and it is independent of charged particle motion (Fig. 2.35). A particle's velocity observed by NTA at any point will therefore be a combination of both solvent electro-osmosis and particle electrophoresis [1].

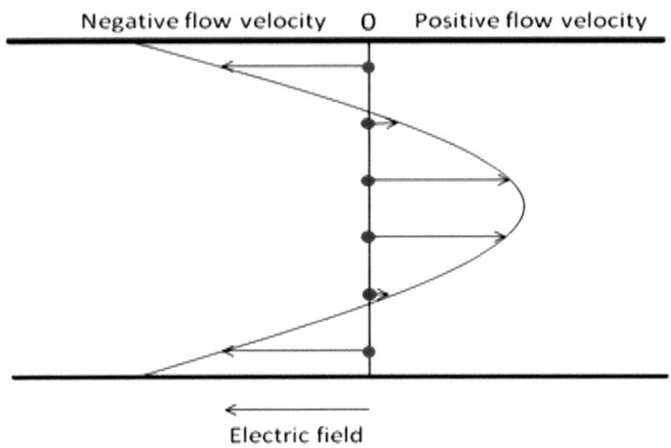

Figure 2.35 The solvent electro-osmotic profile generated within the NTA closed system, which must be measured and subtracted from the total nanoparticle motion detected to allow the nanoparticle electrophoresis to be isolated and from which its zeta potential can then be determined.

2.3.6.3 Experiment and result

To measure the size and number concentration of NBs, NanoSight LM20 (NanoSight Ltd.) was used. NB solution was created by introducing O_2 gas into pure water with NB generator (Kyowakisetsu. Co. Ltd., BUVITAS). To measure ZP of NBs, NanoSight NS500Z (NanoSight Ltd.) was used. NB solution was produced by introducing air gas into pure water with NB generator (IDEC. Co. Ltd., NanoGALF).

NBs containing O_2 gas and blank sample were injected into laser module using the syringe and the light scattering observed. In NBs sample, a lot of light scattering in the view was observed. But in blank sample, nothing was observed. Analysis of NBs was carried out using "Nanoparticle Tracking Analysis (NTA)" software. In the result, the number concentration of bubble was estimated to be 3.0E8 particles/mL, and the size and number concentration of mode peak was 72 nm and 4.6E6 particles/ml, (Figure 2.36) ZP was –53 mV. Also, the particles that have strong intensity in sample liquid could not be seen, but it was assumed that the checked light scattering were not foreign substances, such as metal powder that arose at the time of generation.

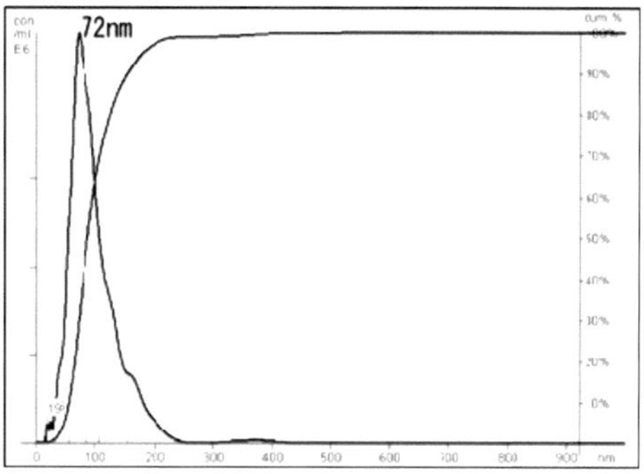

Figure 2.36 Particle size distribution of O_2 NB sample. Mode size and concentration = 72 nm, 4.6E6 particles/mL. Total concentration = 3.0E8 particles/mL.

2.3.6.4 Conclusion

It became clear that initial studies using the NanoSight system would suggest a simple method of measuring the size and concentration of NB.

Moreover, by a ZP measurement result being added, it is assumed that it can be used for optimisation of NB generation, its physical study, etc.

Acknowledgement

This work was supported by NanoSight Ltd., UK. NBs sample was offered by Kowakisetsu, Co., Ltd. Japan and IDEC Co., Ltd. Japan.

References

1. Wright M, Suipa A, Sullivan J et al (2012) Nanoparticle tracking analysis (NTA) for the multiparameter analysis of nanoparticles in liquids, Part 1: Principles and Methodology.

2. M117F Application Note, NanoBubble (2009) NanoSight Ltd., UK.
3. M121 A Application Note, ZetaPotential (2011) NanoSight Ltd., UK.
4. Vasco Filipe et al (2010) Critical evaluation of nanoparticle tracking analysis (NTA) by nanosight for the measurement of nanoparticles and protein aggregates, *Pharm Res*, **27**(5), 796–810.
5. Ayako Irie (2010) The Latest Technology of Microbubbles and Nanobubbles 2. pp. 67–70 (in Japanese).

2.3.7 Particle Image Velocimetry System

Masashi Yasuki
Seika Corporation, Tokyo 100-0005, Japan
myasuki@jp.seika.com

2.3.7.1 Introduction

Particle image velocimetry (PIV) is a technique to measure velocity based on displacement of small particle within small time separation.

As PIV can be used for both gas and liquid flow, it is widely used among many R&D fields.

Originally, it was developed for aerodynamic measurements for aircrafts etc, and now it is widely used from large-scale measurements such as full-scale wind tunnel testing for automobile to microscale measurements such as microfluidic research in bio- and life science field. It has become necessary measurement system in fluid dynamics.

This paper explains the principle of PIV, recent system configuration of PIV system and its performance.

2.3.7.1.1 Principle of PIV

In PIV, small particles are seeded to the flow in advance and visualised by illuminating with so-called laser light sheet, which is generated by expanding the laser beam to planer state.

Taking two photographs with short time separation and applying cross correlation to these two images to get thousands of near instantaneous velocity vectors for whole images. From these velocity vectors, it is possible to calculate several flow parameters, such as vorticity and flow strains. Furthermore, parameters such as turbulent intensity can be obtained by averaging the several tens and several hundreds of these instantaneous velocity fields.

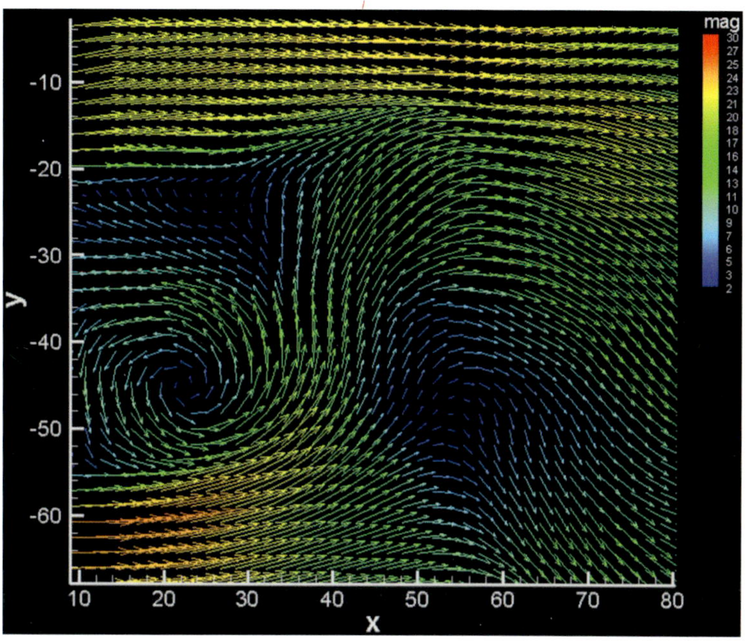

Figure 2.37 Velocity vector map by PIV.

2.3.7.1.2 Image capture by frame straddling

In PIV, the flow is visualised by illuminating with laser light sheet, and particle image is taken by a camera placed normal to the light sheet. At this time, two images with small time separation are captured. To do this, a double pulse laser that can fire two pulses with flexible time separation and a camera that has a double shutter function are used in combination.

A double shatter camera is a digital CCD camera or digital CMOS camera, which can capture two images with very short time separation. Using it in synchronisation with a double pulse laser, it is possible to take two images with minimum time separation of down to about 70 ns (minimum time separations differ according to manufacturers and models). The actual time separation between two images is decided by the timing of firing double pulse, not by the time between two frames. This synchronised imaging is called frame straddling.

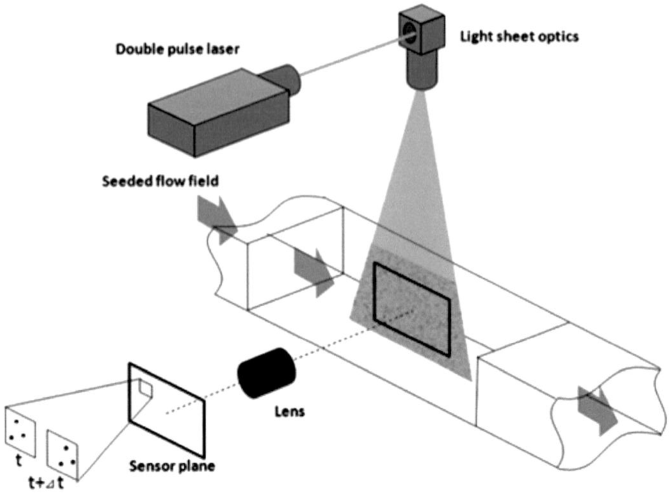

Figure 2.38 Principle of PIV.

It is possible to apply any velocity range by setting the time separation so that the average particle displacement between two images is several pixels apart.

2.3.7.1.3 Seeding

To visualise flows, smokes or small particles are mixed to flows. This is called seeding, and particles used for seeding is called tracer particles.

For air or gas flows, small droplets of about 1 micron in size are used, and a Laskin nozzle-type seeding generator is most useful for seeding. For water or liquid flows, solid particles or microbubbles are used.

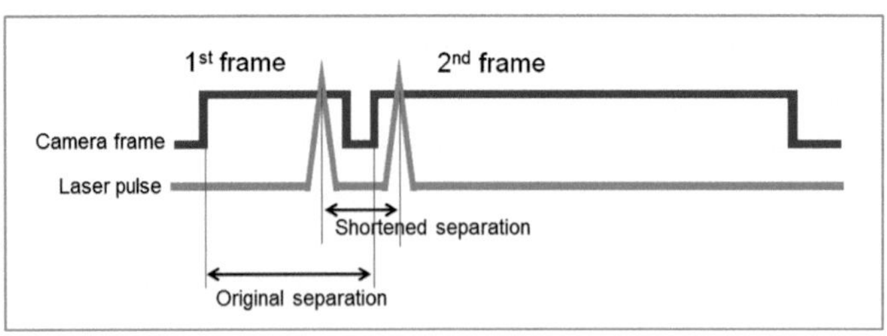

Figure 2.39 Frame straddling.

Various material of solid particle, such as Nylon, Polystyrene and glass particles, is used depending on specific gravity of liquid.

Particle size varies depending on measurement area size, spatial resolution of camera sensors, etc.

It is useful to use microbubbles as a seeding for flow measurements with large quantity such as water tunnels. Microbubble does not require collecting and cleaning after measurements.

In such cases as spray application, seeding is not required because the measuring object itself is small particles.

However, even in the spray application, seeding is required for measuring regions outside of spray or entrainment of the spray.

In any cases, one must be careful that excessive seeding may alter the flow property.

2.3.7.2 PIV system

Here, a SEIKA Koncerto2D-PIV system is explained as an example of standard 2D-PIV system, which is currently most generally used.

2.3.7.2.1 System configuration of 2D-PIV system

2.3.7.2.1.1 Software

Koncerto software is integrated PIV software, which performs all processes required in PIV, such as hardware control of lasers and cameras, digital filters for image processing, pre-processing of background subtraction, etc., velocity evaluation of PIV processing, post-processing of data validation and data visualisation.

2.3.7.2.1.2 Timing controller

A timing controller Seika TT1680 has eight physical output channels, eight logical output channels and eight physical input channels. It synchronises not only PIV system itself but also a more complex synchronisation such as rotating machineries.

2.3.7.2.1.3 Double pulse laser

A double pulse laser consists of two set of independent frequency-doubled pulsed Nd:YAG lasers. Two beams are combined by polarising optics to single beam. Thus, the time separation of two pulses is completely flexible. Typical specification of each laser is as follows. Wave length is 532 nm (green), and pulse duration is 5–10 ns. Repetition rate is 10–15 Hz. Pulse energy is 10–200 mJ.

2.3.7.2.1.4 *PIV camera*

PIV camera is digital CCD of CMOS camera, which must be capable of frame straddling mentioned before.

Typical specification is as follows. Spatial resolution is 1–5 Megapixel. Frame rate is 5–100 frames per second. It can capture two frames with a minimum time separation of 1 ms. Time resolution of PIV is less than a half of frame rate. These specifications are depending on models.

2.3.7.2.1.5 *Light sheet optics*

Light sheet optics is to generate laser light sheet to illuminate flow field with required image area. Seika BZ-60 light sheet optics is adjustable sheet angle and focal length.

2.3.7.2.2 *PIV analysis*

2.3.7.2.2.1 *Cross correlation analysis on Koncerto*

Koncerto employs variety of PIV analysis algorithms based on FFT cross correlation. Here, the concept of these methods is explained.

Standard FFT cross correlation Figure 2.40 is the process of standard FFT cross correlation. An image A in a drawing is the first image and an image B is the second image. Two-dimensional FFT is applied to the interrogation area, which is the small area typically 32 × 32 pixels in both images with same location to obtain a correlation plane. Next, a peak search is performed to correlation plane, and a correlation peak is obtained. Then, the final displacement is obtained by sub-pixel fitting.

Double (multiple) correlation Double (multiple) correlation is a technique to enhance a signal-to-noise ratio of cross correlation. Multiplying two correlation planes, which are obtained from two interrogation windows close to each other, to increase regular signal and decrease irregular noise. Number of windows, rate of overlap and direction (horizontal or vertical) are selectable by the software.

Multiple pass correlation Multiple pass correlation is a technique to increase the accuracy of correlation. It is so-called iterative method.

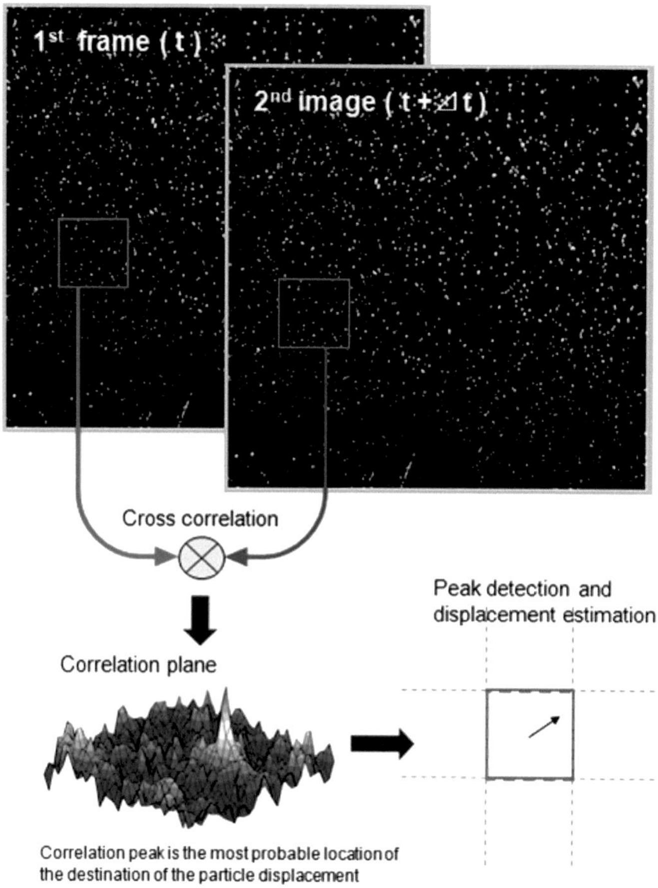

Figure 2.40 Standard FFT correlation.

The result of standard cross correlation is used as a predictor to shift an interrogation window in a second image and apply standard cross correlation again to get higher correlation peak than previous time. Applying this procedure several times iteratively, we get accurate result. Number of iteration is selectable by the software.

Multiple grid correlation Multiple grid correlation is also a technique to increase the accuracy of correlation. It is so-called hierarchical method. Same as multiple pass correlation, the iterative operation is applied, in each iteration, the interrogation window are

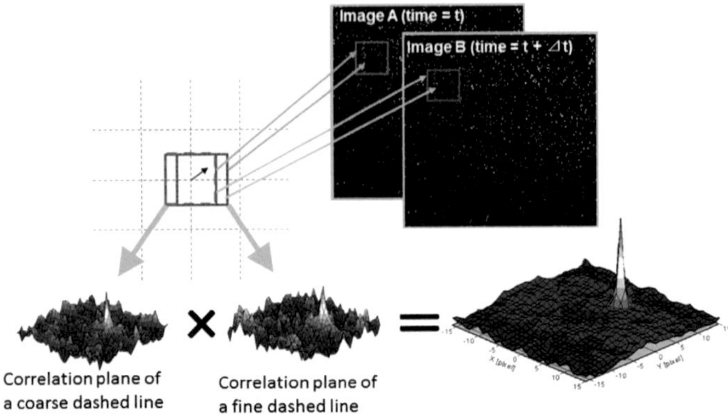

Figure 2.41 Double (multiple) correlation.

divided by four. It has effect of extending dynamic range, increasing spatial resolution and applicable in the case of rotation in an interrogation window.

Image deformation correlation Window shifting by a predictor in the multiple pass correlation and the multiple grid correlation mentioned before is a shift by pixels. In the image deformation correlation, surrounding multiple predictors are used to deform images. Thus, sub-pixel image shift is possible. It is applicable if there is a deformation and/or rotation in an interrogation window, and more accurate result is obtained.

Central difference interrogation The iterative operations, such as multiple pass correlation, multiple grid correlation and image deformation correlation mentioned before, are explained based on forward difference interrogation for the simplicity. Central difference interrogation is a technique to apply window shifting for both images. An interrogation window in second image is shifted by half of a predictor, and an interrogation window in first image is shifted back by half of predictor. The better result can be obtained in such a measurement case that near wall and obstacles in the flow.

2.3.7.2.2.2 Post-processing

Post-processing is after processing of cross correlation analysis, and velocity vector validations, statistical processing and data display are included.

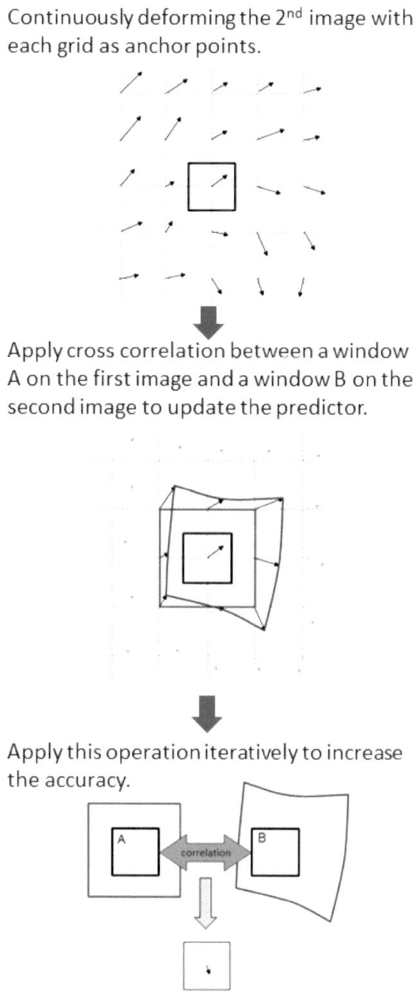

Figure 2.42 Image deformation correlation.

Velocity vector validations In PIV, several thousands of velocity vectors can be obtained from an image pair. However, all these vectors are not always analysed correctly and error vectors are generated by various factors.

A vector validation first determines correct or error of a vector, remove it if it seems erroneous and apply interpolation to fill a hole.

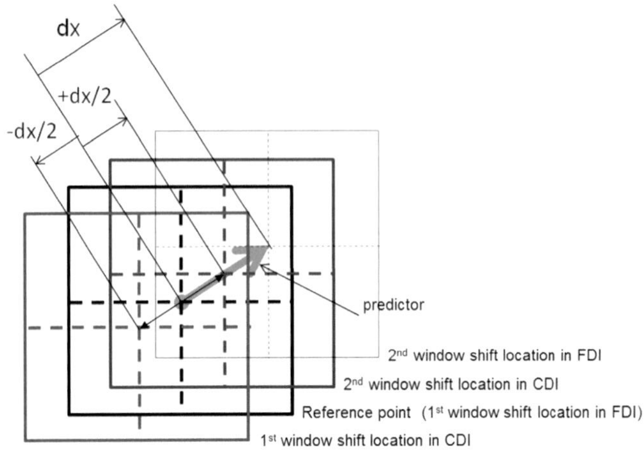

Figure 2.43 Central difference interrogation.

There are two types of validation filters, a global filter and a local filter. Global filters are applied to all vectors such as a standard deviation filter and a max displacement filter. Local filters such as median filter and average filter are applied to local areas typically 3×3 vectors.

Statistical processing and calculation of flow parameters Average vector field and Root mean square (RMS) vector field can be calculated as statistics. Calculation of flow parameters such as vorticity and turbulent intensity is possible.

Statistical processing and calculation of flow parameters Average vector field and Root mean square (RMS) vector field can be calculated as statistics. Calculation of flow parameters such as vorticity and turbulent intensity is possible.

Automated operation Batch processing and script processing are possible as an automated operation to apply various PIV processing to multiple images.

Various PIV systems Lastly, other PIV systems are introduced. A stereo PIV system can measure three velocity components in two-dimensional plane. A time-resolved PIV system is a very high time resolution (up to about 40 kHz) PIV system, which employs a high-speed video camera and a high-repetition rate pulse laser. Micro-PIV can measure microscale velocity field such as microfluidic ships using optical microscope. Volume PIV system is developed to measure three component measurements in a three-dimensional cube. These systems can be combined to integrate complex PIV systems.

References

1. Raffel M, Willert C, Kompenhans J (1998) *Particle Image Velocimetry, A Practical Guide*, Springer, Berlin.
2. The Visualization Society of Japan (2002) PIV hand book, Morikita Publishing Co., Ltd.

Chapter 3

Sonochemistry: Influence of Acoustic Flow and Mechanical Flow on Sonochemical Efficiency

Yoshihiro Kojima[a] **and Shinobu Koda**[b]
[a]*EcoTopia Science Institute, Nagoya University, Aichi 464-8603, Japan*
[b]*Department of Molecular Design and Engineering, Graduate School of Engineering, Nagoya University, Aichi 464-8603, Japan*
ykojima@esi.nagoya-u.ac.jp

3.1 Introduction

The sequential growth and collapsed processes of ultrasonic cavitation bubbles lead to the generation of local fields with enormous pressures and temperatures so that physical and chemical effects emerge at local positions in liquids inside sonochemical reactors. The engineering processes utilising these effects are termed sonochemical processes. The application of the sonochemical process to the decomposition of pollutants and synthesis of chemical compounds in solutions using a laboratory-scale sonochemical reactor have been demonstrated [1,2]. In order to put the sonochemical processes of waste treatment and material processing to practical use, design

Micro- and Nanobubbles: Fundamentals and Applications
Edited by Hideki Tsuge

ISBN 978-981-4463-10-2 (Hardcover), 978-981-4463-11-9 (eBook)
www.panstanford.com

and development of large-scale or pilot-scale sonochemical reactors have been studied by several groups [3–9]. It is well-known that sonochemical reactions are influenced by many factors such as ultrasonic frequency, temperature, dissolved gas species and the physical properties of media [10]. For further development of sonochemical reactors with the high sonochemical efficiency, it is necessary to take account of other factors such as the volume, shape of a reactor and liquid flow. We indicated that the sonochemical efficiency depends on the volume or the height of the liquid [11]. Several researchers have explored the hydrodynamic behaviours and the modelling of velocity in a reactor under sonication [12–16] because ultrasonic irradiation into liquid causes acoustic streaming. The velocity profile is considered to provide important knowledge on the design of a large-scale sonochemical reactor. However, the understanding of the influence of velocity profiles in the reactor on cavitation intensity or the sonochemical efficiency is insufficient. The knowledge of the relationship between the velocity profiles and cavitation intensity or sonochemical efficiency may contribute to the development of a large sonochemical reactor.

Recently, we have investigated the behaviour of visualised bubbles and the velocity profile of liquids in a pilot-scale rectangular parallelepiped sonochemical reactor sonicated at 490 kHz [17]. In addition, we have made an attempt at the combination treatment of sonication with stirring and examined the influence of the mechanical flow involved with revolution of the stirrer on the velocity profile and the sonochemical reaction [17]. In this chapter, we will report and discuss the effect of the electric input power on the velocity profile and the sonochemical reaction field as visualised by sonochemical luminescence emission and the effect of liquid flow on the sonochemical efficiency.

3.2 Flow Profile and Sonochemical Efficiency

A rectangular parallelepiped vessel made of quartz glass with inner dimensions of W200 mm × D200 mm × H650 mm was employed as a reactor. A disc-shaped PZT (piezoelectric zirconate titanate) transducer mounted on a vibration plate of stainless steel was attached at the central position of the reactor bottom. The diameter of the PZT transducer is 50 mm. The liquid under sonication was subjected to a laser light sheet through the side wall of the reactor in order to visualise cavitation bubbles, acoustic streaming and standing waves in our sonochemical reactor. Figure 3.1 shows the photographs of cavitation bubbles under sonication in the electric power range of 5–50 W. The sonochemical luminescence images recorded by a digital camera are also shown in Fig. 3.1. Under sonication at 5 W, a significant number of bubbles are trapped at the antinodal points of the standing wave formed in the cylindrical zone shaped due to the narrow directivity of the 490 kHz ultrasonic beam and located above the transducer, and we will refer to the zone as the cylindrical zone. It is noted that the size of visualised bubbles is above several microns and may be around 1 mm as shown in Fig. 3.1a-1 because the interval of the standing wave is $\lambda/2$ ($\lambda = 3$ mm). Some of the large bubbles ascend towards the liquid surface and move to the reactor wall along the surface. Figure 3.1a-2 demonstrates that chemical luminescence emerges in the region where the standing wave is observed. The acoustic streaming generated by sonication at 10 and 20 W pushes up the visible bubbles towards the liquid surface as shown in Fig. 3.1b-1 and c-1. The chemical luminescence is also observed at an approximately identical location (Fig. 3.1b-2 and c-2), but progressively reshaped. The region expands horizontally and diminishes longitudinally. The standing wave is deformed by the effect of sound wave reflection from the liquid surface and side walls. The increase in electric power laterally widens the region of luminescence observed near the liquid surface. When the electric power exceeds 20 W, large bubbles are expelled from the cylindrical zone and occupies a zone near the side walls of the reactor and the liquid surface, indicating the existence of a diagonally downward liquid flow as described later. A shown in the photographs of Fig. 3.1d-1, e-1 and f-1, under sonication within the electric power range of 30–50 W, the droplets are generated, by ultrasonic atomisation. It is interesting that as shown in the photographs of Fig. 3.1d-2, e-2 and f-2, chemical luminescence can be observed not only in the bulk liquid near the liquid surface but also in the fountain. Moreover, it is found that in the case of 50 W, chemical luminescence also takes place in the region near the transducer (Fig. 3.1f-2). These results suggest that the active bubbles induced by ultrasound are in the fountain and the region near the transducer.

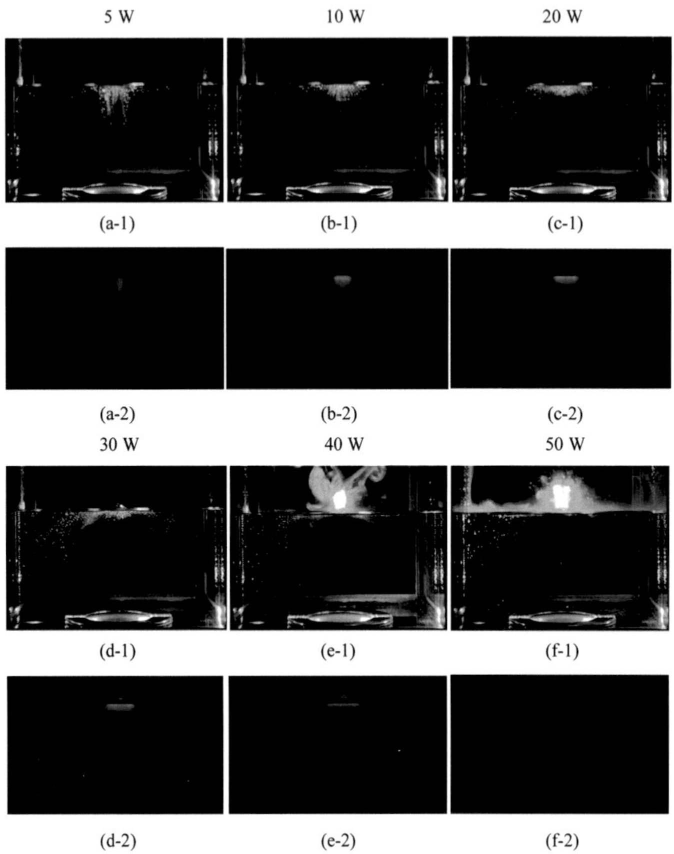

Figure 3.1 Photographs of (1) the flow profiles and (2) the chemical luminescence in a sonochemical reactor.

Further investigation of the distributions of the flow velocity and acoustic pressure contributes to an understanding of reaction fields in the reactor. The flow velocity of liquid was measured according to a laser doppler velocimetre. The nylon powders having the mean particle diameter of 4 μm and specific weight of 1.02 were employed as a tracer of flow. Figure 3.2 shows the distributions of the flow velocity and acoustic pressure in the sonochemical reactor under sonication at the electric powers of 10, 30 and 50 W. The arrow denotes the direction of liquid flow, and the difference in the magnitude of the average flow velocity is indicated by a greyscale. The greyscale in the background of the figure reveals the different intensity of the acoustic pressure. The results demonstrate that the velocity of the upward-moving stream, i.e., the acoustic streaming, changes according to the acoustic pressure gradient. At the low acoustic pressure corresponding to the applied power of 10 W, a slow random flow appears in the neighbourhood of the side wall. At the applied power of 30 W, an up-current of liquid becomes fast and its region widens. A further increase in the applied power, that is the acoustic power, leads to generation of convection currents as shown in Fig. 3.2. The streaming near the oscillator plate directs downwards. The velocity of acoustic streaming for a cylindrical reactor is expressed by the Eckart model [18]. Unfortunately, we cannot reproduce acoustic streaming for a rectangular sonochemical reactor. One of the reasons is that the distribution of flow velocity strongly varies with the position of liquid and input electric power because acoustic pressure depends on the position and input electric power. For example, Fig. 3.3 shows the histograms of the flow velocities measured for sonication with different power values at the height of 8 cm above the bottom of a reactor. The histogram becomes narrower and the mean velocity increases as the applied power increases.

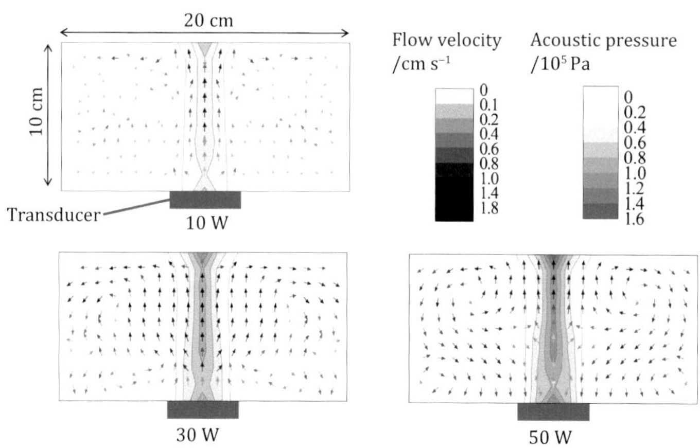

Figure 3.2 Distributions of the flow velocity and the acoustic pressure in a sonochemical reactor. We measured them only on the left side of the solution with respect to the concentric axis of the transducer. The figure involves the mirror image of the left side.

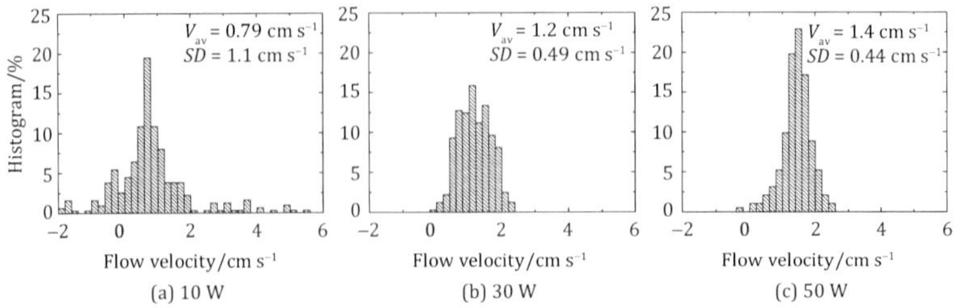

Figure 3.3 Histogram of the flow velocity in a sonochemical reactor ($h = 8$ cm), V_{av} and *SD*.

The sonochemical efficiency (*SE*) corresponds to the ratio of the number of reacted molecules m_I to the ultrasound energy E_{US} absorbed in solution for the irradiation time t_{US}. Thus, *SE* is defined as

$$SE = \frac{m_I}{E_{US}} = \frac{C_I}{P_{US}t_{US}/V_I}, \tag{3.1}$$

where C_I and V_I denote the I_3^- concentration of the 0.1 mol·dm^{-3} KI solution and solution volume, respectively and P_{US} is ultrasound power [19]. To evaluate the sonochemical efficiency under the same sonication conditions as the study of the acoustic flow, the rate of KI oxidation proceeding under sonication and the ultrasound power were measured. The sonochemical efficiency was calculated from the data according to Eq. (3.1). As shown in Fig. 3.4, the sonochemical efficiency decreases with an increase in the electric power but the decrease is slight or it seems to be almost independent of the electric power in the range of 40–50 W. Under sonication at the higher power, the decrease in the sonochemical efficiency is interpreted in term of the decrease in the number of active bubbles per unit energy consumption contributes to sonochemical reactions. As ultrasonic power increases, cavitation becomes active, that is the number of cavitation bubbles increases and the liquid velocity is accelerated. However above a certain acoustic power, some of bubbles coalesce and the active bubbles decrease. This leads to the decrease in sonochemical efficiency. Hatanaka et al. also suggest that sonochemical luminescence is suddenly quenched at excessive ultrasonic power [20]. On the other hand, however, the results shown in Fig. 3.2 indicate that at 50 W, convection currents are formed and flow from the side wall reign toward the inner part of the reactor. It is reasonable to state that this flow may effectively supply the acoustic field with reactants and cavitation nuclei which contribute to the formation of the active bubbles. As a result, in the case of the intensive electric power such as 50 W, the expansion of the luminescence emission region is expected to be observed in the luminol solution above the transducer as indicated in Fig. 3.1f. To enhance the sonochemical efficiency, one needs not only to reduce the chance that bubbles come together but also to widely supply the acoustic field with reactants and cavitation nuclei.

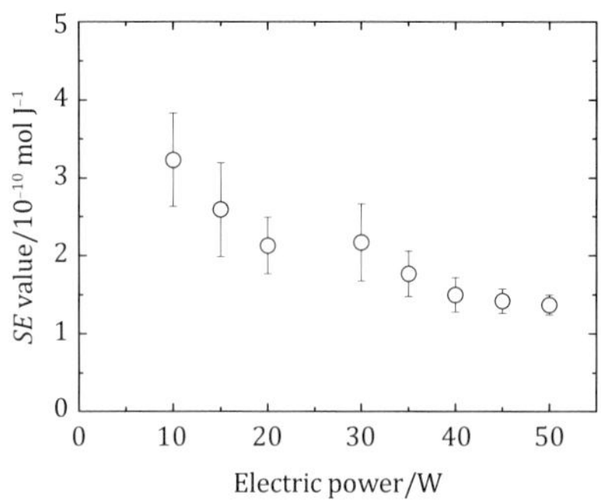

Figure 3.4 Sonochemical efficiencies (*SE*) at different electric powers.

3.3 Combined Treatment of Stirring with Ultrasonication

The results of Figs. 3.1, 3.2, 3.3 and 3.4 suggest that acoustic streaming or the flow profile influences sonochemical reactions or sonochemical efficiency. Therefore, information of visualisation of cavitation and distribution of liquid velocity under sonication with and without an additional mechanical force is expected to be useful to understand the sonochemical reaction field. Figure 3.5 shows the influence of additional stirring on the liquid flow profile under sonication at the electric power of 30 W at different speed values in the range of 0–350 rpm. In addition, when the stirring speed reaches 200 rpm, the increased mechanical flow enhances atomisation of the liquid from the fountain region. Figure 3.5c demonstrates that the trapped bubbles exist in the vicinity of the newly generated flow by the revolution of the stirrer. The number of these trapped bubbles increases as the revolutions increase. When the revolution attains 350 rpm, the fountain begins to incline and the liquid surface oscillates violently.

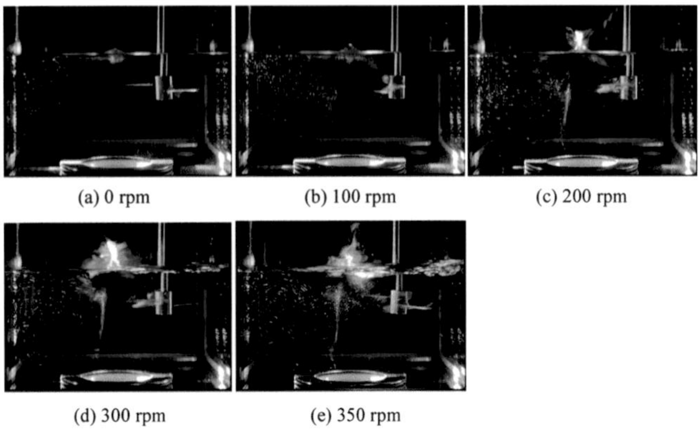

Figure 3.5 Photograph of the flow profiles in a sonochemical reactor by combined treatment of sonication with stirring (30 W).

Figure 3.6 shows the photographs of the chemical luminescence emerging under sonication with stirring in the revolution range of 0–350 rpm. The results demonstrate that the area of the luminescence increases as the revolutions increase. The comparison of Fig. 3.5 with Fig. 3.6 reveals that the newly expanded region of chemical luminescence by the increasing revolutions corresponds approximately to the zone where newly trapped bubbles are observed.

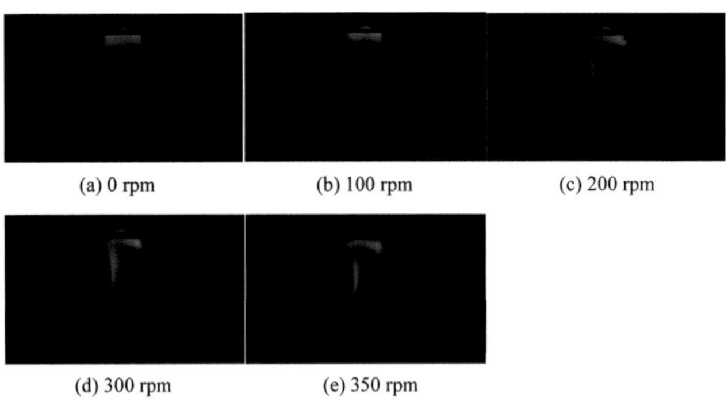

Figure 3.6 Photograph of the chemical luminescence in a sonochemical reactor by combined treatment of sonication with stirring (30 W).

Figure 3.7 shows the distribution of the average flow velocity and the reaction field characterised by chemical luminescence. The mechanical flow induced by the stirrer strongly disturbs the acoustic streaming. The flow velocity under sonication with stirring is about ten times as fast as that without the stirring.

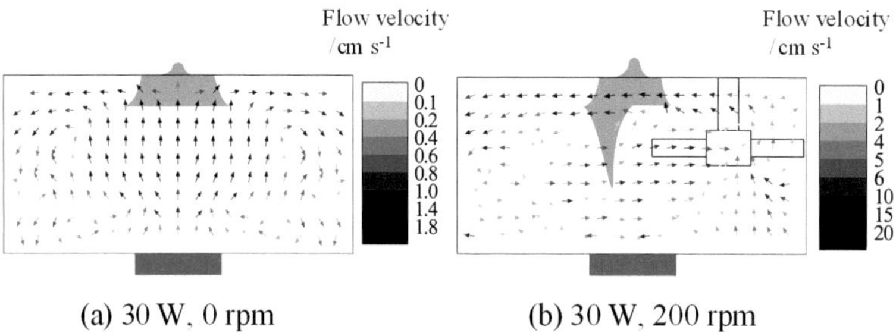

Figure 3.7 Distribution of the flow velocity and the reaction field in a sonochemical reactor by combined treatment of sonication with stirring (30 W, 200 rmp).

Figure 3.8 illustrates the results of the sonochemical efficiency under the different revolutions of the stirrer. The sonochemical efficiency increases with the revolution of the stirrer. The sonochemical efficiency is nearly twice as high at 350 rpm as that without mixing by the stirrer.

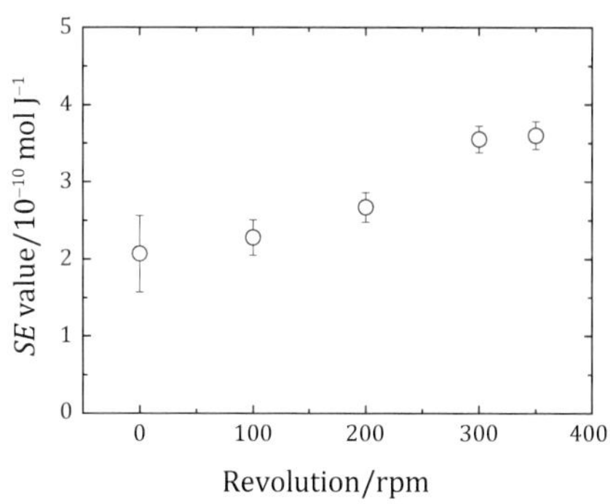

Figure 3.8 Dependence of *SE* value on the revolution of a stirrer in a sonochemical reactor.

The formation of the steady standing wave in liquid under sonication is favourable for the increase in the number of active bubbles which contribute to the sonochemical reactions in the liquid [11]. Hence, in order to enhance the sonochemical reaction rate, it is important to extend the field of the steady standing wave. However, the standing wave of the larger amplitude causes the aggregation of the bubbles, resulting in a reduction of the active bubbles. As a result, the sonochemical efficiency decreases with increasing electric power. On the contrary, the mechanical flow generated by the revolution of a stirrer is responsible for the increase in horizontal flow rate, which prevents the active bubbles from aggregating in the standing wave and additionally supplies the acoustic field with reactant and the cavitation nuclei required for the formation of the active bubbles. An increased liquid velocity might contribute to the increase in sonochemical yield. Therefore, the combination of stirring with sonication enhances the sonochemical reaction rate.

3.4 Conclusions

We have investigated the relation between the flow velocity or the flow profile and the sonochemical efficiency under ultrasonic irradiation with a frequency of 490 kHz in a pilot-scale reactor of a rectangular parallelepiped and can make the following conclusions:

1. The average velocity of liquid flow involved with the acoustic streaming increases with an increase in the electric power applied to a transducer, while the sonochemical efficiency decreases up to a certain electric power due to the decrease in the number of active bubbles per unit energy consumption.
2. The combined treatment of mechanical flow arising from the revolution of a stirrer with acoustic streaming leads to the enhancement of sonochemical efficiency. The sonochemical efficiency is about twice higher compared with that under ultrasonication without stirring.
3. Atomisation is enhanced by increasing the revolution of a stirrer under sonication.

These conclusions suggest that in order to enhance sonochemical efficiency, it is necessary not only to reduce the chance that bubbles come together but also to widely supply the acoustic field with the reactants and cavitation nuclei contributing to generation of active bubbles. Therefore, the flow velocity and flow patterns should be taken into consideration for designing and developing a large-scale sonochemical reactor.

References

1. Kojima Y, Fujita T, Ona EP, et al. (2005) Effects of dissolved gas species on ultrasonic degradation of (4-chloro-2-methylphenoxy) acetic acid (MCPA) in aqueous solution, *Ultrason Sonochem*, **12**, 359.
2. Gedanken A (2004) Using sonochemistry for the fabrication of nanomaterials, *Ultrason Sonochem*, **11**, 47.
3. Mason TJ (1992) Industrial sonochemistry—potential and practicality, *Ultrasonics*, **30**, 192.
4. Dälhnke S, Keil FJ (1998) Modeling of three-dimensional linear pressure fields in sonochemical reactors with homogeneous and inhomogeneous density distributions of cavitation bubbles, *Ind Eng Chem Res*, **37**, 848.
5. Dahlem O, Reisse J, Halloin V (1999) The radially vibrating horn: A scaling-up possibility for sonochemical reactions, *Chem Eng Sci*, **54**, 2829.
6. Gogate PR, Pandit AB (2004) Sonochemical reactors: scale up aspects, *Ultrason Sonochem*, **11**, 105.
7. Asakura Y, Ishio K, Kojima Y, et al. (2007) Effect of liquid height on sonochernical reaction in a large-scale reactor of a rectangular parallelepiped using low frequency ultrasound (SC), *J Chem Eng Japan*, **40**, 1088.
8. Adewuyi YG, Oyenekan BA (2007) Optimization of a sonochemical process using a novel reactor and Taguchi statistical experimental design methodology, *Ind Eng Chem Res*, **46**, 411.
9. Nikitenko SI, Le Naour C, Moisy P (2007) Comparative study of sonochemical reactors with different geometry using thermal and chemical probes, *Ultrason Sonochem*, **14**, 330.
10. Mason TJ, Lorimer JP (2002) *Applied Sonochemistry*, Wiley-VCH, Weinheim.
11. Asakura Y, Nishida T, Matsuoka T, Koda S (2008) Effects of ultrasonic frequency and liquid height on sonochemical efficiency of large-scale sonochemical reactors, *Ultrason Sonochem*, **15**, 244.

12. Gondrexon N, Renaudin V, Petrier C, et al. (1998) Experimental study of the hydrodynamic behaviour of a high frequency ultrasonic reactor, *Ultrason Sonochem*, **5**, 1.

13. Chouvellon M, Largillier A, Fournel T, Boldo P, Gonthier Y (2000) Velocity study in an ultrasonic reactor, *Ultrason Sonochem*, **7**, 207.

14. Servant G, Caltagirone JP, Gérard A, Laborde JL, Hita A (2000) Numerical simulation of cavitation bubble dynamics induced by ultrasound waves in a high frequency reactor, *Ultrason Sonochem,* **7**, 217.

15. Yasui K, Kozuka T, Tuziuti T, et al. (2007) FEM calculation of an acoustic field in a sonochemical reactor, *Ultrason Sonochem*, **14**, 605.

16. Yasuda K, Tachi M, Bando Y, Nakamura M (1999) Effect of liquid mixing on performance of porphyrin decomposition by ultrasonic irradiation, *J Chem Eng Japan,* **32**, 185.

17. Kojima Y, Asakura Y, Sugiyama G, Koda S (2010) The effects of acoustic flow and mechanical flow on the sonochemical efficiency in a rectangular sonochemical reactor, *Ultrason Sonochem*, **17**(5), 978–984.

18. Eckart C (1948) Vortices and streams caused by sound waves, *Phys Rev*, **73**, 68.

19. Koda S, Kimura T, Kondo T, Mitome H (2003) A standard method to calibrate sonochemical efficiency of an individual reaction system, *Ultrason Sonochem*, **10**, 149.

20. Hatanaka S, Mitome H, Yasui K, Hayashi S (2006) Multibubble sonoluminescence enhancement by fluid flow, *Ultrasonics*, **44**, e435.

Part 2

Application of Microbubbles

Chapter 4

Environmental Field

4.1 Applications of Microbubbles in the Field of Water and Wastewater Treatment

Pan Li[a] **and Masayoshi Takahashi**[b]
[a]*School of Environmental Science and Engineering, State Key Laboratory of Control and Resource Reuse, Tongji University, 1239 Siping Road, Shanghai 200092, P.R. China*
[b]*Research Institute for Environmental Management Technology, National Institute of Advanced Industrial Science and Technology, Tsukuba, Ibaraki 305-8569, Japan*
lipan163@hotmail.com, m.taka@aist.go.jp

4.1.1 Background

The most remarkable property of microbubbles is their ability to decrease in size and collapse under water, whereas ordinary macrobubbles rapidly rise and burst at the surface. This phenomenon indicates significant potential for the use of microbubbles in a variety of practical applications.

Along with the shrinkage of microbubbles, the interior pressure increases due to the surface tension, while ions accumulate on the gas-liquid surface. Those phenomena suggest the possibility of

Micro- and Nanobubbles: Fundamentals and Applications
Edited by Hideki Tsuge

ISBN 978-981-4463-10-2 (Hardcover), 978-981-4463-11-9 (eBook)
www.panstanford.com

different mass transfer behaviour or critical reaction field formation [1–3]. Many interesting findings have been demonstrated recently. Crystallisation at the end of microbubble shrinkage was investigated in NaCl solution at below the saturation point and utilised to develop a novel crystalliser [4]. Moreover, it was suggested that microbubbles generated in a deep-flow technique hydroponics-culture system could significantly promote plant growth [5]. Takahashi et al. demonstrated that microbubbles could be useful in the formation of gas hydrates due to their ability to alter the nucleation conditions and their efficient gaseous solubility [1]. They also reported that free radicals could be generated by the collapse of air microbubbles [3]. Furthermore, some researchers indicated that microbubbles could accelerate the formation of hydroxyl radicals (•OH) during the ozonation process [6–8].

The current study demonstrated the effect of catalyst on free radical formation during the collapse of microbubbles. Furthermore, the technique was applied in a model of the decomposition of wastewater containing a refractory contaminant [9,10]. Finally, the mechanism of microbubble collapse will be discussed.

4.1.2 Experimental

4.1.2.1 The effect of copper as a catalyst on the free radical generation during the collapse of microbubbles

4.1.2.1.1 Experimental methods

Figure 4.1 shows a schematic representation of the experimental set-up, including a description of the generator. A commercial microbubble generator (Awawa, Shigen-kaihatu Co., Ltd.) was used in this study. Water in a 3 L beaker was circulated by a pump through a gas-dissolution tank and a microbubble-dispersing nozzle. The gas was sucked in by the pump along with the water, and they were mixed by the turbulent flow. The gas was dissolved in the water at an elevated pressure (0.4 MPa) in the gas-dissolution tank. The highly pressurised water was then injected into water at normal atmospheric pressure through a specially-designed nozzle that comprised an orifice and a net. Microbubbles were formed due to the sudden drop in pressure. The condensed microbubbles gave the water a milky appearance.

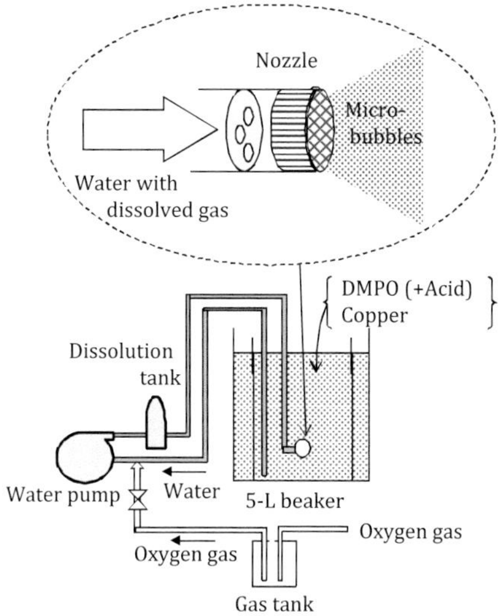

Figure 4.1 Schematic diagram of the experimental set-up.

Free radicals generated by the collapsing microbubbles were detected using the electron-spin resonance (ESR) spin-trap method, which efficiently monitors the presence of short-term free radicals. 5,5-dimethyl-1-pyrroline-N-oxide (DMPO) was chosen as the spin-trap reagent because it has been widely used to identify oxygen-centred radicals such as superoxide and •OH. Oxygen microbubbles were generated for 5 min in the beaker, which was filled with 2.5 L distilled water before the addition of a solution of DMPO+A. Samples containing microbubbles were removed from the beaker 5 min after adding them and transferred to a quartz flat cell for ESR measurements. The test was repeated by varying "A" in three patterns as follows:

A1: nothing
A2: hydrochloric-acid (HCl) solution
A3: copper wires and hydrochloric-acid (HCl) solution

4.1.2.1.2 Results and discussion

Figures 4.2, 4.3 and 4.4 show the ESR spectrums of a sample with added A1, A2 and A3 taken 5 min after the bubbling had started. The ESR parameters in Fig. 4.2 were similar to those of DMPO-R. In Figs. 4.3 and 4.4, four lines (the hyperfine splitting constants

AN = AH = 14.9 G) were observed, and the ESR parameters were similar to those of DMPO-OH, suggesting that •OH had formed. The similar ESR patterns, but lower peaks were obtained in our previous research where microbubbles collapse in the absence of a dynamic stimulus. In the present test, DMPO was directly added into the microbubble generation system, where water with microbubbles was circulated. It is suggested that the harsh pressure variation along the flow path have a great effect on the microbubble collapse. The DMPO-OH peaks in Fig. 4.3, where copper was not added, were much less intense than those shown in Fig. 4.4, where copper was added. This suggested that copper promoted the formation of •OH.

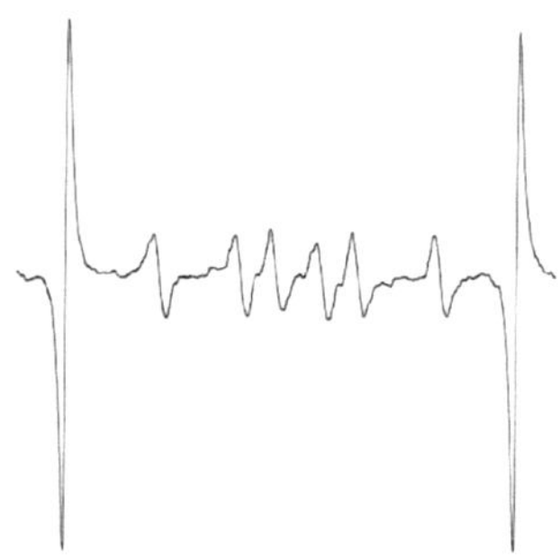

Figure 4.2 ESR spectra of microbubbles in distilled water, using DMPO as a spin-trapping reagent for oxygen microbubbles.

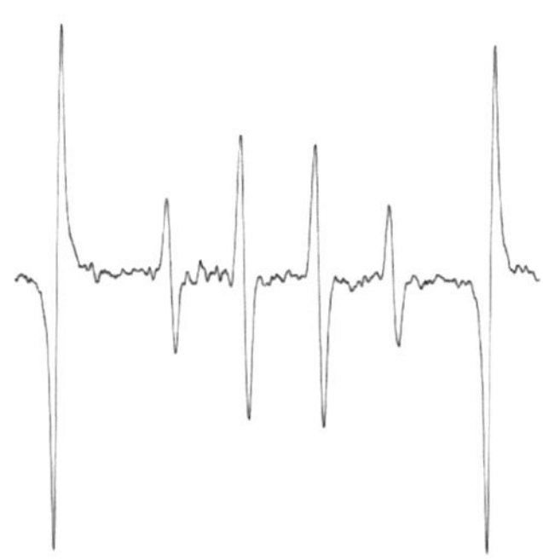

Figure 4.3 ESR spectra of microbubbles in distilled water, using DMPO as a spin-trapping reagent for oxygen microbubbles. HCl was added to the solution.

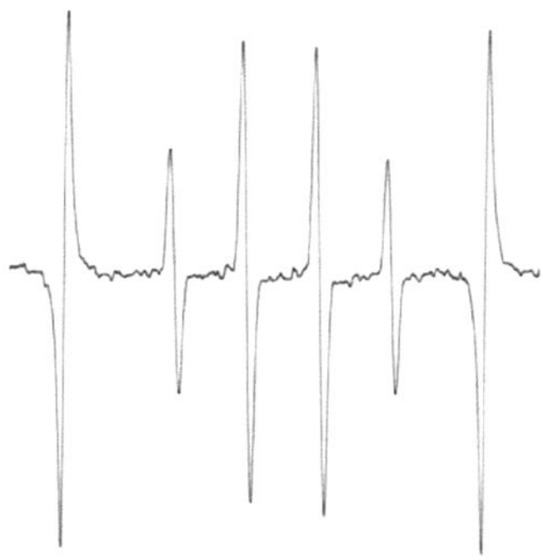

Figure 4.4 ESR spectrum of a distilled water sample containing oxygen microbubbles and DMPO. Copper and HCl was added to the solution.

4.1.2.2 Degradation of phenol

4.1.2.2.1 Experimental methods

We investigated the degradation efficiency of phenol by collapsing microbubbles. The pH of the solution was maintained at 2.3. The water flow rate was maintained at 8.7 L min^{-1} during microbubble generation, while the gas flow rate was 0.5 L min^{-1}. A water bath was used to keep the solution temperature at 35.5°C. Copper wires were added as a catalyst. The type of gas supplied to the microbubble generator was ambient atmosphere, pure nitrogen gas or pure oxygen gas.

4.1.2.2.2 Results and discussion

Figure 4.5 shows the deposition of phenol by collapsing microbubbles when the air was used as the gas source for microbubble generation. During the 2 h test period, reductions of approximately 60% and 20% were observed in the phenol and total organic carbon (TOC) contents, respectively. Catechol and hydroquinone were the main aromatic byproducts of the phenol oxidation, with the former reaching a higher concentration than the latter. The TOC removal rate increased when the catechol and hydroquinone concentrations peaked at 25 min and linearly increased with time thereafter. A control experiment was also conducted in which the supply of gas to the microbubble generator was stopped by closing the valve in the gas line. Neither the phenol concentration nor the TOC content showed any change with time when air microbubbles were not generated.

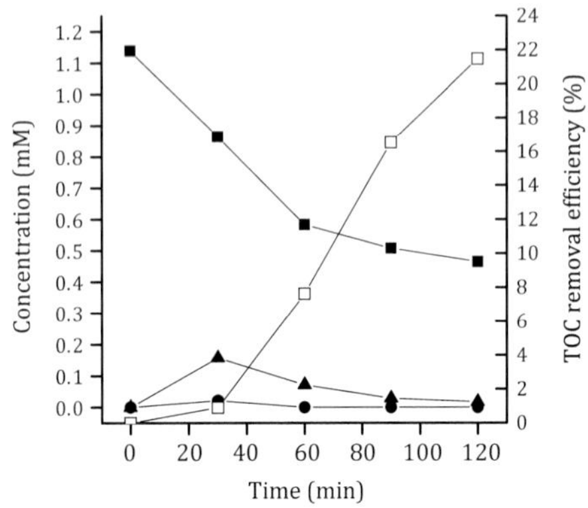

Figure 4.5 Changes in the concentration of phenol (■), primary aromatic byproducts (hydroquinone [●] and catechol [▲]) and TOC removal efficiency (□) as a function of time, with generation of air microbubbles (pH = 2.3).

Figure 4.6 shows the effect on the degradation of phenol by varying the type of gas supplied to the microbubble generator. The phenol decomposition rate increased in the order nitrogen < air < oxygen, with 36%, 59% and 83% of the phenol being decomposed during the 2 h treatment, respectively. It is suggested that degradation efficiency increases with the partial pressure of oxygen within the microbubbles.

We next investigated the process in the presence of tert-butyl alcohol (t-BuOH; 5×10^{-2} M), which is a known scavenger of •OH radicals. Using nitrogen microbubbles, the changes in phenol concentration with and without added t-BuOH are shown in Fig. 4.7. Almost no phenol decomposition occurred when t-BuOH was added. This suggests that when nitrogen microbubbles are used, •OH radicals are only generated under acidic conditions. Figure 4.7 also shows the changes in phenol concentration that occurred with and without added t-BuOH when oxygen microbubbles were generated. The conversion of phenol was significantly suppressed by the addition of t-BuOH. Therefore, •OH radicals are also mainly responsible for the decomposition of phenol in the case of oxygen microbubbles.

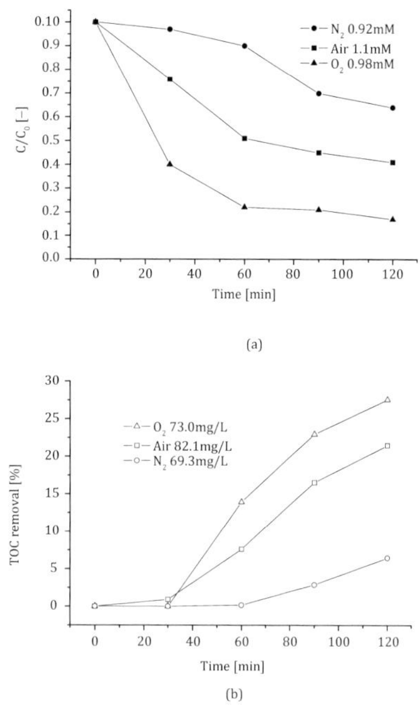

Figure 4.6 Effect of the type of gas supplied to the microbubble generator on the degradation of phenol (pH = 2.3): filled symbols denote phenol content and open symbols denote TOC content; ● and ○ indicate nitrogen microbubbles; ■ and □ indicate air microbubbles and ▲ and Δ indicate oxygen microbubbles.

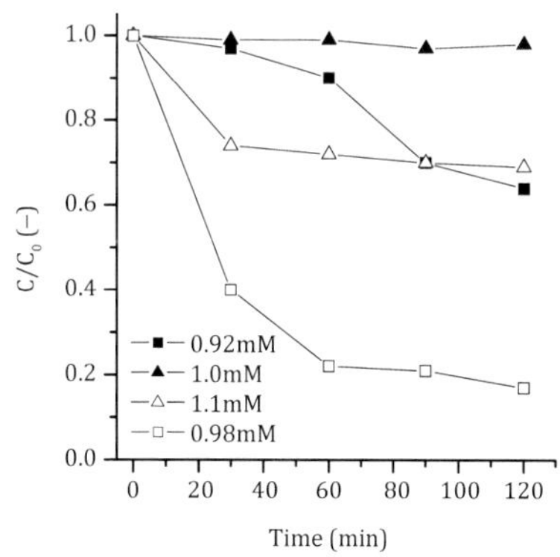

Figure 4.7 Degradation of phenol with (▲ Δ) and without (■ □) t-BuOH (pH = 2.3) by nitrogen microbubbles (▲ ■) and oxygen microbubbles (Δ □). C_0 is the intial concentration of phenol.

4.1.3 Discussions on the Mechanism of the Collapse of Microbubbles

Takahashi et al. demonstrated free radical generation from collapsing microbubbles in aqueous solutions in the absence of a dynamic stimulus [3]. In contrast to ultrasound-induced cavitation bubbles, which collapse violently after microseconds, the microbubbles collapsed softly under water after several minutes. It was therefore unlikely that the mechanism of free-radical generation by microbubbles was similar to that in cavitation bubbles—the latter being related to the extremely high temperatures caused by adiabatic compression during violent collapse. The mechanism might have been related to ionic accumulation around the collapsing microbubbles. The increase in surface charges (zeta potentials) of the microbubbles, which were measured during their collapse, supported the hypothesis that the significant increase in ion concentration around the shrinking gas–water interface provided the mechanism for radical generation.

We demonstrated the catalytic performance of copper in this process under milder conditions (pH = 2.2). The mechanism was likely to be similar to that observed in another advanced oxidation process, wet-air oxidation (WAO), in which copper or copper oxide has been used as a catalyst [9]. WAO involves the liquid-phase oxidation of organic or oxidisable inorganic components at elevated temperatures (T = 453–588 K) and oxygen pressures (P_{02} = 2–15 MPa) using a gaseous source of oxygen (usually air). The enhanced solubility of oxygen in aqueous solutions at elevated temperatures and pressures provides a strong driving force for oxidation. Free radicals play an important role in WAO [11,12]. Copper-based materials have been found to have catalytic activity and create milder oxidation conditions for WAO [11,13]. Typical conditions applied in catalytic wet air oxidation (CWAO) are T = 373–473 K and P_{02} = 0.3–3.5 MPa. The pH of solution and the partial pressure of oxygen are import factors in CWAO systems. Imamura et al. investigated the effect of pH on the wet oxidation of acetic acid in the presence of Co/Bi catalyst [14]. They found that the oxidation rate greatly decreased when varying the pH of the solution from 3.4 to 10.0. The wet oxidation is improved by increasing the partial pressure of oxygen because its mass transfer rate from the gas phase to the liquid phase is proportional to its partial pressure.

The mechanism of generation of •OH from the collapse of oxygen (or air) microbubbles appeared to be similar to that seen in CWAO and might have been related to environmental changes inside the collapsing microbubbles. A characteristic property of microbubbles is a high internal pressure, which results from surface tension at the gas–liquid interface. The Young–Laplace equation expresses the relationship between pressure and diameter

$$P = P_1 + \frac{4\sigma}{d_b}, \tag{4.1}$$

where P is the interior gas pressure, P_1 is the liquid pressure outside of the bubble, σ is the surface tension of the liquid and d_b is the diameter of the bubble. One example of the relationship between the interior gas pressure and the bubble diameter is given in Fig. 4.8. The interior gas pressure increases as the bubble becomes smaller; moreover, the rate of increase is inversely proportional to the bubble size. Therefore, a high-pressure spot might be formed during the final stage of the collapse process. Moreover, even though it is unlikely that a hot spot with an extremely high temperature, as in cavitation bubbles, could be formed during the microbubble collapse process, the temperature inside the bubbles might increase to a level that is sufficiently high to have an effect similar to that seen in WAO. The catalytic activity of copper during the microbubble collapse process could be similar to that observed in CWAO.

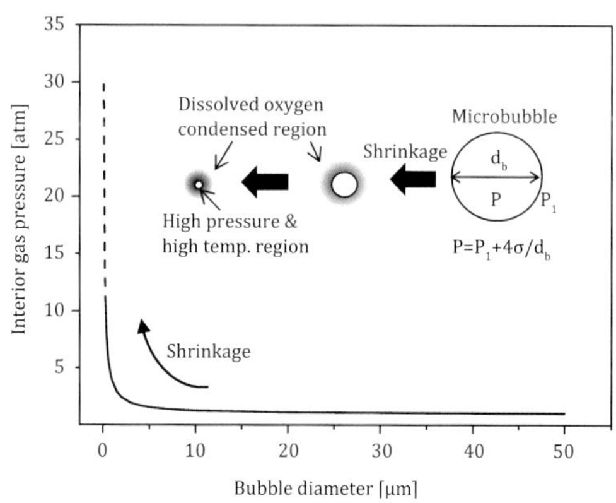

Figure 4.8 Increase in the interior gas pressure of microbubbles during shrinkage (at T = 298 K, P_1 = 0.1 MPa).

4.1.4 Conclusions

The generation of •OH from the collapse of oxygen (or air) microbubbles was markedly enhanced by a copper catalyst. The mechanism appeared to be similar to that seen in CWAO and might have been related to the environmental changes inside the collapsing microbubbles. The decomposition of phenol by air microbubbles in the presence of copper indicated that microbubbles are promising candidates for practical applications in the field of water and wastewater treatment. More research is needed to understand the mechanism of the catalytic collapse of microbubbles and evaluate its efficiency in the treatment of real wastewater.

References

1. Takahashi M, Kawamura T, Yamamoto Y, Ohnari H, Himuro S, Shakutsui H (2003) Effect of shrinking microbubble on gas hydrate formation, *J. Phys. Chem. B*, **107**, 2171–2173.
2. Takahashi M (2005) ζ potential of microbubbles in aqueous solutions: electrical properties of the gas-water, *J. Phys. Chem. B*, **109**, 21858–21864.
3. Takahashi MM, Chiba K, Li P (2007) Free-radical generation from collapsing microbubbles in the absence of a dynamic stimulus, *J. Phys. Chem. B*, **111**, 1343–1347.
4. Ohkubo M, Sato Y, Terasaka K, Obayashi D (19–21 Sepember 2008) *Development of Novel Crystallization Using Micro-bubble Shrinkage*, 40th Autumn Meeting of the Society of Chemical Engineers Japan, Sendai, Japan, A215 (in Japanese).
5. Park JS, Kurata K (2009) Application of microbubbles to hydroponics solution promotes lettuce growth, *Horttechnology*, **19**, 212–215.
6. Chu LB, Yan ST, Xing XH, Yu AF, Sun XL, Jurcik B (2007) Enhanced sludge solubilization by microbubble ozonation, *Chemosphere*, **72**, 205–212.
7. Chu LB, Xing XH, Yu AF, Zhou YN, Sun XL, Jurcik B (2007) Enhanced ozonation of simulated dyestuff wastewater by microbubbles, *Chemosphere*, **68**, 1854–1860.
8. Bando Y, Takahashi Y, Luo W, Wang Y, Yasuda K, Nakamura M, Funato Y, Oshima M (2008) Flow characteristics in concurrent upflow bubble column dispersed with micro-bubbles, *J. Chem. Eng. Jpn.*, **41**, 562–567.
9. Li P, Takahashi M, Kaneo C (2009) Enhanced free-radical generation by shrinking microbubbles using a copper catalyst, *Chemosphere*, **77**, 1157–1160.

10. Li P, Takahashi M, Kaneo C (2009) Degradation of phenol by the collapse of microbubbles, *Chemosphere*, **75**, 1371–1375.
11. Suresh KB, James T, Jaidev P, Karl F, Deepak BA, Stephen CG (2006) Wet oxidation and catalytic wet oxidation, *Ind. Eng. Chem. Res.* **45**, 1221–1258.
12. Robert R, Barbati S, Ricq N (2002) Intermediates in wet oxidation of cellulose: identification of hydroxyl radical and characterization of hydrogen peroxide, *Wat. Res.*, **36**, 4821–4829.
13. Vedprakash SM, Vijaykumar VM, Jyeshtharaj BJ (1995) Wet air oxidation, *Ind. Eng. Chem. Res.* **34**, 2–48.
14. Imamura S (1982) Wet oxidation of acetic acid catalyzed by Co-Bi complex oxide, *Ind. Eng. Chem. Prod. Res. Dev.*, **21**, 570–575.

4.2 Purification of Water Quality

Koji Michioku
Graduate School of Engineering, Kobe University, Kobe-shi, Hyogo Prefecture, Japan
michioku@kobe-u.ac.jp

Significant pollutant loads have been yielded from catchments due to extensive human activities. This produces large amounts of nutrient and oxygen demanding substances, such as organic matter, and brings various negative impacts to the natural environment of rivers and oceans. In this section, two examples of microbubble aeration technology have being reported as a countermeasure against water quality troubles in a river catchment. The first topic is microbubble aeration to prevent deoxidation in reservoirs. The second is a microbubble technology for treatment of leachate from garbage-dumped landfill.

4.2.1 Microbubble Aeration of Anoxic Hypolimnion in Reservoirs

4.2.1.1 Eutrophication and water quality problems in reservoirs

Reservoirs with long retention times generally suffer from water quality troubles due to eutrophication. This mechanism is illustrated in Fig. 4.9. During warm seasons, nutrients running off the catchment induce algae blooming in the photic zone. Organic compounds decompose in the course of settling, consuming a large amount of dissolved oxygen in the aphotic zone and making the hypolimnion anoxic. Here, hypolimnion refers to the layer below the thermocline interface. Because the system is thermally stratified due to solar radiation, the anoxic lower layer is isolated from ambient water, and therefore, oxygen is scarcely supplied from the aerobic surface layer. When the benthic layer is exposed to the anoxic water, dissolved substances such as ionic metals, organic compounds and nutrients are released with hydrogen sulfide and methane also being generated from sediments through reduction. This is how the impounded water is contaminated. The key to water purification is how to feed oxygen into the anoxic layer beyond a certain level.

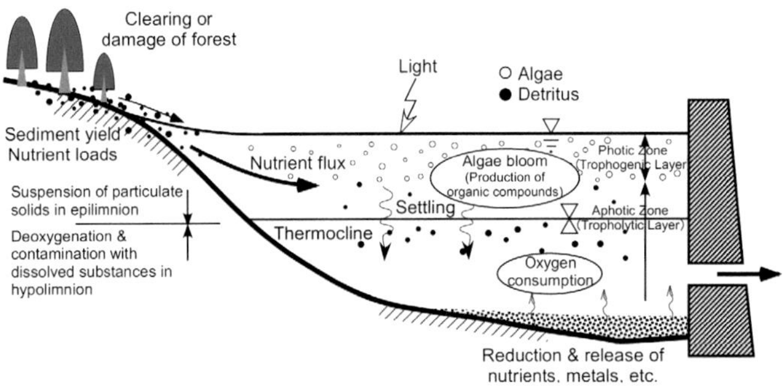

Figure 4.9 Mechanism of reservoir eutrophication and oxygen consumption.

There are two ways of feeding oxygen and therefore aerating the hypolimnion. The first method is the artificial overturning of thermal stratification by using air bubblers, air bubble jets or mechanical mixer [4]. In doing so the anoxic hypolimnetic water is vertically exchanged and mixed with the aerobic epilimnetic water. The other method is through hypolimnetic aeration to discharge air bubbles or aerobic water into the anoxic hypolimnion without generating any vertical mixing [1], [2], [3], [8], [9]. One of the conventional systems for hypolimnetic aeration is shown in Fig. 4.10. Air bubbles are generated and lifted up through a vertical cylinder with entraining anoxic water from the cylinder's lower end inlet. In the course of lifting up the cylinder, the water is oxidized and then the aerated water is returned to the hypolimnion.

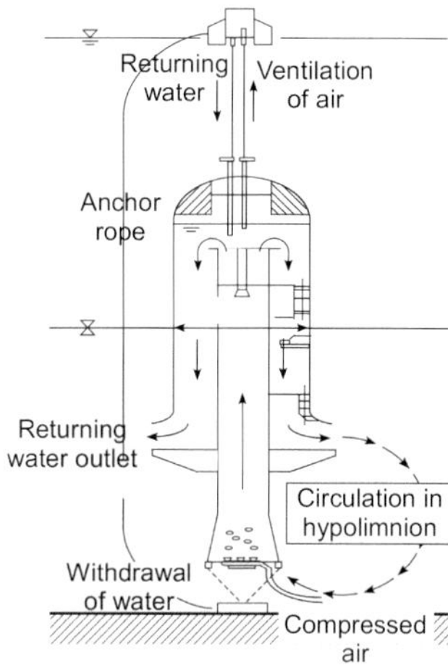

Figure 4.10 A conventional reservoir.

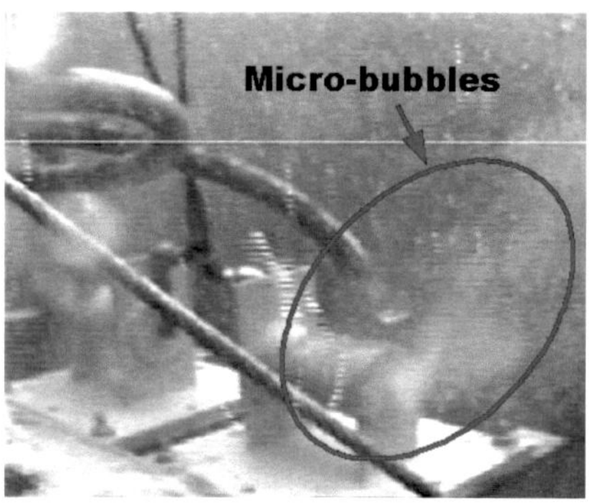

Photo 4.1 Microbubbles generated in a hypolimnetic aerator.

When the hypolimnion is seriously anoxic and contaminated with toxic matter such as heavy metals, vertical mixing must be avoided, and the first type of aeration is not recommended. In this case, the only solution is through hypolimnetic aeration. Although various types of hypolimnetic aeration systems have been proposed so far, most of them were to feed fairly coarse air bubbles with mean diameters greater than several millimetres. Because high solubility of oxygen cannot be expected under these conditions, aerators with quite a large capacity are required for obtaining a desirable accumulation rate of oxygen. In addition, systems with complicated structures such as in Fig. 4.10 are not easily able to maintain stable performance. Based on this background, the author has developed a microbubble aeration system in order to efficiently feed oxygen into the hypolimnion with a minimum of cost and energy [5], [6]. Because the system is able to produce very fine microbubbles with diameters ranging around 50–100 μm as shown in Photo 4.1, this enables high oxygen solubility and significant performance in water purification.

4.2.1.2 Microbubble aeration system for reservoir hypolimnion

In order to minimise water quality troubles, an experiment on hypolimnetic aeration was carried out by using an aeration system as shown in Fig. 4.11. A microbubble generator, YJ-II (Enviro Vision Co. Ltd.), shown in Photo 4.2 was equipped in the hypolimnion [7]. Instead of natural air, oxygen was used so that higher oxygen solubility could be obtained with less energy. Microbubbles were generated by taking the anoxic water with a discharge of $Q_L = 300$ [l/min] by pump and mixing it with oxygen that was supplied from a compressor. In this manner, a liquid-oxygen multi-phase jet is flushed out to the ambient water from the aerator's nozzle.

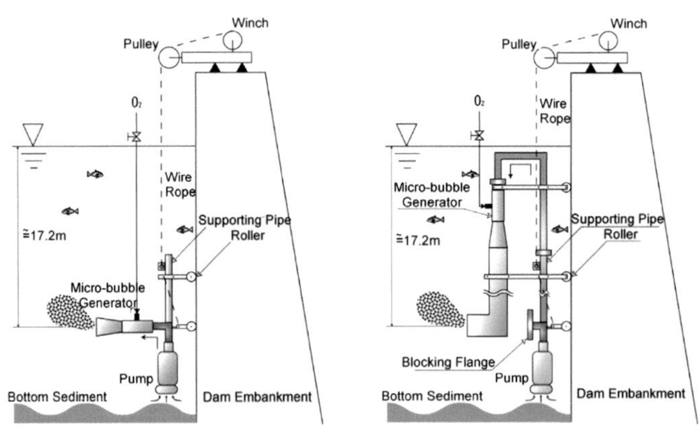

(a) Test-I (b) Test-II

Figure 4.11 Schematic of the experimental aeration systems.

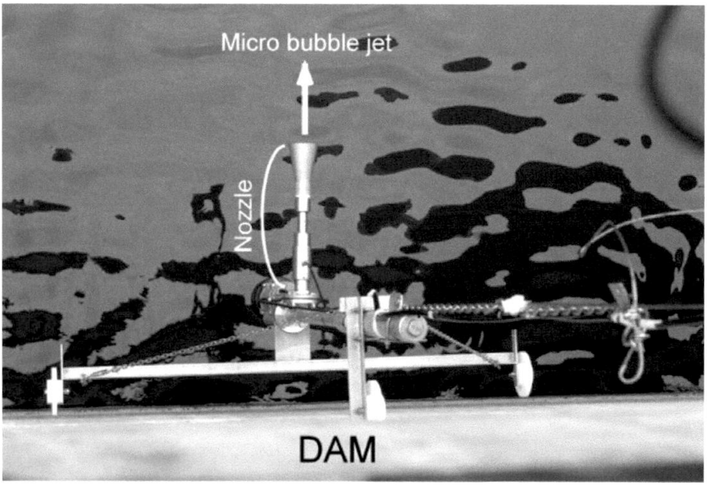

Photo 4.2 YJ-II microbubble aerator.

Two types of aeration systems were devised. Test I was to directly inject the microbubble mixture into the ambient water. It was conducted from 30 May to 24 July 2007. Test II was to feed oxidised water that was produced by an aerator inserted into a pipe as shown in Fig. 4.11b to the ambient water. Test II was conducted from 23rd August to 23rd October in 2007. Because distance between the generator and the nozzle outlet is longer in Test II than in Test I, more oxygen is expected to be dissolved through the confined pipe while running Test II.

It was found from a preliminary experiment that continuous operation of the present aeration system supplies so much Dissolved Oxygen, DO that water tends to become oversaturated with DO. Therefore, the system was intermittently operated by monitoring DO concentration in the hypolimnion at 10:00 AM every morning. The aerator was switched on and off so as to keep DO concentration within a suitable range between 5–10 mg/L in both Tests I and II. Oxygen discharge was kept to 0.6 m^3/h except during Term (3) of 0.4 m^3/h. Aeration was absent and construction work was carried out to change the system from Test I to Test II between 24 July and 23 August 2007. At the beginning of Test I, the bubble's diameter was a little larger than we expected; so, the nozzle was exchanged to another model on 5 July in order to decrease bubble size.

In Test II, microbubbles were not well generated at first because of the sudden contraction of the pipe connected to the aerator caused macrobubbles. It was replaced with a gradual contraction module, and thereafter, fine microbubbles were successfully generated.

Table 4.1 Operation schedule

TEST	Term	Operation period	Discharge (m^3/hour)	Operated (days)	Absent (days)
TEST-I	(0)	5/30–6/1	0.6	2	12
	(1)	6/13–6/22	0.6	9	14
	(2)*	7/6–7/24	0.6	18	30
TEST-II	(3)	8/24–9/4	0.4	12	14
	(4)**	9/18–10/3	0.6	15	9
	(5)	10/12–10/23	0.6	11	

* Nozzle was exchanged to decrease bubble's diameter on 5 July.
** Pipe contraction was improved to gradual contraction on 18 September.

The operation schedule is listed in Table 4.1. Water quality measurement was carried out 21 times from 16 May to 30 November 2007. Vertical profiles of temperature, dissolved oxygen, electric conductivity, Chemical Oxygen Demand, COD etc., were measured at the deepest point, about 15 m distance from the aeration point.

4.2.1.3 Temperature and dissolved oxygen

Temperature isopleths observed during the aeration are shown in Fig. 4.12. A sharp thermocline had already developed at EL.172 m when aeration started on 16 May 2007. The stable thermal stratification was kept until Fall, overturning only in late November. Shaded stripes with numbers correspond to periods during which the system was being operated with the other periods being durations of no aeration. Because the microbubble plume generates a weak turbulence, the epilimnetic water was entrained into the hypolimnion and the temperature of the hypolimnetic water slightly increased during system operation. Despite the turbulence, full-depth scale vertical mixing never occurred, and the epilimnetic and hypolimnetic waters were well separated.

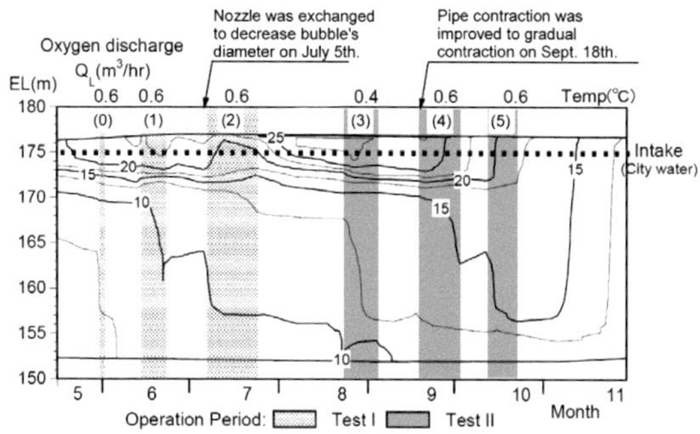

Figure 4.12 Time dependency of vertical temperature profile.

A time history of dissolved oxygen concentration is shown in Fig. 4.13. While the hypolimnion was completely anoxic before starting the aeration, immense recovery of DO was achieved through microbubble aeration. However, a small amount of the bottom layer below EL.156 m remained anoxic because only the water around EL.256 m was withdrawn by the pump, and the bottom layer lying below was left without being entrained. The figure documents well the increase in DO concentration during aeration especially in the layer between EL.156 and EL.169 m, while it rapidly decreased when aeration was stopped. Most of the DO demanding materials are considered to be organic matters, ammonia, ionic metals and other reductive components. It was confirmed that a high amount of manganese was also removed due to oxidation, which is an advantage for water works.

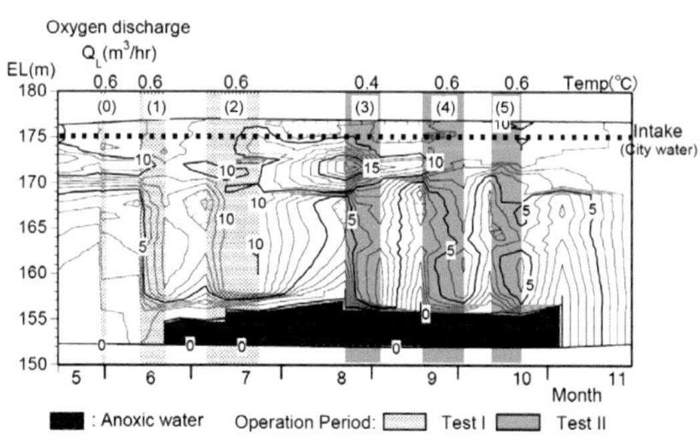

Figure 4.13 Seasonal variation of dissolved oxygen concentration.

4.2.1.4 Performance of hypolimnetic aeration

Aeration performance is evaluated by DO solubility obtained from the measurement of DO concentration. An increasing rate of DO is described by the following equation:

$$\frac{d[\mathrm{DO}]}{dt} = k_{\mathrm{La}}\left([\mathrm{DO}^*]-[\mathrm{DO}]\right), \tag{4.1}$$

where [DO] is DO concentration in mg/L, [DO*] is a saturation concentration for a given water temperature, k_{La} is an oxygen transfer capacity coefficient in s^{-1} and t is elapsed time per second.

By integrating Eq. (4.1) under the initial condition that [DO] = 0.0 mg/L at $t = t_0$, a solution for time-dependency of DO concentration is obtained as follows:

$$[\mathrm{DO}] = [\mathrm{DO}^*]\left(1-\exp\{-k_{\mathrm{La}}\cdot(t-t_0)\}\right). \tag{4.2}$$

The increasing behaviours of DO concentration observed at EL.160.6 m during aeration are compared with the solution Eq. (4.2) in Fig. 4.14. As a result, k_{La} is identified as listed in Table 4.2, which shows that Test II gives a slightly higher oxygen transfer rate than Test I.

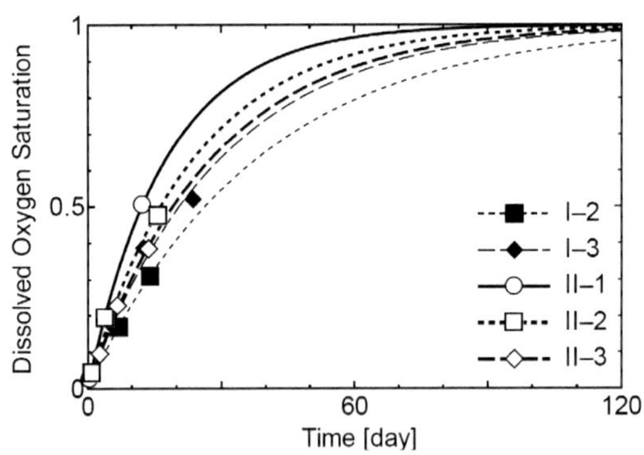

Figure 4.14 Plot of dissolved oxygen concentration as a function of elapsed time compared with the theory.

Table 4.2 Operation schedule

Number of period	Oxygen gas flow rate [N1/min]	Water flow rate [1/min]	k_{La} [1/s]
I–2	10.0	400	0.026 3
I–3	10.0	400	0.033 1
II–1	6.67	350	0.056 4
II–2	10.0	300	0.041 7
II–3	10.0	300	0.035 9

4.2.2 Microbubble Aeration to Purify Leachate from Waste-Dumped Landfills

4.2.2.1 Background of leachate problem

In Japan, a huge amount of municipal solid waste has been directly dumped in landfills prior to issuance of the waste management and public cleansing law in 1970. Leachate from waste-dumped landfills shows very different water quality from ordinary groundwater runoff as well as from municipal sewage waters. Most of the organic matter and phosphorous are filtered out and oxygen is completely consumed in the course of retention, infiltration and travelling underground. On the other hand, ammonium nitrogen is the dominant pollutant component with extremely significant loads.

The field site is a municipal waste landfill located in a mountainous area of 65×10^6 m^2 in Kobe City, where municipal solid waste was dumped in the 1960s and 1970s. Photo 4.3 shows an aerial view of the study field. In the last decade, the landfill has released highly contaminated leachate with nitrogenous and carbon compounds, etc. At the moment, leachate is tentatively impounded in a reservoir with a volume capacity of about 20,000 m^3 as shown in Photo 4.4 and released after purification in a leachate treatment plant. However, due to ageing plant equipment, the plant needs to be closed as soon as possible as more than 25 years have passed since completion of the landfill dumping. It is necessary to develop an alternative water treatment system that must be more economic and requires minimum maintenance.

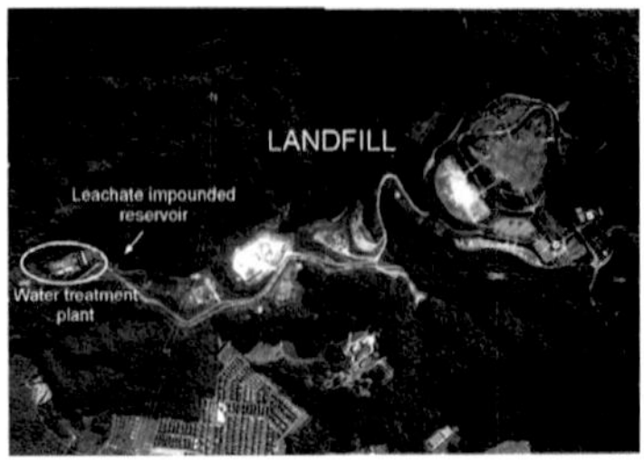

Photo 4.3 Aerial view of garbage dumped landfill and facilities.

Photo 4.4 Leachate impounded reservoir.

4.2.2.2 Microbubble aeration for nitrification of leachate: a laboratory experiment [10]

Nitrogen removal techniques consist of two stages: the nitrification and denitrification processes. The focus here is placed on the former. Microbubble aeration is expected to be a new technology to efficiently nitrify leachate in an energy-saving manner.

A preliminary experiment was carried out in a laboratory by using a test tank of 1 m^3 capacity in order to fundamentally examine the aerator's performance in leachate nitrification. The experimental set-up is schematically shown in Fig. 4.15. The M-2 type aerator (see Photo 4.5) developed by the Nanoplanet Research Institute Co. Ltd. was used in the experiment.

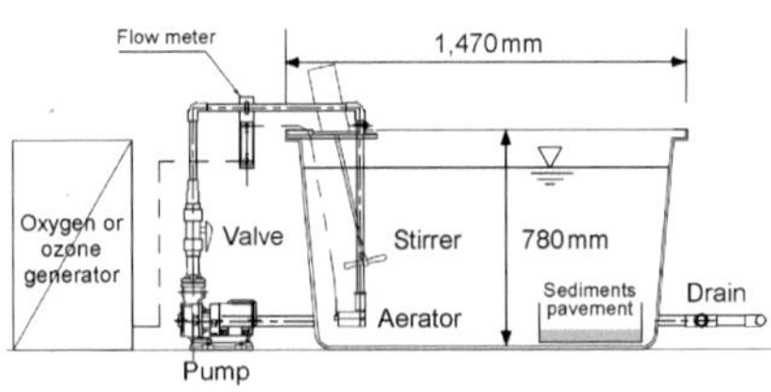

Figure 4.15 Experimental setup for leachate nitrification.

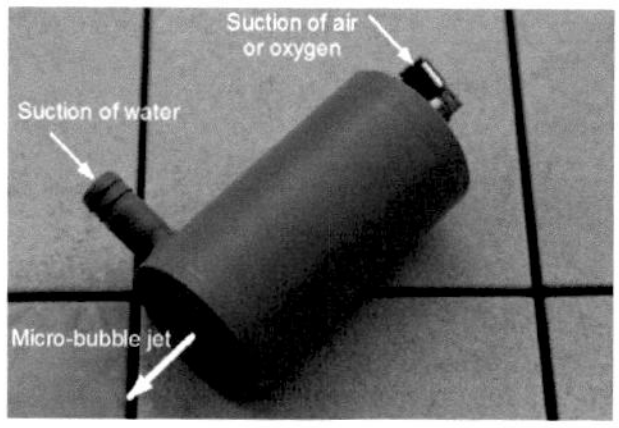

Photo 4.5 M2 type aerator.

Three types of gases: (i) natural air (Case-G1), (ii) oxygen (Case-G2) and (iii) ozone (Case-G3), were used for the aeration in order to examine water purification performance of each gas. A reference case (Case-G0) with no aeration was also examined for comparison. Sediments collected from the reservoir bed were paved at the bottom of the tank as a source of nitrification bacteria.

The aeration discharge for each gas varied depending on gas solubility allowing the leachate to be oxidised at approximately the same speed in each test tank. Nitrification performance of each gas was analyzed by plotting the concentration of nitrogenous compounds against a cumulative gas discharge $V(t)$ defined by

$$V(t) = \int_0^t Q(t)dt, \tag{4.3}$$

where t is the elapsed time and $Q(t)$ the discharge.

Concentrations of nitrogen compounds are plotted as a function of V in Figs. 4.16, 4.17 and 4.18. At the first stage, ammonia NH_4-N was oxidised into nitrite NO_2-N. After consumption of NH_4-N, nitrification from NO_2-N nitrate NO_3-N is accelerated in each case. The best performance of nitrogen oxidation is found in oxygen aeration (Case-G2). The second best performance is found in natural air (Case-G1) with ozone (Case-G3) showing the least performance.

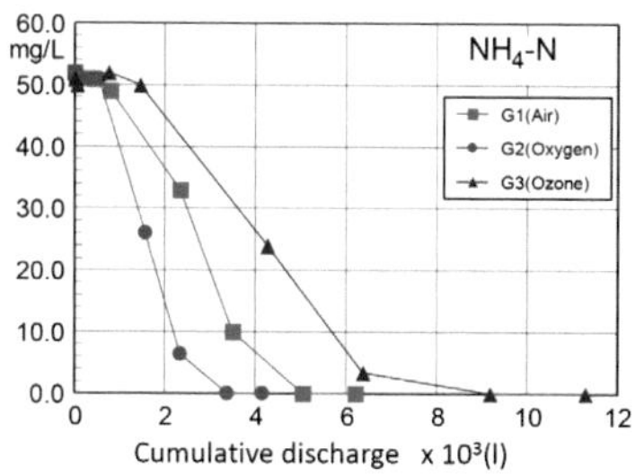

Figure 4.16 Time-dependent concentration of ammonia NH_4-N plotting against cumulative gas discharge, $V(t)$.

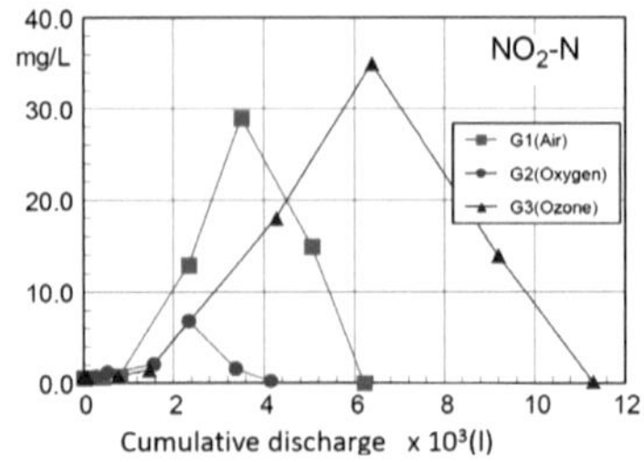

Figure 4.17 Time-dependent concentration of nitrite NO_2-N plotting against cumulative gas discharge, $V(t)$.

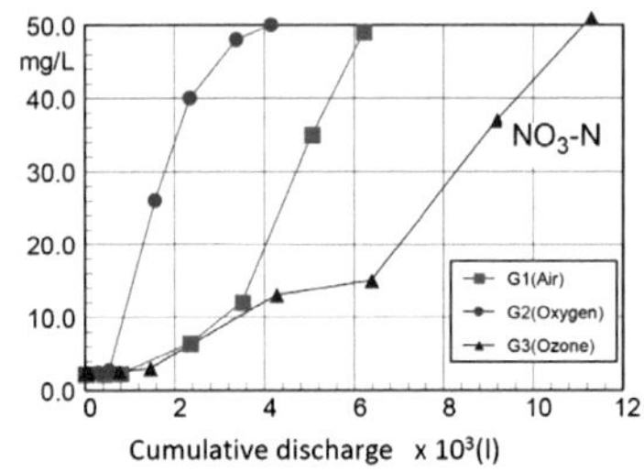

Figure 4.18 Time-dependent concentration of nitrate NO_3-N plotting against cumulative gas discharge, $V(t)$.

4.2.2.3 Microbubble aeration for nitrification of leachate: an in situ experiment

Capacity and dimensions of prototype aeration system were determined based on the laboratory experiment. Fifty units of M-2 type aerator were finally equipped in the leachate impoundment as shown in Fig. 4.19 and Photo 4.6. Oxygen was supplied instead of natural air to the aerators by using an oxygen generator so that higher oxygen solubility could be achieved to promote nitrification with minimum energy.

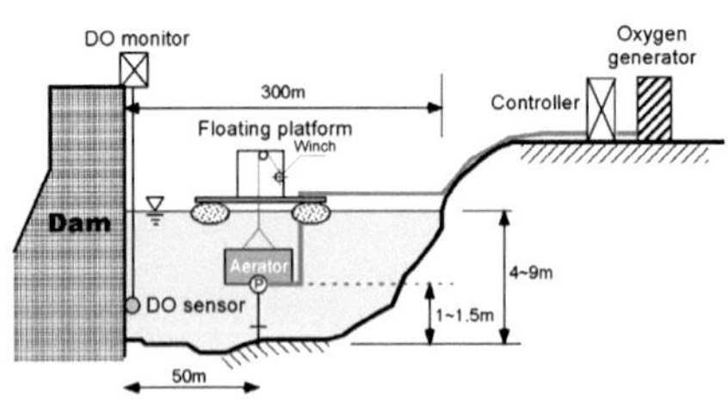

Figure 4.19 A prototype aeration system for leachate impoundment.

Photo 4.6 Aerators equipped in the leachate impoundment.

Figure 4.20 shows the time-dependent concentration of dissolved oxygen DO in the impoundment. The aeration system was operated intermittently depending on the DO concentration. The shaded strips in the figure correspond to durations of aeration system operation. An aerobic condition with DO concentration ranging between 10 and 20 mg/L was kept by microbubble aeration, except for a period in the summer of 2008 during which the system did not function due to the aerator nozzle clogging with suspended solids.

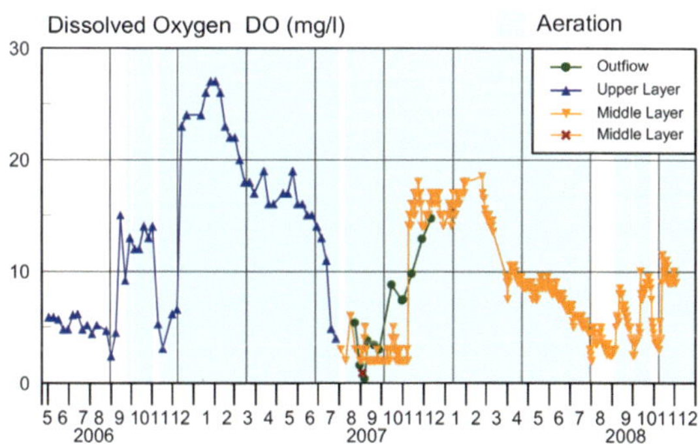

Figure 4.20 Time-dependent DO concentrations in the upper, middle and lower layers in the impoundment and in the outflow water.

Figures 4.21 and 4.22 show time histories of ammonium nitrogen NH_4-N and nitrate nitrogen NO_3-N which was measured in samples collected from the original leachate, the upper layer of impoundment and the outflow water, where the outflow water has approximately equivalent quality to the impounded water. Concentration difference between the leachate and the impounded water in the figures corresponds to the oxidised nitrogen amount from ammonia to nitrate. The nitrification performance can be evaluated from the decrease in NH_4-N or from the increase in NO_3-N. The figures show that the leachate was fairly well nitrified during the aeration. On the other hand, efficient nitrification did not occur during an absence of aeration.

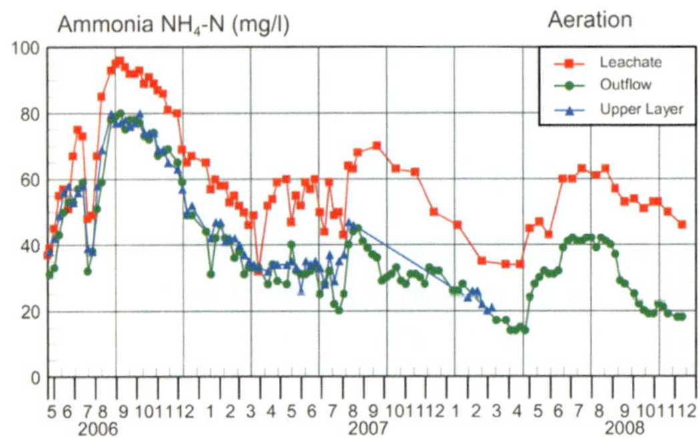

Figure 4.21 Concentration of ammonium nitrogen NH_4-N in the leachate, the upper layer of impoundment and the outflow water.

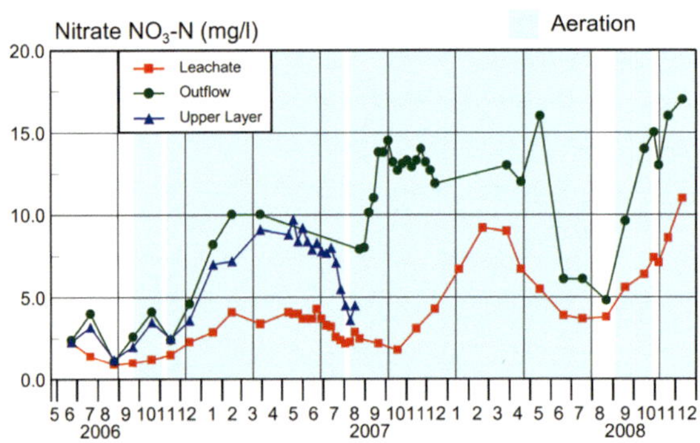

Figure 4.22 Concentration of nitrate nitrogen NO_3-N in the leachate, the upper layer of impoundment and the outflow water.

Because bacterial activity is one of the most important parameters in nitrification, the system's performance also depends on natural conditions such as water temperature, precipitation, leachate runoff discharge, etc. In order to achieve a higher performance of nitrification, ecological aspects such as installment of microbe-bound carriers, adding nutrients and organic compounds for nourishing bacteria, artificial control of discharge and temperature of leachate and so on, should be taken into consideration in the design and operation of the leachate treatment system. The lifecycle cost is also an important parameter for sustainable water quality management.

References

1. Ashley KI (1985) Hypolimnetic aeration: Practical design and application, *Wat. Res.*, **19** (6), 735–740.
2. Ashley KI, Hay S, Scholten GH (1987) Hypolimnetic aeration: Field test of the empirical sizing method, *Wat. Res.*, **21** (2), 223–227.
3. McQueen DJ, Lean DRS, Charlton MN (1986) The effects of hypolimnetic aeration on iron-phosphorus interactions, *Wat. Res.*, **20** (9), 1129–1135.
4. Stephens R, Imberger J (1993) Reservoir destratification via mechanical mixers, *J. Hydraulic Eng., Proc. ASCE*, **119** (HY4), 438–457.
5. Michioku K, Kanda T (11–14 December 2006) *Thermosolutal Stratification and Water Quality Behaviors in a Reservoir Hypolimnion Before and After Micro-bubble Aeration*, paper presented at the Proc. 6th International Symposium on Stratified Flows, Perth, Australia.
6. Michioku K, Sakatani Y, Matsuo K, Oda T, Hara Y (10–13 September 2006) *Reservoir Purification by Using Micro-bubble Aerator*, in Proc. 7th International Conference on HydroScience and Engineering, Philadelphia, PA.
7. Michioku K, Minagawa H, Yamada S, Ikeguchi T and Ohkawa N. (2012). An in-situ experiment on oxygen solubility of mcro-bubble aeration in a eutrophic dam-reservoir, *Proc. Intnl. Sympo. Dams for a Changing World*, ICOLD 2012, pp. 2-99–2-104.
8. Moore BC, Chen PH, Funk WH, Yonge D (1996) A model for predicting lake sediment oxygen demand following hypolimnetic aeration, *Wat. Resour. Bull.*, American Water Resources Association, **32** (4), 723–731.
9. Smith SA, Knauer DR, Wirth TL (1975) Aeration as a lake management technique, *Tech. Bull.*, **87**, Wisconsin Dept. Natural Resources, 39.
10. Michioku K, (9–14 August 2009) *Micro-bubble Aeration of Polluted Leachate from a Landfill*, in *Proc. 33rd IAHR Congress*, Vancouver, BC, Canada.

4.3 Applicability of Soil Improvement Using Microbubbles Water Against Soil Liquefaction

Koichi Nagao,[a] Naoaki Suemasa,[b] and Tatsuo Akashi[c]

[a]*Sato Kogyo Co. Ltd., Atsugi, Kanagawa, Japan*

[b]*Tokyo City University Setagaya, Tokyo, Japan*

[c]*NILIM, Government of Japan, Tsukuba, Ibaraki, Japan*

nagao@satokogyo.co.jp, nsuemasa@tcu.ac.jp, akashi-t2fw@nilim.go.jp

4.3.1 Introduction

Soil liquefaction disaster is caused by seismic shocks which are added to the loose sandy ground under the groundwater level. And then the ground structure which is made of sand particles is fractured and sand particles become as though they were just floating in the water as the sign of the liquefaction phenomenon of the ground.

Soil liquefaction is a major engineering interest for the geotechnical engineers in earthquake-prone countries. In Japan, because most built-up areas are concentrated on alluvial plain and soft ground, soil liquefaction is one of the latent threats.

Though various kinds of liquefaction countermeasures have been developed and performed until now, they have some mutual problems. Most countermeasures are difficult to use in narrow places where existing structures such as housings stand close together. Also, those countermeasures are quite expensive because they are required to perfectly prevent from the soil liquefaction.

Soil improvement against soil liquefaction, which utilises the principle of bubble shrinkage, is partially saturated using air injections into the ground, and the air cushion made of bubbles prevents pore water pressure from rising during earthquakes [1].

Though this method has not become practical yet, it can be prospected as an effective anti-liquefaction method to solve problems such as high cost and space shortage due to closely-spaced houses.

First, because just simple equipment is necessary for this method without removal of existing structures, it could be executed in narrow spaces. Because only air and water are used, it is economical as well as friendly to natural environment.

Therefore, this brand new method using microbubble water proposed in this chapter is expected to be one of the most practical measures against soil liquefaction with low cost and suitable even for build-up areas.

The outline of this method of construction is as follows: inject pressurised microbubble water into the suspected liquefaction ground by the bore hole. Figure 4.23 shows the process of decreasing saturation degree of ground by injecting the microbubble water and the mechanism of how the bubbles absorb the increased water pressure by ground shakes. When the microbubble water is injected into sand particles of the saturated ground, some of microbubble may pass through the element by the water flowing because the diameter of the bubbles is in the order of 10 μm and less than the size of voids among sand particles. The other microbubble, however, may be trapped in small air pockets among sand particles and both of them are united to grow up to millimetre size in diameter. As a result, the trapped microbubble decreases its surface tension and the air pressure in the bubble. Because water surrounding the bubble dissolves the air, the air precipitated into the bubble may help enlarge the bubble. Though excess pore water pressure tends to increase by repeated seismic stress, the increased pressure is absorbed by shrinkage of the bubbles.

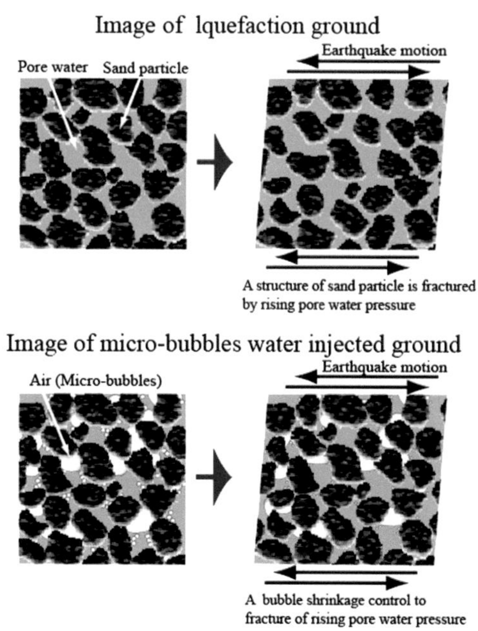

Figure 4.23 Mechanism of controlling liquefaction by microbubbles.

It is cleared from the laboratory test results that a saturation level of the ground decreases to about 80% by injecting micro-bubble water into loose sand [2].

Figure 4.24 shows a representative example of a microbubble water injection test that inspects the saturation of column about 74 mm in diameter and 300 mm tall. At first, the saturation degree decreases rapidly to about 90% (at the 30% injection of the volume of void), then decreases more gradually after that and finally reaches about 80%. The liquefaction resistance of the specimen de-saturated by injecting microbubble water, in which the degree of saturation was 90% and the relative density was 60%, was about two times larger than that of completely saturated sand.

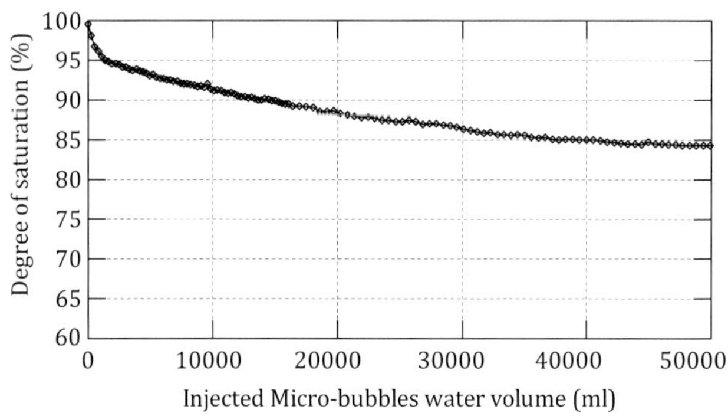

Figure 4.24 Decreasing of saturation by injecting microbubbles water.

Followings are demonstrations of the seismic vibration tests using a large scale flexible shaking box which clarify the effectiveness of the micro-bubble water injected into the ground as a countermeasure against soil liquefaction [3].

In addition, this experiment was conducted by National Institute for Land Infrastructure Management of Japan in December 2008 and January 2009.

4.3.2 Instance of Liquefaction Countermeasure (Large Scale Seismic Vibration Test)

4.3.2.1 Outline of tests

In the experiment, a large scale shaking table apparatus with a laminated shear box was used, which is owned by the Building Research Institute in Japan (Photo 4.7).

Photo 4.7 Scene of the experiment.

An overview of the large scale flexible shaking box is shown in Fig. 4.25. The shaking table is equipped with two actuators which have the ability to activate an inertial load of 450 tonf/tf at a maximum velocity of 20 cm/s in the horizontal direction.

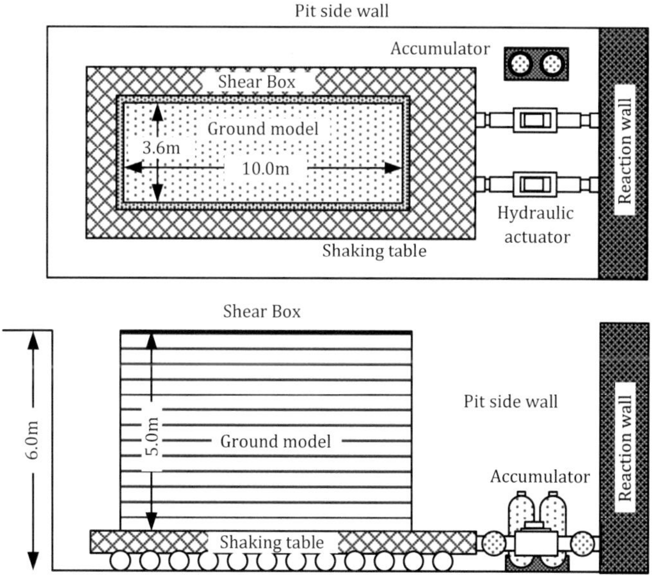

Figure 4.25 Large scale flexible shaking box.

The laminated shear box, mounted on the shaking table, consists of 17 layers of frames. Each layer with 300 mm height is connected through bearing rollers to the others in order to be independently oscillated. A model ground set up in the box is 3.6 m wide, 10 m long and 5 m high.

For the test, Nikko silica sand was used because of its tendency to liquefaction. The model ground was made of wet sand which was filled up until the ground reached 4800 mm in height. The density of ground was loose, and an N-value was about 6–7.

In order to confirm the effect as a countermeasure against a residential building, a model made of concrete block was put on the centre of the ground surface, which is 1.3 m by 1.3 m in area, 0.5 m in height and weighs 2.5 tonf/tf so that a contact pressure of the block was set to be approximately as large as that of an average house which was 15 kN/m^2.

4.3.2.2 Microbubbles water inject into the test ground

An outline of the microbubble water and the degassed water injection system are shown in Fig. 4.26. The system consists of a microbubble generator produced by Nikuni Corporation Ltd. and a pipe line system to feed the water to the model ground. The microbubble generator is mainly composed of a vortex flow turbine pump and an excess air disjoin tank. In the pump, water and air are self-fed by suction of the pump and mixed while the air dissolving in the water. More air dissolution is enhanced under the pressure of the excess air disjoin tank. In the tank, excess air bubbles not to be dissolved are ejected from the tank. The pressurised and air-dissolved water from the tank is spurted from injection slits of pipes put on the bottom of the model ground. Then microbubble is precipitated again out of the air dissolved water at the slits.

In the test, the pressure under which the microbubble water was generated and pneumatically transported was about 0.6 MPa. An inflow of air volume was 5 L/min, and the flow rate of water was 50 L/min. A concentration of dissolved oxygen of the air dissolved water was about 14 mg/L.

Schematics of arrangements of pipelines and measurement apparatus are described in Fig. 4.27. Six injection pipes in each of which has three injection slits (check valve) were laid at 1.6 m intervals on the floor of the box. Before filling the sand, a flow rate of each pipe was adjusted to be almost the same as the others' by using a controlled valve of each pipe.

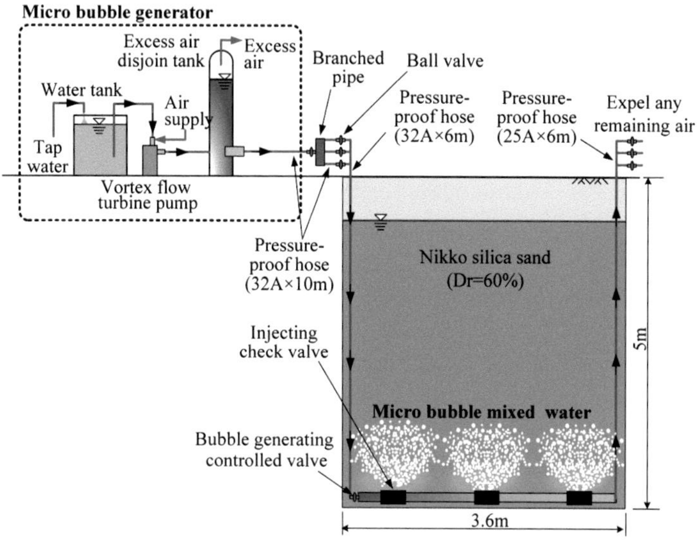

Figure 4.26 Outline of microbubble water injection and degassed water injection.

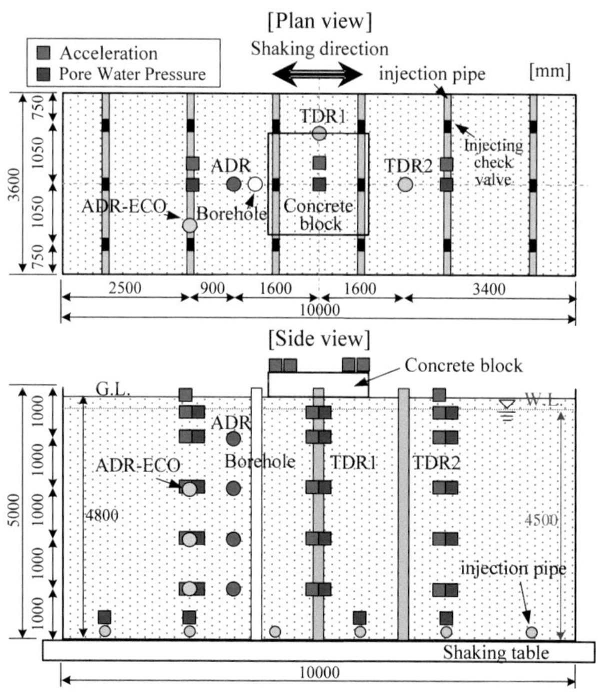

Figure 4.27 Arrangement plan of experimental ground.

On the other hand, degassed water, defined as Dissolved Oxygen value less than 3 ppm, was made using a degassed pump joined to the micro-bubble system in the condition not to feed the air. Except for this point, the other system for degassed water was the same as that of microbubble water.

Amplitude domain reflectmetry and time domain reflectmetry were used for measurement of saturation. These sensors were dielectric methods to measure soil moisture content and the degree of saturation measured to convert a volume water content measured by these sensors.

Figure 4.28 shows depth distributions of the saturation degree of the model grounds after injecting water. The saturation degree of the ground in the case of microbubble water injection became less than that in using degassed water injection with a difference of 2%–8%.

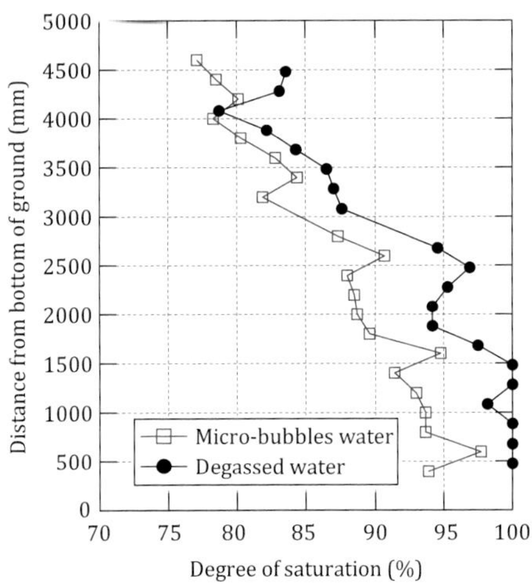

Figure 4.28 Distributions of saturation degree.

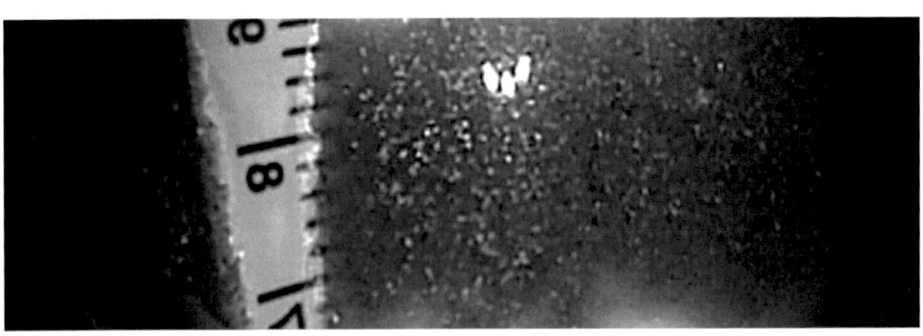

Photo 4.8 Existence of air bubbles in the experimental ground.

The existence of air bubbles was observed in Photo 4.8, taken at a height of around 1.0 m before the shaking test. Dots shining white in the photographs are bubbles. The same scenes were observed in other different depths as well. Therefore, it was confirmed that microbubbles fed in the ground remained as millimetre-sized bubbles among soil particles.

4.3.2.3 Seismic vibration tests

The following conditions of oscillation were set up: 20 waves of sinusoidal were used with 2 Hz of frequency for one stage and three stages of oscillations executed in which maximum acceleration of waves were 50, 100 and 150 Gal in order.

Horizontal acceleration records for each case are indicated in Fig. 4.29. In the microbubbles' case, the maximum acceleration at the ground surface was amplified to 180 Gal to the input acceleration of 100 and 210 Gal to the 150 Gal input. It was pointed out that considering those acceleration levels, liquefaction of the model ground was not so much observed than usually expected in these amplified levels of accelerations.

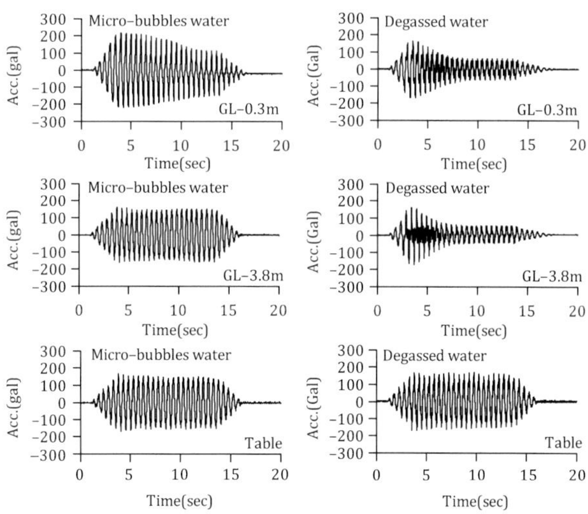

Figure 4.29 Comparison acceleration result of the ground.

The records of the excess pore water pressure are showed in Fig. 4.30. The maximum pore water pressures in microbubble water injected sand ground were smaller than that in the degassed water injected one at any depth and shaking level. Also, the increasing rate of the pore pressure in the microbubble water was smaller compared with the degassed water's results. According to the data, it was also confirmed that the microbubble water injection could suppress the increase of excess pore water pressure.

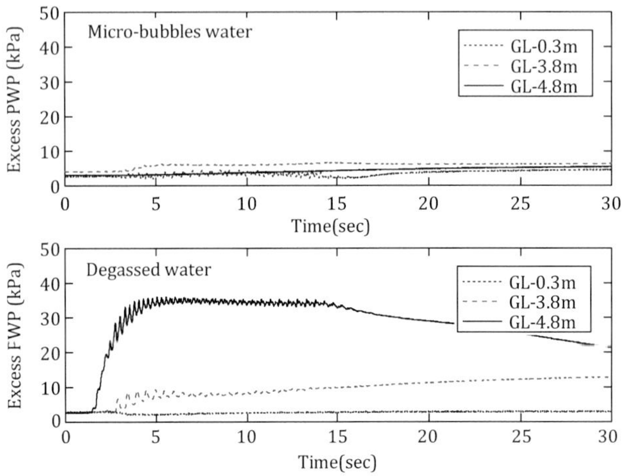

Figure 4.30 Comparison excess pwp result of the ground.

4.3.3 Conclusion

According to the result of the seismic vibration tests, the applicability of the microbubble water injection method was examined. It was observed that liquefaction did not occur in the condition of maximum acceleration of 200 Gal to the lower saturated sand ground of 80% with N-value of around 7.

If a degree of saturation of the ground can be lowered more homogeneously to about 80% with the efficiency of microbubble water injection, it can expect to gain enough strength against liquefaction too.

The brand new method of injecting water containing microbubbles into the ground will not only be simple and cost-effective but also friendly to environment countermeasure technology in the near future.

In addition, the method of injecting air into the ground is examined to be used for an application of environment field such as soil polluted purification as well as carbon dioxide emissions reduction.

It is also desirable that this technology would be applied to the various fields for a strategic position in near future.

References

1. Yoshimi Y, Tanaka K, Tokimatsu K (1998) Liquefaction resistance of a partially saturated sand, *Soils Found.*, **29**(3), 157–600.
2. Nagao K, Azegami Y, Yamada S, Suemasa N, Katada T (2007) A micro-bubble injection method for a countermeasure against liquefaction, in *4th International Conference on Earthquake Geotechnical Engineering* (Pitilakis, Kyriazis D, ed.), Springer, Thessaloniki, Greece.
3. National Institute for Land and Infrastructure Management. *Development of Planning and Management Technologies for the Ultra-long-life Houses*, Internet website, http://www.nilim.go.jp/engineer/index.html.

Chapter 5

Industrial Field

5.1 Environment Friendly Cleaning Technology using Microbubbles [1–4]

Makoto Miyamoto
Samsung Yokohama Research Institute, Minoh City, Osaka 562-0036, Japan
m.miyamoto@samsung.com

5.1.1 Introduction

As there have been an increasing number of different abnormal weathers reported, more responsible exertions are required in all industrial activities to improve the environment. Japan is facing an even higher challenge for environmental issues as an advanced country in order to meet the carbon dioxide emissions reduction targets that are based on the Kyoto Protocol and the government's statement of 25% reduction of carbon dioxide emissions in 1990 by 2020. Several legislations are underway, such as Pollutant Release and Transfer Register (PRTR) and revised Energy Conservation Law. Also, several systems are under discussion such as domestic trading for carbon dioxide emission credits, introduction of fixed price purchasing system of renewable energy and prevention tax of greenhouse gas (environment tax). In corporate management, efficient

Micro- and Nanobubbles: Fundamentals and Applications
Edited by Hideki Tsuge

ISBN 978-981-4463-10-2 (Hardcover), 978-981-4463-11-9 (eBook)
www.panstanford.com

management to intensified international competition as well as the environmentally friendly corporate activities such as the process development involving environment, safety and health [5,6] are becoming more important. Likewise, the most important issue in the manufacturer's future is the environment issue. Therefore, building manufacturing engineering systems to reduce energy consumption, CO_2 emissions and water consumption are highly demanded.

In electronics manufacturing for semi-conductors, liquid crystal displays and fine-structured equipment, the cleaning process is one of the most important processes to determine the qualities of the materials and products, in some cases, occupies a third of the total process. In fine-structured equipment manufacturing, in order to remove industrial grease (degreasing) used for cutting metals, the annual amount of 100,000 tonnes of volatile organic compounds, acidic and alkaline solvents, are consumed domestically and are also to be reduced in consumption. However, the solvent consumption is not reduced because the reduction has direct effects on the product qualities. On the other hand, environmentally friendly cleaning technologies are developed and commercialised, such as ionised water cleaning, functional water cleaning and ultrasonic cleaning.

Under the circumstances mentioned above, a new cleaning technology utilising microbubbles with diameters smaller than 0.1 mm is developed as an environmentally friendly cleaning technology. Different cleaning schemes using gas-liquid boundary control methods are summarised in the Table 5.1 in terms of the characteristics and control factors. The functional water is in the state of completely dissolved gas and ozonised water cleaning is a well-known example. The method utilising fine droplets and/or vapour in the gas is called mist cleaning or vapour cleaning. The method utilising impacts of high-speed water supplied to the objects to be cleaned is called jet cleaning. The microbubble cleaning utilises gas bubbles in the liquid to clean the object. Therefore, the microbubble cleaning method is different from the other methods.

The microbubble cleaning method has the following characteristics: the fine particles are expected to be removed by the electrostatic charge interaction effects to the bubble surfaces, and the organic solvent is expected to be removed by the hydrophobic interaction effects on the bubble surfaces. Other characteristics are

Table 5.1 Air/water interface control technologies in industrial cleaning

		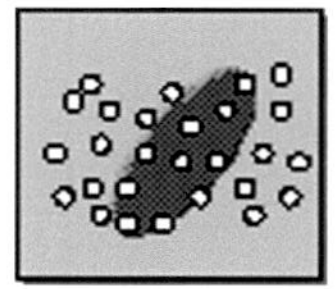	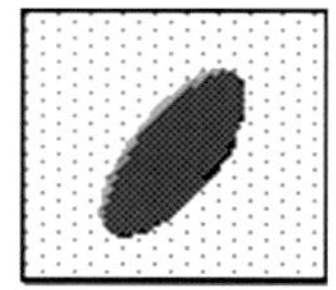	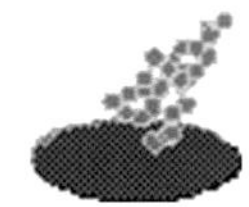
Method	**Dissolution/Dissolved**	**Bubbles**	**Mist/Vapor**	**Droplets**
Cleaning method	Functional water	Microbubble	HF Vapor Ozone Vapor	Dual-flow Jet Aerosols
Major physical factors	Henry constant (Solubility, ppm)	Void ratio (Surface area) (%)	Vaporization pressure (Surface area)(Pa, %)	Droplet diameters
Major control factors	Oxidation-reduction (Ozone, Hydrogen, pH) Mega-sonic	Bubble diameters, Densities	Ozone, Organic solvent, Temperature/Pressure	Flow velocities, Surface tensions
Typical removal target	Organic compounds, Particles, Metals	Organic compounds, (Particles)	Organic compounds, Particles	Particles
Challenges in commercialization	High solubility	**Denser bubbles**	Higher concentration of gas	Large area treatment

- the large surface area in the unit volume,
- the hydrophobic property on the bubble surface,
- the electrostatic charge in either positive or negative,
- the bubble collapse under a certain pressure or force,
- the increased interaction efficiency due to a long retention time in the liquid,
- the adaptability for complex structured objects due to the micro-sizes.

For the number of cleaning processes in the electronics manufacturing, the commercialisation of the microbubble method is reported to apply the degreasing process that consumes a large amount of solvent [1]. It is realised with the development of an environmentally friendly anti-bubble coalescence that enables high density bubbles. It is also a method with high cleaning abilities and separation abilities based on the large surface area. In this chapter, the main topic is the microbubble degreasing technology, and the outline of the technology and the effects on the mass-production processes are described.

5.1.2 The Problems of the Conventional Cleaning Methods and the Aim for Microbubble Cleaning Method

The principle of the cleaning methods using solvents and microbubbles are shown in Fig. 5.1. The fundamental cleaning principle of the solvent for general uses (used with ultrasonic in most cases) is based on solubility and physical stripping. The fundamental challenge is the disability in separating the pollutant from the solvent due to the high affinity of the solvent and removed substances. Thus, more removed substances retained in the solvent in larger processed amounts decrease the cleaning ability. There is an estimate of 50 billion yen annually consumed in domestic solvent use [6].

Miyamoto et al. have found out that the key characteristics of bubbles having hydrophobic surface and buoyancy overcome the fundamental challenge mentioned above. Adsorption of the grease to the bubble surfaces is expected in the water with the bubbles. Also, the grease is to be separated from the water after bubbles reach the water surface by the buoyancy. The cleaning processes in the mass-production lines require cleaning hundreds of objects of different sizes and complex shapes. The major requirements are the following:

- to reach in minutes the cleanliness at the same level as the solvent used,
- to ensure the uniform cleanliness on the object surfaces,
- to maintain the cleaning ability for a long term.

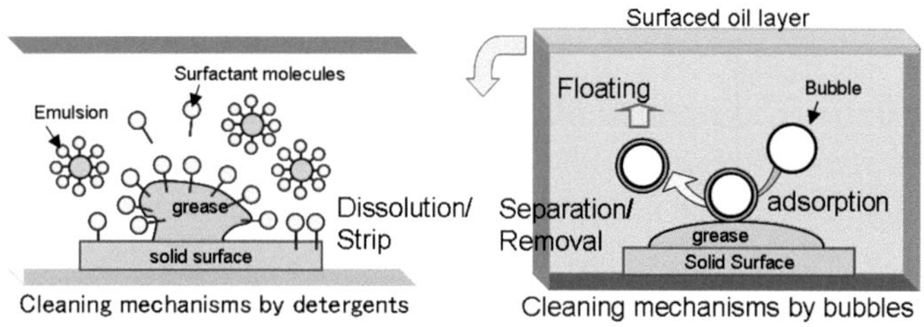

Figure 5.1 Comparing of degreasing principles in case of detergents (conventional) and bubbles (developments).

In order to reach the same cleanliness, the surface area of bubbles has to be increased to enable a large amount of degreasing. The volume of a bubble with diameter d is $\pi d^3/6$, and the surface area S is πd^2; thus, the surface area in unit volume S/V is $6/d$. The relation states that the surface area in unit volume is inversely proportional to the bubble diameter. Therefore, it is important to decrease the bubble diameters and increase the bubble number density for ensuring productivity. In addition, decreasing the bubble diameter is required to clean the object thoroughly and to ensure the uniformity of the cleanliness. An efficient levitation of the pollutant using the bubble buoyancy is important to maintain the cleaning ability for a long term.

On the other hand, in the case of particle removal, the electrostatic charges on the bubble surfaces in the water and the charges on the particle surfaces [7] are utilised. Terasaka et al. have investigated the adsorption of iron oxide particles onto microbubble surfaces and found that adsorption occurs in the pH region of negative microbubble surface charge and positive iron oxide surface charge, and the adsorption efficiency has strong dependencies to both surface charges [8]. In the case of fine particle removal, it is important to control the electrostatic charge interaction between the bubbles and the particles.

5.1.3 Generation of High-Density Microbubbles Using Environmentally Friendly Additives

Microbubbles are generated in the shearing forces applied between gas and liquid flows, breaking a large bubble into smaller ones with physical forces for example and in ejecting gas into liquid through fine holes. The major challenge for higher bubble number density is prevention of coalescing bubbles. Otherwise, bubble sizes become larger no matter what methods are used to generate bubbles in pure water. The selected photographs of coalescing bubbles using a high-speed camera are shown in Fig. 5.2. As shown in the Fig. 5.2, the bubbles become larger whenever the bubbles collide with each other in the pure water. Therefore, it is difficult to reach the required bubble number density for degreasing when the bubbles are generated in the pure water. However, certain chemical compounds have been reported to exhibit the effect of preventing coalescing bubbles [9]. Miyamoto et al. have investigated several additives as well as factors such as molecule mass, polarities and kinds and numbers of functional groups in terms of the environmental stress, costs, safety and usability under the following five conditions [10]:

Chemical property viewpoints

- soluble to water,
- safe chemical compounds (excluding poison deleterious substance, PRTR), effective in low density.

Cleaning viewpoints

- low forming,
- stable effect in warm water (higher than 50 °C).

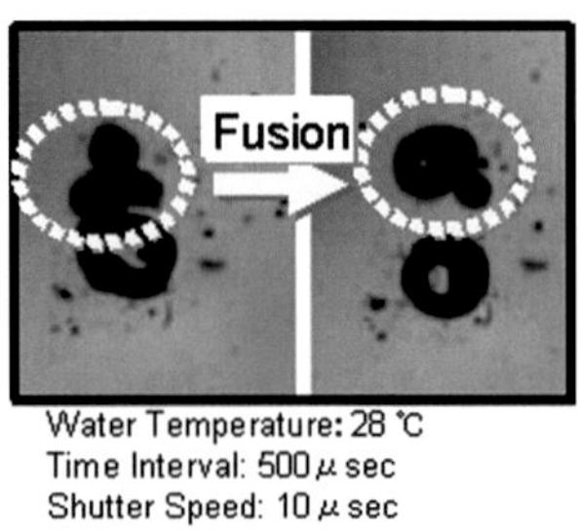

Figure 5.2 Photographs of bubble fusion obtained with high-speed microscopic video camera.

Photographs of the comparison of microbubbles with a smaller amount of developed additives and without any additives (both apparatus are identical) are shown in Fig. 5.3. The chamber of the microbubbles with the additives exhibits white turbidity like milk due to the high density bubbles. Thus, the preventive effect of the additives to bubble coalescence is significant.

Figure 5.3 Photographs of comparing high-density microbubbles generated in pure water with a trace amount of additive (left side) and bubbles generated in pure water without the additive (right side).

5.1.4 Principle of Oil Removal Using Microbubbles

One of the most important functions of microbubbles is to adsorb grease compounds such as machine oil and cutting oil onto the bubble surfaces. Typical images of visualised such phenomena using high-speed camera are shown in Fig. 5.4. It shows the details of the bubbles generated at the tip of the needle located below the slide glass that is covered with a thin cutting oil layer. The bubble that is floating from the syringe stays on the oil surface; at the same time, oil spreads to the entire bubble surface. Later, the bubble leaves the oil surface ③. The bubble surface is seen to have lost the shine that it originally had and formed a thin oil layer on the surface. The oil layer on the slide oil was removed within a few seconds ④. The oil exits the water in an unstable state due to its hydrophobic property, shown in the right in the Fig. 5.4. By supplying the hydrophobic field with bubbles, the oil molecules adsorb onto the bubble surfaces due to the affinity of the hydrophobic group. Although the amount of oil adsorption onto a single bubble is small, it is possible for the entire oil layer to be removed by repeated bubble interactions. This sequence of the reactions is the principle of oil removal using microbubbles.

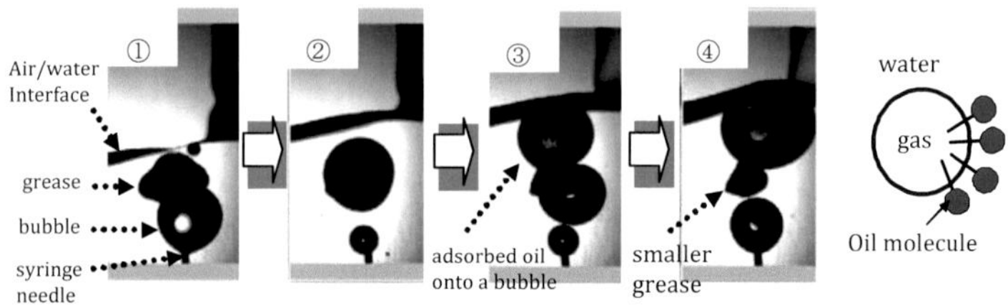

Figure 5.4 Images obtained from the video to show degreasing process using bubbles.

Although the amount of oil adsorption onto a single bubble is limited, oil removal from a solid surface is possible by repeated bubble interactions. Thus, a larger surface area is required to remove a large amount of oil in a realistic duration of operation. In other words, generating higher densities of bubbles is important. Also, smaller sizes of bubbles are required, proven by the experiment, in order to obtain uniform cleaning on the complex and uneven surface profiles.

The principle of the degreasing system is shown in Fig. 5.5. The microbubbles are generated and supplied to the groups of the target parts with high densities. Then, the buoyed oil is separated from the water of the cleaning bath by overflowing into the next chamber (overflow chamber).

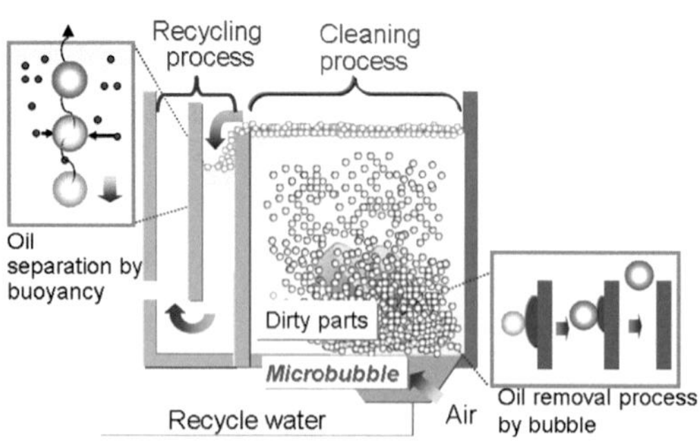

Figure 5.5 Images of cleaning mechanisms of microbubble.

5.1.5 Microbubble Cleaning System and the Characteristics

As mentioned in the above sections, the high density microbubble generations using a small amount of additives and the principle of oil removal using the microbubble are described. Microbubble cleaning systems are developed utilising those advantages. An image of a developed microbubble cleaning system is shown in Fig. 5.6. The system's major function is to realise highly efficient interactions between the microbubbles from the generator and the target parts in the basket. Also, in the case of microbubble cleaning, an oil layer removal mechanism and a separated overflow chamber are implemented due to the property of the oil–water separation. The advantages of the system are the following:

- Low cost: low running cost is achieved using water-based cleaning liquid.
- Low environmental load: waste can be drained due to water-based cleaning liquid.
- Safe: small amount of additives is the only chemical compound and poisonless.

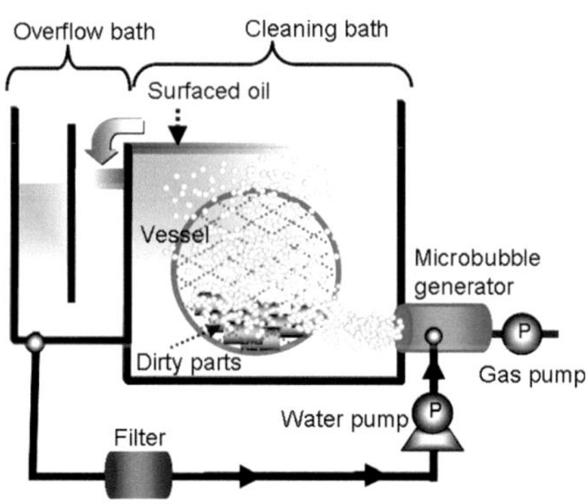

Figure 5.6 Schematic illustrations of microbubble cleaning system.

5.1.6 Effect of Microbubble Cleaning

The comparison of a mechanical part before and after microbubble cleaning are illustrated in Fig. 5.7, originally contaminated with water-based cutting oil that is difficult to remove. The oil adhered to the surface before cleaning is completely removed by microbubble cleaning and rinsed with water that partly remained on the surface, showing that the surface is now seen to be hydrophilic.

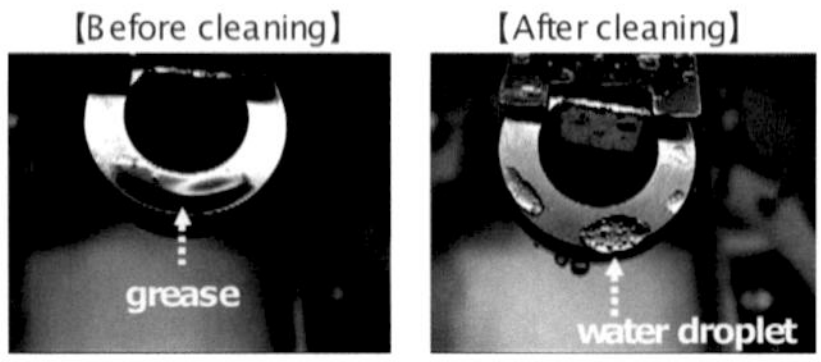

Figure 5.7 Photographs of a test piece stained with hydrophobic grease (left side) before and (right side) after 10 s microbubble cleaning.

The quantitative analysis on the microbubble cleaning performance is shown in Fig. 5.8. The analysis is based on the following cleaning procedures and the oil densities (μg/cm^2) are calculated. First, cylindrical parts are contaminated with hydrophobic oil on the entire surfaces. Second, microbubble cleaning is performed for 2 min. Third, the parts are rinsed with pure water and dried with nitrogen gas spray. Finally, the remaining oil amounts are measured. The remaining oil density after cleaning only with water is approximately half of the original, 1280 μg/cm^2, and the density after bubbles without the additives, where the bubble diameters are in mm, is 180 μg/cm^2. In comparison to those, the oil density after microbubble cleaning is 4.8 μg/cm^2; thus, the performance is 100 times greater than cleaning only with water and 30 times greater than normal bubble cleanings. The performance is comparable to alkaline cleaning liquids available in the market. Microbubble cleaning is proven to be able to achieve the high performance that has been achieved only with the solvents as the results of the surface area increased due to the smaller sizes of the bubbles and uniform distribution to the complex surfaces of the targets.

The most important outcome of microbubble cleaning is to reuse the cleaning water repeatedly. A system with cleaning liquid that has to be replaced often would not be commercialised. The results of the repeated cleaning with the developed cleaning system are shown in Fig. 5.9 that shows the remaining oil density after hundreds of parts were cleaned. The cleanliness of the parts remains constant in the developed system compared to that using alkaline cleaning liquid showing that cleanliness decreased with the number of the batches. Therefore, repeated cleaning performances are confirmed based on microbubble cleaning having the high oil removal abilities and oil–water separating properties.

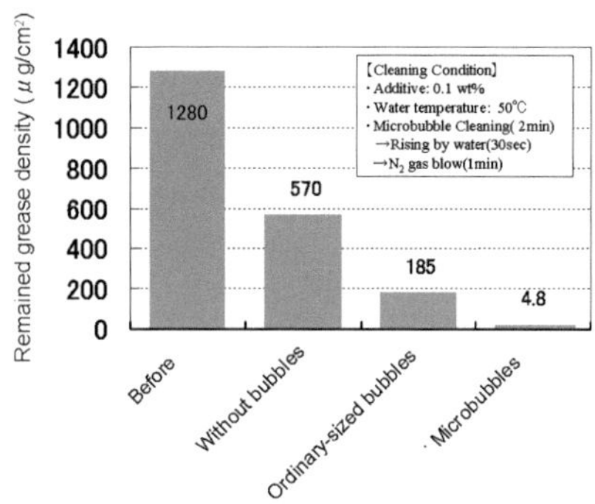

Figure 5.8 Efficiency of removal of persistent hydrophobic grease.

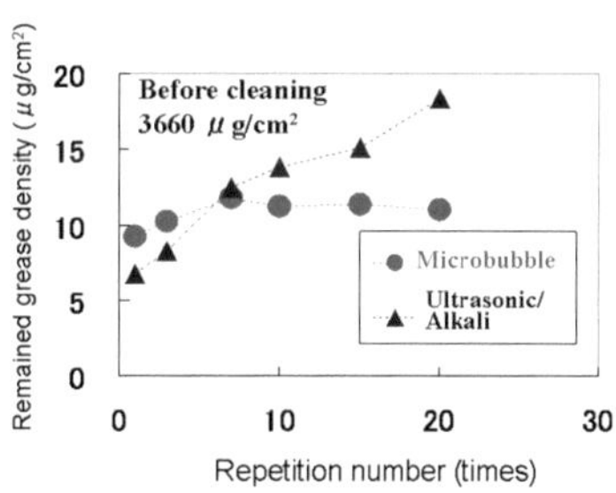

Figure 5.9 Grease removal efficiency in repeated batch treatments cleaning for multiple test pieces.

5.1.7 Commercialisation Verifications

The cleaning level variations in microbubble cleaning systems implemented on a mass-production line are described in this section in comparison to that for the currently operated system that utilises alkaline solvents. The cleanliness of the microbubble system is less than 10 μg/cm^2, which shows no variations over a month of operations. The right hand side image of Fig. 5.10 shows the cleaning water and the removed oil sampled after 6 month operations. The cleaning water exhibits complete transparency after cleaning tens of thousands of parts due to the oil separating properties of microbubbles.

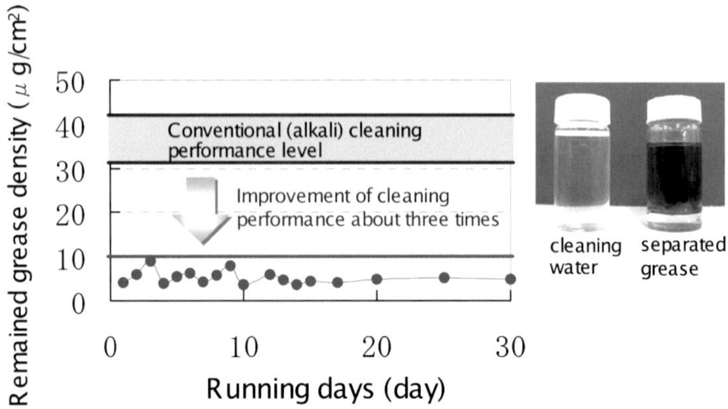

Figure 5.10 Cleaning performance by microbubbles in mass-production line and photographs of cleaning water and separated oil obtained after 180 days.

In the case of the developed system replacing the current system, the estimated CO_2 emission is less than one thousandth, and one twelfth is estimated for alkaline solvent replacement. If all the solvent cleaning systems in Japan were replaced, the estimated CO_2 emission reduction of 400,000 tonnes would be possible. In addition to the large environmental load reduction, a running cost reduction to less than one tenth the conventional solvent cleaning systems is reported. Also, an improved product quality is realised using the stable cleaning process developed.

5.1.8 Conclusions

The 21st century is called the environment century, and the activities for reducing the environmental load are more important than ever in electronic fabrication processes. Especially in the cleaning processes, a huge amount of solvent is consumed to remove large amounts of organic contaminants. The high density microbubble cleaning method is expected to reduce the environmental load and the operational cost due to the increased cleaning performance with increased surface area and the advantage of the oil separating properties.

As described above, the microbubble cleaning method is completely new and different from the conventional methods, it is said to be based on a new principle of a boundary controlled cleaning method. More research activities in the near future are expected to widen the applied field with high environmental loads and high cleanliness requirement, such as semi-conductor and liquid crystal panel fabrications, beyond the degreasing. In conclusion, the microbubble cleaning method is believed to become the standard method to contribute to environmental conservation in the future.

References

1. Miyamoto M, Ueyama S, Matsui H, Kuzumoto M (2008) *Mitsubishi Electric Technical Report*, **82** (8), 489 (in Japanese).
2. Miyamoto M, Ueyama S, Hinomoto N, Saitoh T, Maekawa S, Hirotsuji J (2007) *JpnJ Appl Phys*, **46**(3A), 1236.
3. Miyamoto M (2007) *Clean Technology*, **17**, 24 (in Japanese).
4. Miyamoto M, Ueyama S (2005) *Handbook of Microscale and Nanoscale Heat and Fluid Flow*, NTS, Japan, 340 (in Japanese).
5. Handbook of Industrial Cleaning Technology, *Realize*, 3 (1994) (in Japanese).
6. Search Report for Industrial detergents in Japan, Japan Industrial Conference on Cleaning, p6 (1993) (in Japanese).
7. Takahashi M (2005), *J Phys Chem B*, **109**, 21858.
8. http://www.applc.keio.ac.jp/~terasaka/
9. Takagi S (2004) *Nagare*, **23**, 17 (in Japanese).
10. Ueyama S, Japanese Patent No. 4210536 (in Japanese).

5.2 Characteristics of Gas Oil Containing Micro- and Nano Air-Bubbles

Yasuhito Nakatake, Takashi Watanabe, and Toshihiko Eguchi
Department of Mechanical Engineering, Kurume National College of Technology, Fukuoka, 830-8555, Japan
nakatake@kurume-nct.ac.jp

5.2.1 Introduction

The authors are conducting research on the reduction of fuel consumption, smoke and harmful emissions including PM of diesel engines using micro- and nano air-bubbles mixed in gas oil [1–3].

The contents of the research that have been presented were that of a performance test of a diesel engine which used gas oil mixed with micro- and nano air-bubbles within the fuel tank, and as a result, fuel consumption and smoke were reduced by 2.5% to 13.7% on average against the partial load change observed in an industrial turbo-diesel engine with a four-cylinder [1] and a single-cylinder marine engine [2,3].

In this paper, an on-board–type micro- and nano air-bubble mixture device that generates micro- and nano air-bubbles and mixes them into gas oil within a fuel supply line was used as a method maximising the effect of mixed micro- and nano air-bubbles, with the mounting of the device on a real vehicle taken into consideration, and the physicochemical characteristics of the gas oil mixed with micro- and nano air-bubbles generated in this device and the effect of decreasing fuel consumption of a diesel engine will be described. The effects of the two generating methods of ejector-type (EMB) and pressurised dissolution-type (PMB) on fuel properties and engine performance were examined.

What is called "micro- and nano air-bubble mixed gas oil" is the gas oil mixed with nano air-bubbles within the fuel tank or mixing tank, which does not mean that the fuel is mixed with micro air-bubbles of several microns to several dozen micros is led to the engine as it is, and presumably, the micro-sized air-bubbles emerge on the surface within the tank, and only the air-bubbles of a nano-size level are led to the engine because of the terminal velocity of air-bubbles, the size of tank and the engine fuel consumption rate. The term "micro bubble gas oil" has caused misunderstanding; so, it is called "micro- and nano air-bubble mixed gas oil" in this study.

5.2.2 Physicochemical Characteristics of Micro- and Nano Air-Bubble Mixed with Gas Oil

The characteristics of micro- and nano air-bubble mixed gas oil become different according to not only the kinds of mixing device and operating conditions, but also the time they are mixed (that is the discharge rate (circulation) against container size and the lapse of time (after the mixing stopped). Therefore, an in-line type micro- and nano air-bubble mixture device to be installed in the fuel line was experimentally produced for mounting on a real vehicle, and the physicochemical characteristics of the micro- and nano air-bubble mixed gas oil produced by the device were examined.

Figure 5.11 shows an outline of an on-board type micro- and nano air-bubble mixture device. Figure 5.11a shows appearance photographs of the mixture device, and Fig. 5.11b is its schematic diagram. As shown in Fig. 5.11b, the gas oil sent from the fuel tank is pressured through a micro- and nano air-bubble mixture device and absorbs air by itself in the ejector component, and the gas oil with micro- and nano air-bubbles is sent to a mixing tank. When the gas oil surface in the mixing tank rises and the switch of level indicator comes on, the switch shifts to a circulation line in the mixing tank. This circulation plays a role in increasing the gas oil passage through the ejector component and enhances the effect of mixing micro- and nano air-bubbles.

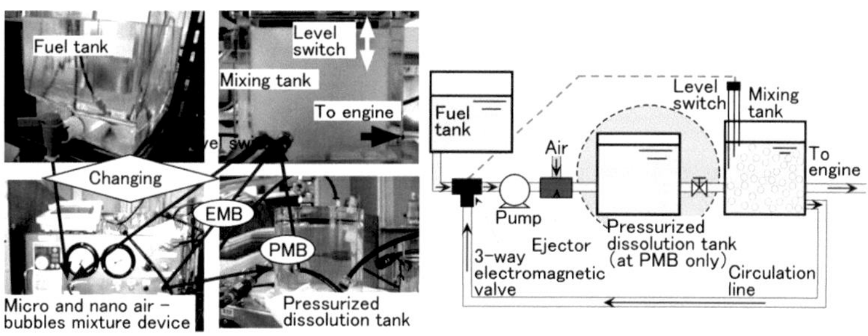

(a) (b)

Figure 5.11 On-board type micro- and nano air-bubbles mixture device. (a) Appearance photograph. (b) Schematic diagram.

As shown in Fig. 5.12, an EMB [1–3] was fundamentally employed for micro- and nano air-bubble mixture, and the method can be changed to a pressurised dissolution-type too, by placing it upstream in the mixing tank. The PMB is to once dissolve air by increasing the pressure in a pressurised dissolution tank by narrowing down a discharging port of the tank using a valve and then releasing the pressure in a mixing tank downstream. As a result, micro- and nano air-bubbles can be generated. The characteristics of micro- and nano air-bubble mixed gas oil were measured using the gas oil in the mixing tank or extracted from it. The details of the mixing condition cannot be described. However, naturally, it is impossible to establish the same conditions of the pressure of the pump, flow and the pressure of air absorbed by the gas oil for both the EMB and PMB. Therefore, the conditions of mixing operation for both methods were established so that the changes of characteristics of fuel before and after being mixed with micro- and nano air-bubbles were remarkable.

Figure 5.13 shows a droplet distribution after atomisation measured by an emersion method atomising against a water tank at a fuel injection pressure of 20 MPa under atmospheric pressure using an EMB generator. The average diameter of the droplets of sprayed liquid after injection decreased from 20.0 to 17.5 μm caused by being mixed with nano air-bubbles, which suggests that the fuel was further atomised by being mixed with nano air-bubbles.

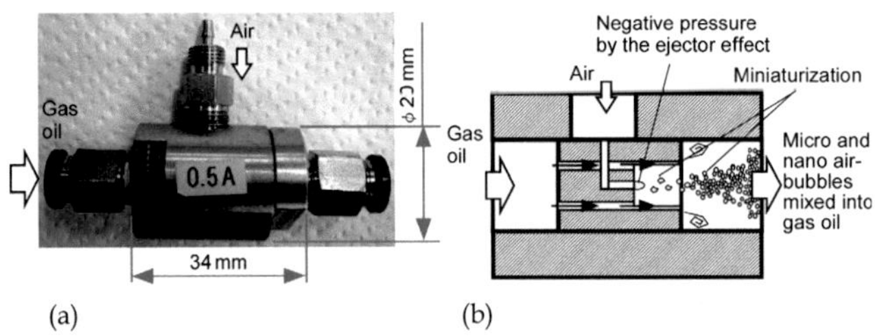

Figure 5.12 Ejector-type micro- and nano air-bubbles mixture device. (a) Appearance photograph. (b) Internal structure.

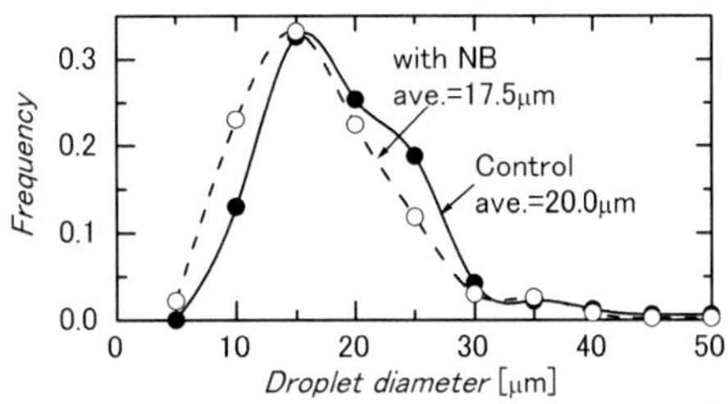

Figure 5.13 Droplet diameter distribution.

Aniline points of gas oil mixed with micro- and nano air-bubbles by each method of EMB and PMB were measured, and the calculated results of cetane numbers are shown in Fig. 5.14. In the case of EMB, the cetane number increased by 0.9% compared to the control gas oil, and in the case of PMB, it increased by 1.9%, which is equivalent to approximately twice as large as the number of EMB.

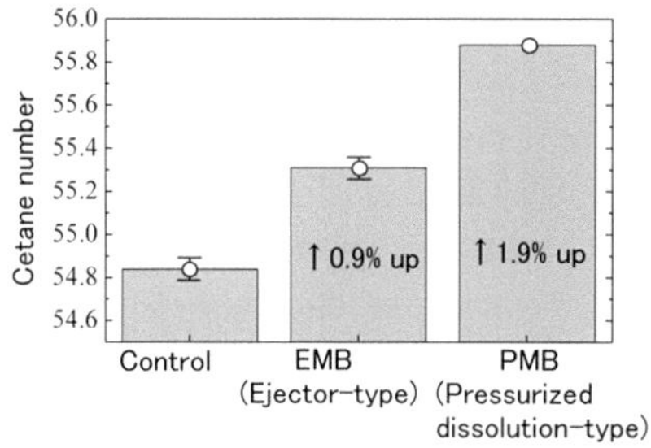

Figure 5.14 Comparison of the cetane number by the difference of the mixture method.

Figure 5.15 shows the changes with the passage of time of the characteristics of fuel when it was mixed with micro- and nano air-bubbles for 1 h by each method. The generation methods were (a) EMB and (b) PMB. The left vertical axes of both graphs show pH (concentration of hydrogen ions) and fuel temperature [°C], and the right vertical axes show relative dissolved oxygen (DO_R). DO_R is a non-dimensionalisation with saturated-dissolved oxygen of the water of the same temper as that of gas oil because the temperature of gas oil increases by being mixed with micro- and nano air-bubbles; however, the saturated-dissolved oxygen of gas oil is unknown.

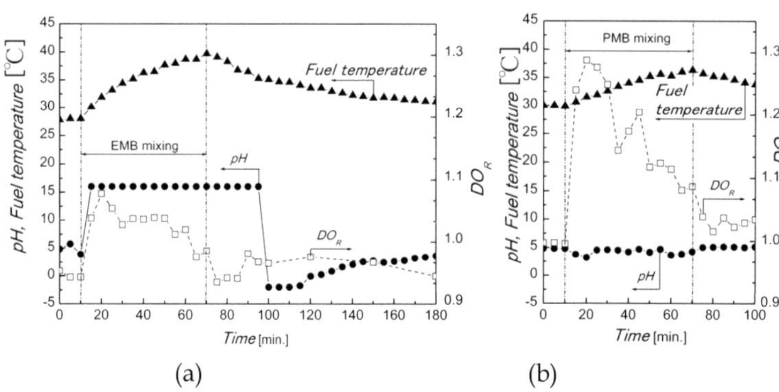

Figure 5.15 Change of the minute air bubbles light oil property by the difference of the mixture method. (a) EMB; (b) PMB.

As shown in Fig. 5.15, DO_R increased immediately after the start of mixture by both methods (a) and (b), and then, it gradually decreased along with the increase in the temperature of the fuel. The width of the increase was larger in (b) PMB more than in (a) EMB. It increased from 1.0 to 1.3, approximately by 30%, in the case of (b) PMB. Meanwhile, it increased from 0.95 to 1.08, approximately by 14%, in the case of (a) EMB, which means the increase in the EMB is about half of that in the PMB. The increases in DO_R are similar to those of the cetane numbers as shown in Fig. 5.14. After stopping mixing, DO_R returned to the original value steadily but swiftly.

The pH was around 4.8 before bubble mixing. In the case of (a) EMB, the pH drastically changed as soon as the bubble mixture started, and it increased up to 16.0 of the maximum limited value of the gauge. After bubble mixture stopped, the same condition was kept for 30 min. After that, it drastically decreased to –2.0 of the minimum limited value, and later, although it gradually increased, it kept a low value for a longer time than the control gas oil. The large fluctuation in the pH caused by micro- and nano air-bubble mixture is considered to be the effect of the electrical potential around bubbles. Meanwhile, in the case of (b) PMB, the pH slightly decreased to approximately 3.2 during bubble mixture, and when the bubble mixture was stopped, it increased to approximately 5.0, which was slightly higher than the original value and 0.2 higher than the control gas oil.

The difference in the principle of generation caused by different generation methods is considered greatly to affect the difference in these changes. In the case of PMB, the micro- and nano air-bubbles are generated by the fuel to be mixed with bubbles being pressured in a pressure tank and being released to atmospheric pressure. In this method, because air is dissolved into fuel by the fuel with bubbles being pressurised, plenty of oxygen is dissolved into the fuel. As a result, DO_R becomes high. Meanwhile, in the case of EMB, it is considered that although plenty of oxygen cannot be dissolved into the fuel like in the case of PMB, many chemical bonds are broken because the bubbles are sheared more rapidly when micro- and nano air-bubbles are generated. As a result, many ions are generated, and the pH rapidly changes. In the case of EMB, it is considered that shearing actions occur between molecules and weaken molecular attractions while micro- and nano air-bubbles are generated, and furthermore, when micro- and nano air-bubbles burst, because it generates a large energy, the combinations between molecules are broken, and as a result, the molecules decompose into low molecules. Due to these phenomena, fuel performance is expected to improve.

The decreasing rate of surface tension of EMB was larger than that of PMB. The decrease in surface tension is an important factor for the improvement of fuel performance because it leads to miniaturisation of the diameter of atomised particles when the fuel is atomised into the cylinders of the engine.

Furthermore, it is known from Fig. 5.15a that the duration of pH change caused by micro- and nano air-bubbles is longer than that of PMB (Fig. 5.15b). In the experiment device and real engine used in this experiment, fuel was sent to the engine through the fuel pipe and filters. It takes some time to generate micro- and nano air-bubbles in a mixing tank and use the fuel in an engine, and it is also important that the duration of chemical changes of EMB is long.

In addition to the difference in the abovementioned principles of generation, the differences in the amount of air intake and pressure caused by fuel flow passing through the ejector component are considered to have a great effect on the main cause of the differences in the changes in properties between both methods. In the case of PMB, the fuel was pressurised in the pressurised tank by the fuel flow channel being narrowed down and the flow channel resistance being made larger, the fuel flow passing through the ejector component became less than that of EMB. Meanwhile, in the case of EMB, the fuel flow and flow velocity passing through the ejector component became larger, and as a result, it was possible to mix gas oil with micro nano air-bubbles having a high effect on reforming fuel with great shearing action and constrictive action by the raising vacuum degree of the ejector component.

5.2.3 Reduction of Fuel Consumption of a Diesel Engine

A performance test of a diesel engine was conducted using control gas oil and fuels generated by the two kinds of methods mixing with micro- and nano air-bubbles.

A single-cylinder air-cooled diesel engine, product of Fuji Heavy Industries Ltd, Model No. Robin DY27-2DS (Direct

Injection System, bore × stroke: ϕ75 × 60, engine displacement: 265 cc, compression ratio: 21, the maximum output power: 3.9 kW/3000 rpm, the maximum torque: 12.8 Nm/2400 rpm) was used as a test engine.

A partial load experiment was conducted using the control gas oil and the fuels generated by the EMB and PMB methods, with the load changed from 50 to 650 kPa and engine speed from 1500 to 2100/100 rpm.

In Fig. 5.16a, the Δ symbol and the broken line show the experimental results of the fuel generated by EMB; in Fig. 5.16b, the □ symbol and the broken line show those of PMB; and in both figures, the • symbol and the broken line show those of the control gas oil.

The horizontal axes of both graphs show the corrected brake mean effective pressure, P_{mec} [kPa], which is the brake average effective pressure presenting the load of the engine adjusted by climatic conditions. The vertical axes show brake specific fuel consumption [g/kWh] presenting fuel consumption rate per unit of time and per unit of output.

First, as shown in Fig. 5.16a of EMB, the decreasing rate of fuel consumption was 12% at load average, and a maximum decreasing rate of 20% was observed in the neighbourhood of $P_{\mathrm{mec}} = 650$ kPa. As shown in Fig. 5.16b of PMB, the decreasing rate of fuel consumption was 7% at load average, and a maximum decreasing rate of 13% was observed in the neighbourhood of $P_{\mathrm{mec}} = 200$ kPa. As a result of the engine performance test, EMB results in a larger effect of reduction of energy consumption in a more extensive range than PMB.

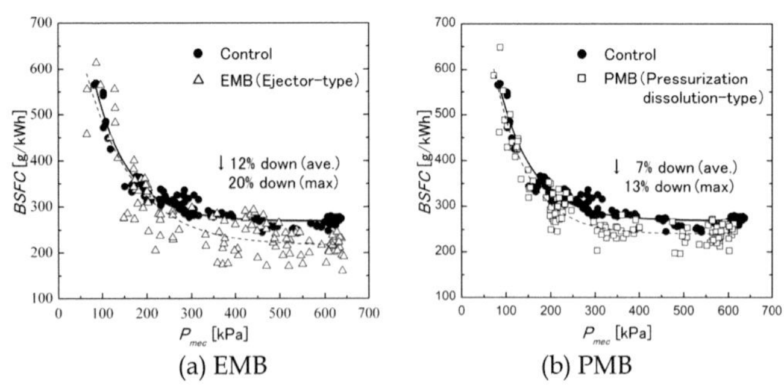

Figure 5.16 Mileage reduction of the diesel engine by the difference of the mixture method.

5.2.4 Summary

The effect of decreasing fuel consumption of diesel engines differs by the methods of mixing gas oil with micro nano air-bubbles. The PMB that can keep the cetane number and DO_R value high resulted in decreasing fuel consumption rate by 7% at load average and a maximum decreasing rate of 13%. Meanwhile, the EMB that is superior in the change of pH and decreasing effect of surface tension resulted in decreasing fuel consumption rate by 12% at load average and a maximum decreasing rate of 20%. The reductions of fuel consumption achieved by these methods are considered to have been caused not only by the physicochemical effects of reforming fuel, but also the effect of nano air-bubbles mixed in the fuel and led to the engine.

Mixing fuel with micro- and nano air-bubbles does not increase the amount of heat generation, but it increases useful work output by shortening combustion time, and the micro- and nano air-bubble fuel presents its effectiveness for internal-combustion engines.

References

1. Nakatake Y, Watanabe T, Harada S, Sugai M, Sezaki T, Eguchi T (2008) Characteritics of diesel combustion with ejector-type micro-bubble mixed into gas oil, *Proceeding of JSAE*, **115-08**, 155-20085669, 17–20 (in Japanese)
2. Nakatake Y, Watanabe T, Eguchi T (2007) Combustion improvement for diesel engines with ejector-type micro-bubble mixed fuel, *JSME Series B*, **73–735**, 2368–2374 (in Japanese).
3. Eguchi T, Watanabe T, Nakatake Y (2007) Improvment of diesel combustion with ejector-type micro-bubble mixed into fuel, in *The Latest Technology on Microbubbles and Nanobubbles* (Tsuge H, ed), CMC publication, Japan, pp 149–156 (in Japanese).

5.3 Suppression Technique for Aggregation of Nanometre-Sized Diamond Particles Using Acoustic Cavitation Bubbles

Shinichi Takeuchi[a] and Takeyoshi Uchida[b]

[a]*Department of Clinical Engineering, Faculty of Biomedical Engineering, Toin University of Yokohama, Yokohama, Kanagawa, 225-8503, Japan*

[b]*National Metrology Institute of Japan, National Institute of Advanced Industrial Science and Technology, Tsukuba, Ibaraki, 305-8568v, Japan*

shin1@cc.toin.ac.jp, takeyoshi.uchida@aist.go.jp

5.3.1 Introduction

The phenomenon of acoustic cavitation occurs by ultrasound irradiation of a large amplitude in water [1–6]. This phenomenon is observed by cavitation bubbles generated in water with negative pressure, which is produced by ultrasound stronger than inner directional forces due to viscosity and surface tension of surrounding liquid. The studies for application of the acoustic cavitation for industrial and medical fields are accomplished actively in our laboratory.

New ultrasound diagnostic systems tend to irradiate larger intensity ultrasound than the conventional systems to improve the quality of diagnostic image and to obtain new diagnostic information. On the other hand, tissue or cells may be damaged by acoustic cavitation. It is also important to study tissue damage due to acoustic cavitation or risk management of acoustic cavitation. Acoustic cavitation is thought of as a two-edged sword because it acts as a convenient tool and a dangerous weapon.

Study on surface modification for disaggregation or avoidance from re-aggregation of nanometre-sized diamond particles for polishing abrasive agents or solid lubricants by using acoustic cavitation bubbles is introduced in this section. Tiny diamond particles are very easy to aggregate. Surface modification is required for disaggregation or avoidance from re-aggregation of nanometre-sized particle polishing abrasive agents. Aggregated nanometre-sized diamond particles were disaggregated by the mechanical method such as ball milling, and re-aggregation was avoided by adding surfactant in conventional methods. However, there were problems such as the polishing surface being contaminated by broken pieces of milling ball and surfactant. We use acoustic cavitation for disaggregation of aggregated nanometre-sized diamond particles and avoid re-aggregation by irradiation of ultrasound into the slurry (suspension) of nanometre-sized diamond particles. There is no risk of contamination like the conventional polishing method using ball mill and surfactant.

5.3.2 Ultrasound, Acoustic Cavitation and Cavitation Bubbles

The phenomenon of acoustic cavitation occurs by generation of ultrasound of a large amplitude in water [1–6]. The acoustic cavitation is the phenomenon where bubbles are generated when tension from ambient negative pressure by ultrasound irradiation in water is greater than surface tension and viscosity of atmospheric liquid. Our normal atmospheric pressure is 1 atm, negative pressure. The behaviour of acoustic cavitation is shown in Fig. 5.17.

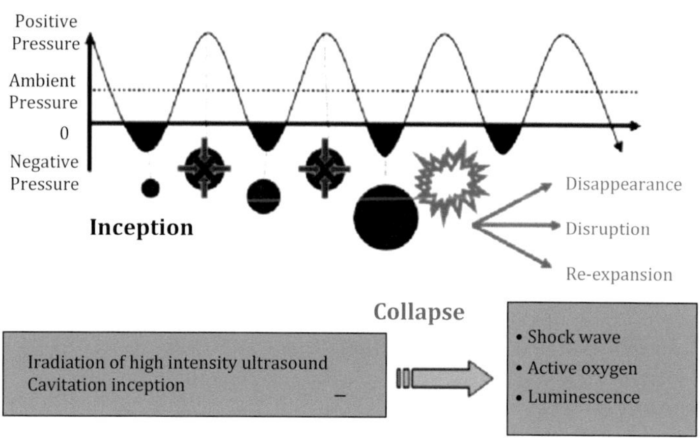

Figure 5.17 Mechanism of acoustic cavitation, generation and collapse of cavitation bubbles in water by irradiation of ultrasound with large amplitude.

The cavitation bubbles expand and shrink repeatedly due to change in sound pressure by ultrasound and finally collapse at positive sound pressure of irradiated ultrasound. There are two types of acoustic cavitation: "stable cavitation" and "inertial cavitation". The size of cavitation bubbles change (expand and shrink) cyclically in the "stable cavitation". The cavitation bubbles collapse at positive sound pressure of irradiated ultrasound in the inertial cavitation. The cavitation bubbles expand and shrink repeatedly at the same frequency as irradiated ultrasound frequency. Acoustic cavitation is a high speed phenomenon compared with the time required for heat transfer between the inside and outside cavitation bubbles. Therefore, the insides cavitation bubbles are in an adiabatic state. Hot spots with high temperature and high pressure (higher than a few thousands degree Celsius and a few thousands atmospheric pressure) was formed in the cavitation bubbles.

It is well-known that water molecules are dissociated, and the various activated oxygen species (free radicals) [4,7] like hydroxyl radical, singlet oxygen, superoxide and so on are generated by effect of shock waves and hot spots with high temperature and high pressure, due to acoustic cavitation. Sono-chemical reactions by acoustic cavitation are shown in Eq. 5.1.

$$\begin{aligned}
&H_2O \rightarrow H\cdot + OH\cdot \\
&H\cdot + OH\cdot \rightarrow H_2O \\
&2H\cdot \rightarrow H_2 \\
&2OH\cdot \rightarrow H_2O_2 \\
&2OH\cdot \rightarrow O\cdot + H_2O \\
&2O\cdot \rightarrow O_2 \\
&\tfrac{1}{2}O\cdot + 2H\cdot \rightarrow H_2O \\
&O\cdot + H_2O \rightarrow H_2O_2
\end{aligned} \tag{5.1}$$

5.3.3 Nanometre-Sized Diamond Particles

Diamond particles with sizes smaller than sub-microns are used for precise polishing abrasive agents and solid lubricants. Particularly, sub-micron diamond particles are required for precise polishing abrasive agents for the hard disk device with high recording density and magnetic head due to evolution of small-sized laptop computer (mini-note book) and high performance cell phones like smart phone. The gap between the hard disk surface and magnetic head is smaller than about 0.2 μm in miniature hard disk devices with high recording density [8–10]. Thus, rough hard disk surface causes failure of head crash. It is important to avoid failure to polish hard disk surfaces precisely before coat of magnetic media.

Nanometre-sized diamond particles with primary sizes smaller than sub-microns are expected as abrasive agents to polish disk surfaces of such hard disk devices precisely [8]. Cluster diamond particles are a kind of nanometre-sized diamond particles with primary sizes smaller than about 5 nm.

Cluster diamond particles can be obtained in the soot resulting from the incomplete combustion of exploded trinitrotoluene (TNT) [9–12]. Such tiny diamond particles have very small primary sizes, but aggregated with sizes greater than 10 microns immediately after explosion. They should be disaggregated to the primary size and should be kept at their primary sizes for long periods.

Examples of photographs of nanometre-sized diamond particles are shown in Fig. 5.18. Figure 5.18a shows the outer view of nanometre-sized diamond particles, Fig. 5.18b is the transmission electron microscope (TEM) image which shows primary particles of nanometre-sized diamond particles and Fig. 5.18c is the Scanning Electron Microscope (SEM) image which shows aggregated nanometre-sized diamond particles.

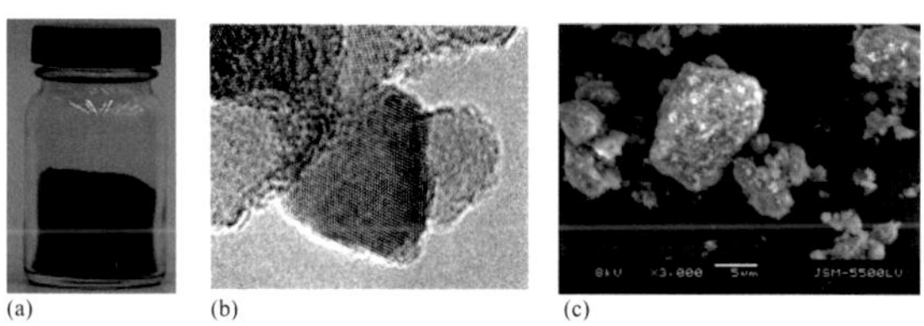

Figure 5.18 Example images of nanometre-sized diamond particles. (a) Out of view. (b) Primary particles observed. (c) Aggregated diamond particles by TEM observed by SEM. TEM, Transmission Electron Microscope, SEM, Scanning Electron Microscope.

Aggregated nanometre-sized diamond particles were disaggregated by the mechanical method such as ball milling, and re-aggregation was avoided by adding surfactant in conventional methods for fabrication of nanometre-sized diamond particle polishing abrasive agents. However, there are problems of inefficiency of disaggregation using ball mill and contamination of surfactant.

Therefore, we disaggregate the aggregated diamond particles using shockwaves from acoustic cavitation [13–15] by irradiation of ultrasound into the slurry, in which nanometre-sized diamond particles are suspended in distilled water. The magnitude of surface potential (Zeta potential) increased by forming OH group on the surface of disaggregated diamond particles because activated oxygen species (hydroxyl radical, OH radical) are generated by acoustic cavitation. We proposed a method to avoid re-aggregation using electrostatic repulsion between diamond particles, which are generated in this time.

5.3.4 Avoidance from Aggregation of Nanometre-Sized Diamond Particles by Ultrasound

5.3.4.1 Apparatus for ultrasound irradiation

There are two types of ultrasound irradiation apparatuses for dispersion of such particles (disaggregation of aggregated particles and suspending the disaggregated particles homogeneously into the medium) in general. One is the ultrasound irradiation method by insertion of ultrasound horn driven with Langevin-type transducer into the particle slurry (suspension) in a vessel shown in Fig. 5.19a. Another is an ultrasound irradiation method by setting a beaker with particle slurry in a vessel of an ultrasound cleaner shown in Fig. 5.19b.

It was thought formerly that the method using an ultrasound horn, shown in Fig. 5.19a, is advantageous compared to the method using an ultrasonic cleaner shown in Fig. 5.19b. Cavitation bubble jets are radiated together with high power ultrasound waves travelling from the ultrasound horn in the method shown in Fig. 5.19a. However, it had a problem that the effective area to the dispersion was confined to the vicinity of the ultrasound horn because ultrasound propagation to the entire slurry was inhibited due to reflection by cavitation bubble jets with extremely low specific acoustic impedance. On the other hand, there was a problem that ultrasound energy could not propagate enough into the particle slurry in the glass beaker because ultrasound was reflected at the surface of the glass beaker with specific acoustic impedance extremely different from that of particle slurry in the method of ultrasound irradiation by setting a beaker with particle slurry in a vessel of an ultrasound cleaner shown in Fig. 5.19b. It was thought that the method shown in Fig. 5.19a has higher efficiency than the method shown in Fig. 5.19b. However, when ultrasound was irradiated into the particle slurry in the water directly without the use of the glass beaker, standing waves were formed by irradiated ultrasound from the bottom of the vessel and reflected waves from the liquid surface. Then, it was found that particles and cavitation bubbles are trapped at loops of sound pressure standing waves, and particles could be disaggregated and their surface could be modified stably using shockwaves and sono-chemical reactions from collapse of cavitation bubbles. Thus, we put particle slurry into water directly for disaggregation and surface modification of diamond particles without the glass beaker to avoid reflection of ultrasound at the surface of glass beaker as shown in Fig. 5.20.

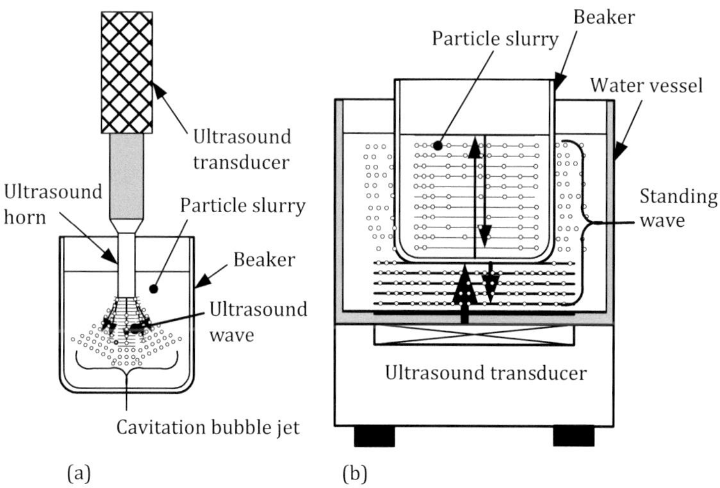

Figure 5.19 Ultrasound irradiation apparatus for dispersion by disaggregation and surface modification of particles in slurry (suspension). (a) Method with ultrasound horn. (b) Method by formation of standing wave sound field with ultrasound cleaner.

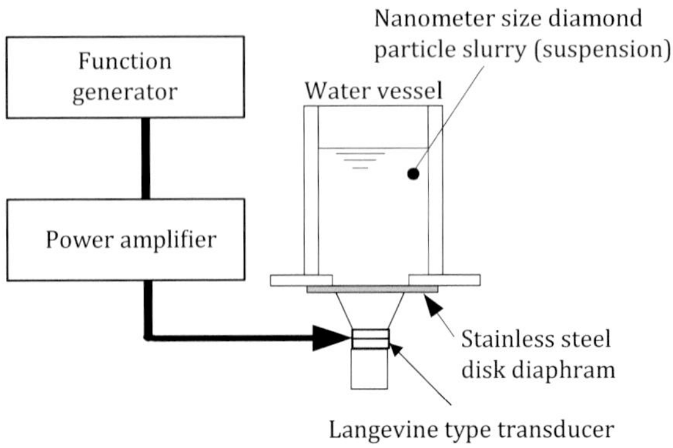

Figure 5.20 Basic configuration of our ultrasound irradiation system for disaggregation and surface modification of nanometre-sized diamond particles.

We use the ultrasound irradiation with stainless steel disk diaphragm driven with Langevin-type transducer (Type HEC-4502 HONDA ELECTRONICS CO., LTD.) on the bottom of the vessel by using O-ring. The system was designed such that the O-ring contacted the node of vibration mode on the stainless steel diaphragm in order to avoid the effect of the vibration mode on the stainless steel diaphragm. The photographs of this ultrasound irradiation system are shown in Fig. 5.21.

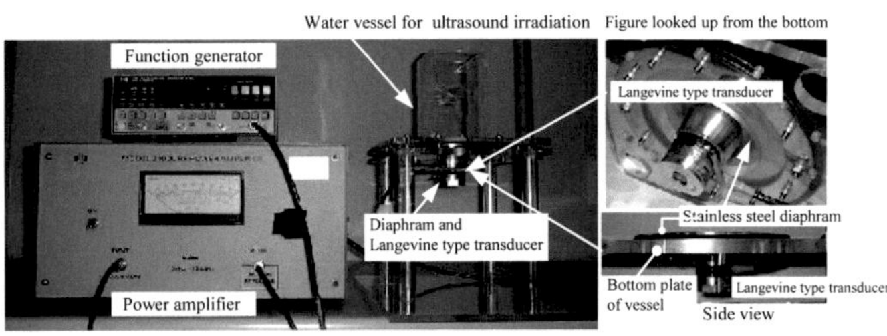

Figure 5.21 Photographs of our ultrasound irradiation system for disaggregation and surface modification of nanometre-sized diamond particles.

Output signals from a function generator (8116A, HP) were amplified by a power amplifier (2100L, ENI) with a gain of 50 dB and applied to the Langevin-type transducer fixed on the bottom of water vessel in order to irradiate ultrasound. The stainless steel diaphragm has the shape of disk with a thickness of 2 mm and diameter of

180 mm and a Langevin-type transducer is equipped at the centre of the stainless steel diaphragm. Langevin-type transducer is a type of ultrasound transducer with piezoelectric vibrators sandwiched between metal blocks in order to vibrate at low frequency which is difficult to vibrate with only piezoelectric vibrators.

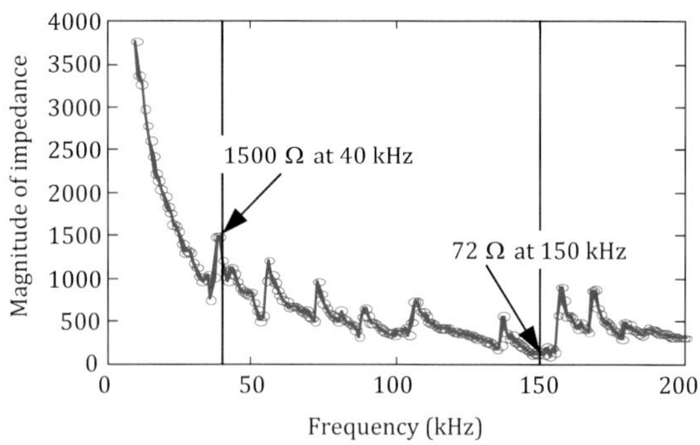

Figure 5.22 Frequency characteristics of input impedance of our ultrasound irradiation system.

We employed the Langevin-type transducer of 40 kHz (Type HEC-4502 HONDA ELECTRONICS CO., LTD.) in this study. This transducer showed minimum input impedance value (72 Ω) at 150 kHz. The input impedance of this transducer is best matched to the output impedance of the power amplifier with nominal impedance of 50 Ω and maximum electric power can be input at 150 kHz. This frequency of 150 kHz coincides with the resonance frequency of piezoelectric vibrators in the Langevin-type transducer. The internal dimensions of the water vessel are 70 mm (width) × 70 mm (depth) × 150 mm (height). The slurry suspended with nanometre-sized diamond particles of 30 mg into the distilled water of 500 mL was poured into this water vessel. The depth of the nanometre-sized diamond slurry was 100 mm in this situation. This depth was almost 10 times the wavelength in water at 150 kHz.

Frequency characteristics of sono-chemiluminescence intensity in water in a water vessel of our ultrasound irradiation system were measured. Sono-chemiluminescence is a sono-chemical reaction by luminal anion and activated oxygen species like hydroxyl radicals which are generated together in acoustic cavitation. Sono-chemiluminescence is the phenomenon which emits weak luminescence in water by acoustic cavitation. When more activated oxygen species are generated, higher intensity of sono-chemiluminescence is observed. Therefore, sono-chemiluminescence can be thought as an effective index to estimate occurrence of acoustic cavitation and acoustic field. Measured frequency characteristics of sono-chemiluminescence are shown in Fig. 5.23. Because it could be confirmed that the highest sono-chemiluminescence could be observed at a frequency of 150 kHz, we decided to measure sono-chemiluminescence at 150 kHz in this study. This frequency coincides with the resonance frequency of the piezoelectric vibrator in the Langevin-type transducer.

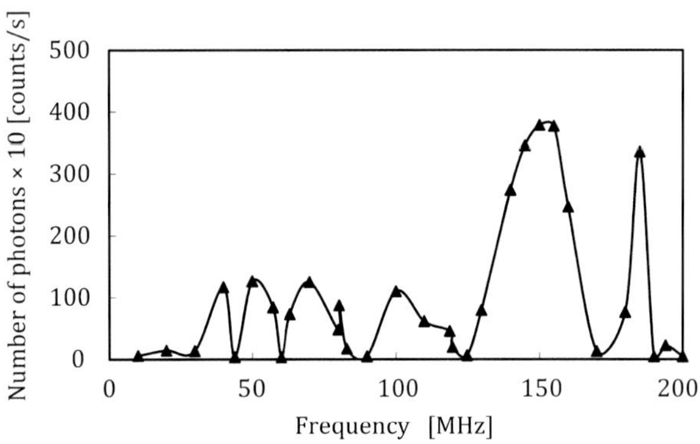

Figure 5.23 Frequency characteristics of intensity of sono-chemiluminescence measured in our water vessel.

5.3.4.2 Acoustic Field and Acoustic Streaming [16]

Ultrasound was irradiated into water in the vessel by applying continuous sinusoidal waves with amplitudes of about 61 V_{op} to the Langevin-type transducer equipped on the bottom of the vessel of our ultrasound irradiation system. Acoustic fields (Spatial distributions of sound pressure) in the direction of water depth of the water vessel and acoustic field on the surface parallel to the stainless steel diaphragm at the height of the loop of the standing wave were measured using a needle type hydrophone. Average sound pressure was about 70 kPa and spatial averaged acoustic intensity I_{SA} was about 3000 W/cm^2 in this situation. It was confirmed that sono-chemiluminescence could be observed in water in this water vessel with averaged acoustic intensity higher than 800 W/m^2. An example of acoustic field (spatial distribution of sound pressure) in the horizontal plane parallel to the stainless steel diaphragm is shown in Fig. 5.24.

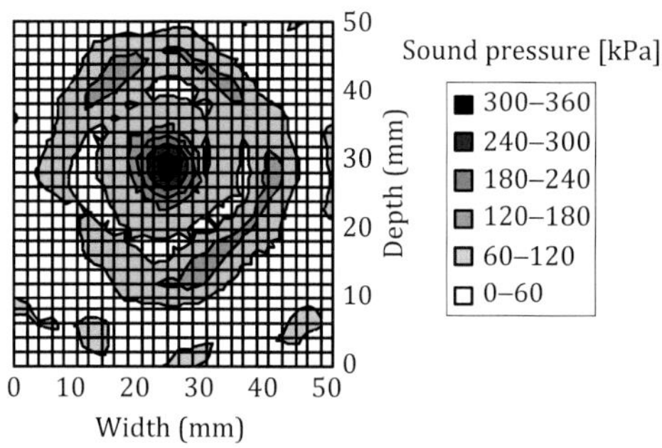

Figure 5.24 An example of measured acoustic fields in parallel plane to the stainless steel diaphragm.

Figure 5.25a shows the measured spatial distribution of sound pressure in the vertical plane in water in the vessel of our ultrasound irradiation system by using a hydrophone. Figure 5.25b shows the sono-chemiluminescence pattern measured in the same vertical plane as Fig. 5.25a. Figure 5.25c was measured by using a high speed video camera. This figure shows the spatial distribution of water streaming (acoustic streaming) in the same vertical plane as Fig. 5.25a. Aluminium powders were suspended as the tracer for the measurement of streaming shown in Fig. 5.25c. The obvious standing wave can be observed in the upper half area of the water vessel shown in Fig. 5.25a. The horizontal stripe pattern of sono-chemiluminescence can be observed at the positions of the loops of standing waves in the spatial distribution of sono-chemiluminescence (sono-chemiluminescence pattern) in Fig. 5.25b. On the other hand, the region with extreme high acoustic pressure can be observed in the lower centre area (the area near the bottom) of water in the water vessel shown in Fig. 5.25a. The dark area with low intensity of sono-chmiluminescence is observed in the area corresponding to above mentioned area in Fig. 5.25b. Furthermore, Fig. 5.25c shows that high speed acoustic streaming occurred in this region in Fig. 5.25c. We think that the sono-chemiluminescence has become weak due to washing away of the cavitation bubbles by this acoustic streaming in this lower centre region. It can be thought that the effect of disaggregation and surface modification of nanometre-sized diamond particles is small in this region because the sono-chemical reaction is weak in this region.

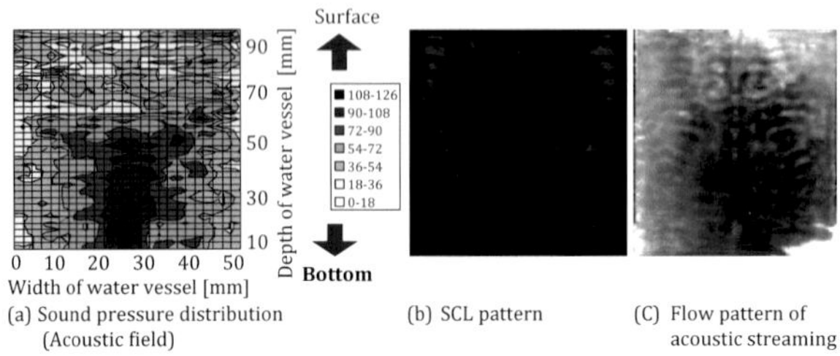

Figure 5.25 Comparison of distribution pattern sound pressure (acoustic field), sono-chemiluminescence pattern and spatial distribution of flow in water in water vessel of our ultrasound irradiation system used in this study. (a) Sound pressure distribution (acoustic field). (b) SCL pattern. (c) Flow pattern of acoustic streaming.

5.3.4.3 Suppression of Aggregation of Nanometre-Sized Diamond Particles (Disaggregation and Surface Modification)

Continuous ultrasound waves were irradiated into the diamond particle slurry in the water vessel by driving our ultrasound irradiation system in this study. The measured relationships between ultrasound irradiation time and size distribution of sonicated diamond particles are shown in Fig. 5.26 and the measured relationship between ultrasound irradiation time and averaged size of sonicated diamond particles are shown in Fig. 5.27. It was found in experimental results that the average size of diamond particles decreased to about 4 μm after ultrasound exposure from 30 s to 60 s, decreased to about 220 μm after 220 s exposure and decreased to the size smaller than about 80 nm rapidly after ultrasound exposure longer than 300 s.

The measured relationship between ultrasound exposure time and zeta potential on the surface of sonicated diamond particles is shown in Fig. 5.28, the measured results of the relationship between ultrasound exposure time and yield of OIH radicals generated in the water by the sono-chemical reaction of acoustic cavitation are shown in Fig. 5.29. Yield of OH radicals were measured with KI solution dosimetry. The higher the light absorption from I^{3-} generated by reaction with OH radicals in KI solution dosimetry means the larger the amount of OH radicals generated. Extreme change which was observed in such data of the average particle size and size distribution of particles exposed to acoustic cavitation between 220 s and 300 s could not be observed in these data.

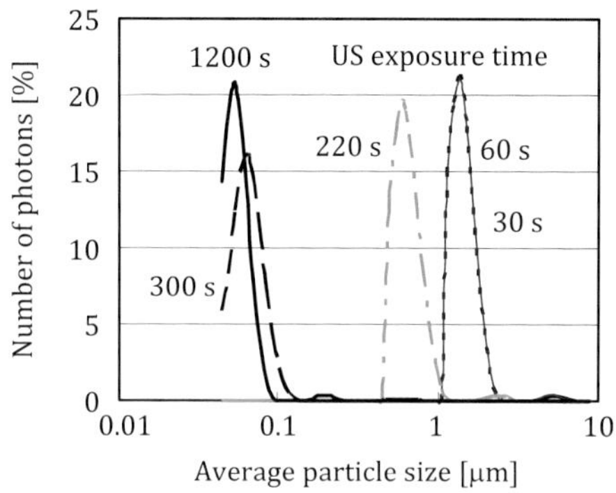

Figure 5.26 Relationship between ultrasound exposure time and size distribution of sonicated diamond particles.

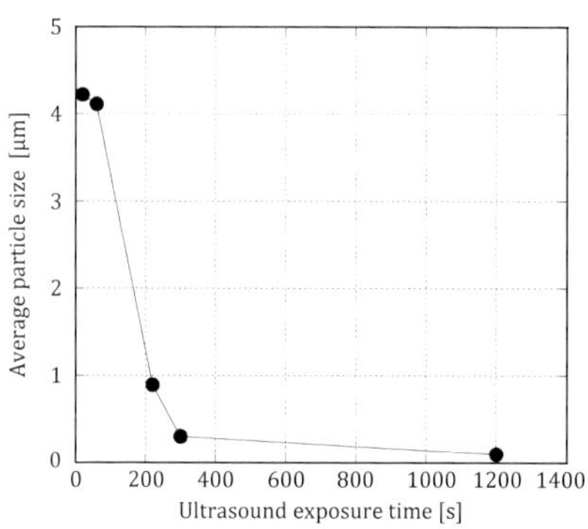

Figure 5.27 Relationship between ultrasound exposure time and averaged sizes of sonicated diamond particles.

We discussed the mechanism of disaggregation of aggregated diamond particles and that of avoidance from re-aggregation of disaggregated diamond particle formerly. The aggregated diamond particles were disaggregated by ultrasound and by shock waves from acoustic cavitation. Immediately after disaggregation, surfaces of disaggregated diamond particles are covered with OH groups which are formed by OH radicals generated in water by acoustic cavitation. Then, OH groups covering the surfaces of diamond particles increased their zeta potential. Disaggregated diamond particles get electrostatic repulsion between each other by covered with OH groups. Disaggregated diamond particles avoid re-aggregation if electrostatic repulsions between the diamond particles overcome the cohesive forces from the van der Waals force. However, significant differences are not identified between yield of OH radicals with ultrasound exposure times of 220 s and 300 s in Fig. 5.28, and significant differences are not identified also between zeta potentials with ultrasound exposure times of 220 s and 300 s in Fig. 5.29. Therefore, it was difficult to explain the cause of difference between averaged sizes of diamond particles exposed to ultrasound for 220 s and for 300 s by using DLVO theory(derived by Derjaguin and Landau, Verwey and Overbeek) [17,18] and by using electrostatic repulsion from OH groups covering the surfaces of diamond particles. Although energy difference of irradiated ultrasound might be involved, unfortunately it is difficult to explain this point clearly enough at the moment. We would like to make this mechanism obvious by more precise experiments in the future.

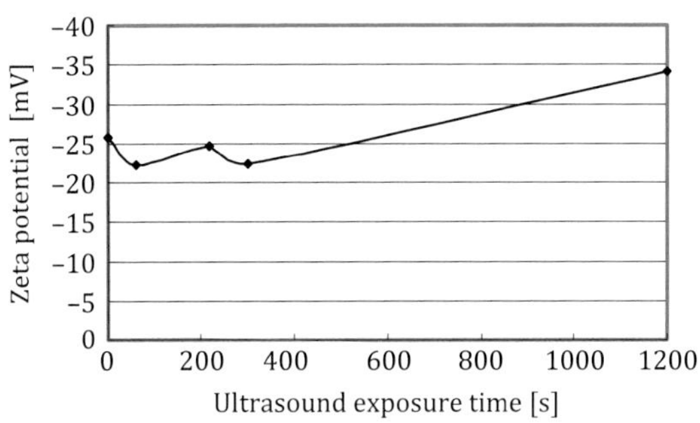

Figure 5.28 Relationship between ultrasound irradiation time and zeta potential on the surface of diamond particles.

Yields of OH radicals were measured using KI solution dosimetry. The higher the light absorption by I^{3-} ions which are generated in the chemical reaction with OH radicals, the larger the amount of OH radicals generation.

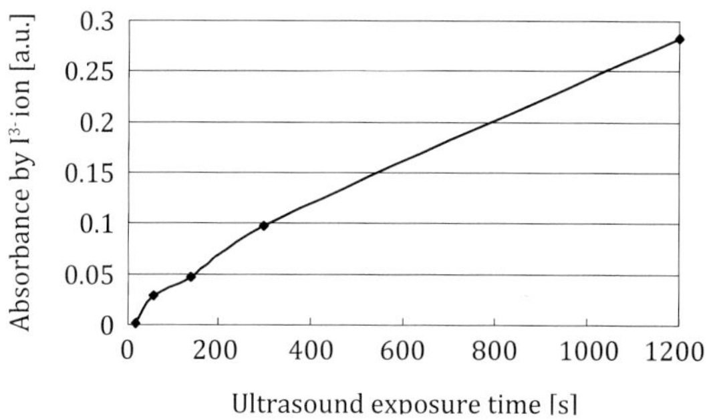

Figure 5.29 Relationship between ultrasound irradiation time and yields of OH radical generated by acoustic cavitation.

5.3.5 Conclusions

We described in this paper a study on improving dispersion of a slurry of diamond particles by disaggregation of aggregated diamond particles with nanometre-sized primary particles and surface modification to avoid re-aggregation by using ultrasound irradiation and acoustic cavitation. The nanometre-sized diamond particles are expected to act as abrasive agents for ultra precise polishing of ultra miniature hard disk devices with high recording densities or as solid lubricants. Furthermore, the nanometre-sized diamond particles also seem to be expected to be applied in the medical field. We succeeded in disaggregating the aggregated diamond particles and modifying their surfaces to avoid re-aggregation by forming standing wave acoustic field in the diamond particle slurry in the water vessel of our ultrasound irradiation system and by trapping them at the loops of sound pressure standing waves together with acoustic cavitation bubbles in the diamond particle slurry.

We considered the relationships between ultrasound exposure time and particle size of diamond particles and those between ultrasound exposure time and zeta potential on their surface by using our ultrasound exposure system. It was found as the results that the average particle size of diamond particles decreased extremely, but zeta potential on the diamond particles did not change so extremely from about 0.8 μm to about 80 nm between ultrasound exposure times of 220 and 300 s. Although energy difference of irradiated ultrasound might be involved, unfortunately it is difficult to explain this point clearly enough at the moment. We will make this mechanism obvious by more precise experiments in the future.

References

1. Suslick KS (1990) *Science*, **247**, 1439.
2. Doktycz SJ, Suslick KS (1990) *Science*, **247**, 1067.
3. Suslick KS (1995) *Mrs Bulletin*, **20**, 29.
4. Suslick KS (1999) *Annu Rev Mater Sci*, **29**, 295.
5. Suslick KS, Didenko Y, Fang M, et al. (1999) *Philos Trans R Soc London Ser A*, **357**, 335.
6. Koda S, Kimura T, Kondo T, Mitome H (2003) *Ultrason Sonochem*, **10**, 149.
7. Murakami T, Takahashi M, Kawashıma N (2000) *Chem Lett*, 413–414.
8. Uchida T, Takatera T, Sato T, Takeuchi S, Kuramochi N, Kawashima N (7–10 October 2001), in *Proc. IEEE Ultrasonics Symp.* (Yuhans DE, Schneider SC, eds), IEEE, Atlanta, GA, p 431.
9. Uchida T, Hamano A, Kawashima N, Takeuchi S (18–21 September 2005), paper presented at the Proc. IEEE Ultrasonic Symposium, Rotterdam, The Netherlands.
10. Uchida T, Hamano A, Kawashima N, Takeuch S (2005) *Jpn J Appl Phys*, **44**, Part 1, 6B.
11. Gubarevich AV, Usuba S, Kakudate Y, Tanaka A, Odawara O (2004) *Jpn J Appl Phys*, **43**, 920.
12. Baidakova MV, Vul Ya A, Siklitskii VI, Faleev NN (1998) *Phy Solid State*, **40**, 715.
13. Uchida T, Sato T, Takeuchi S, Kuramochi N, Kawasima N (2003) *Jpn J Appl Phys*, **42**, 2967.
14. Mason TJ (2003) *Ultrason Sonochem*, **10**, 175.
15. Uchida T, Hamano A, Sawada Y, Kawashima N, Takeuchi S (2004) Improvement of dispersion of nano-sized diamonds by sonochemical reaction—Relationships among acoustic intensity, disaggregation and surface modification-, Autumn Meeting Acoustical Society of Japan.
16. Uchida T, Kikuchi T, Sato S, Kawashima N, Takeuchi S (2006) Effects of sound field and acoustic streaming on nano-meter sized

diamond particles dispersion system using ultrasound, in *2006 IEEE International Ultrasonics Symposium* (Yuhas M, ed), Vancouver.

17. Uchida T, Kikuchi T, Aoki T, Kawashima N, Takeuchi S (2009) Effects of ultrasound exposure time on nano-meter-sized diamond particles dispersion, *Jpn J Appl Phys*, **48**(7), 07GH03–07GH03.
18. Kitahara A, Furusawa K, Ozaki M, Ohshima H (1997) *Zeta Potential*, Scientist-Sha, Tokyo.

Chapter 6

Engineering Field

6.1 Introduction

Keiji Yasuda[a] **and Yoshiyuki Bando**[b]
[a]*Department of Chemical Engineering, Nagoya University, Nagoya-shi, Aichi 464-8603, Japan*
[b]*Adviser, Morimatsu Industry Co., Ltd., Motosu-shi, Gifu, Japan*
yasuda@nuce.nagoya-u.ac.jp, Yoshiyuki_Bando@morimatsu.co.jp

The application-related material production, gas absorption and gas–liquid reaction were introduced. The flow characteristics of micro-bubbles in reactor were examined by using the cocurrent upflow bubble column. When millibubbles coexist with dispersed microbubbles, microbubbles disappear. The volumetric mass transfer coefficient is higher for microbubbles than for millbubbles, but the mass transfer coefficient is much lower for the former than for the latter. The installation of draft tube enhanced the mass transfer and decomposition performance. The design guides of bubble column-type reactor for the enhancement in mass transfer and decomposition performance are decided.

6.1.1 Introduction

Micro- and nanobubbles are applied in the industrial field such as material production, gas absorption and gas–liquid reaction. The gas–liquid reactors optimised for micro- and nanobubbles are developed.

Micro- and Nanobubbles: Fundamentals and Applications
Edited by Hideki Tsuge

ISBN 978-981-4463-10-2 (Hardcover), 978-981-4463-11-9 (eBook)
www.panstanford.com

In this section, the application-related material production, gas absorption and gas–liquid reaction are briefly introduced. The flow characteristics of microbubbles in reactor are described, and the development of airlift bubble column for microbubbles is introduced.

6.1.2 Application of Microbubbles for Industrial Area

The microbubbles are used in some material productions. By mixing with air microbubbles, foamed polyurethane with a pore diameter of 1 μm or less is obtained. This polyurethane has a light reflectivity of 90% or larger [1]. The plastic bumper containing microbubbles has low density and high heat resistance [2]. The novolac-type phenolic resin is obtained high yield using aldehyde microbubbles [3]. The rod-like particles made from ferric hydroxide are synthesised by oxygen microbubbles [4]. The size distribution of these rod-like particles is small. The microcapsule containing microbubbles was produced [5], and the preparation and property of microcapsule will be explained in following section.

In the recent studies related to gas absorption and gas–liquid reaction, a carbon dioxide was often used as microbubbles. The compact rector for neutralisation of alkaline wastewater was developed [6]. The system retaining carbon dioxide in the ground was also studied [7].

Since the rising velocity of microbubbles is very low, in many cases the microbubbles flow together with liquid. Accordingly, the flow pattern of microbubbles in the reactor governs the performances of chemical process. In next section, the flow characteristics of microbubbles in bubble column are described.

6.1.3 Flow Characteristics in Cocurrent Upflow Bubble Column Dispersed with Microbubbles

The microbubble generator (MBG) was installed in bubble column. The flow characteristics of microbubbles were measured, and the mass transfer coefficient of microbubbles was estimated.

Figure 6.1 shows the outline of the experimental apparatus [8]. The diameter of bubble column was 100 mm, and the height was changed in the range of 1.2–4.2 m. The MBG(Nagoya Oshima

Machinery Co., Ltd., Japan), which consisted of a liquid pump and a special line mixer, was installed at the bottom of the column. For the comparison, a sintered glass gas sparger was set at the lower part of the column. It generated bubbles with 2–3 mm in diameter. As the gas and liquid phases, air and tap water were used.

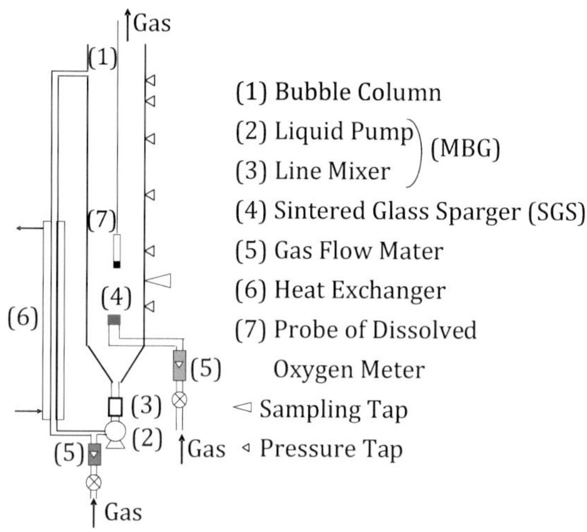

Figure 6.1 Outline of experimental apparatus.

In MBG used in this study, the peak diameter of microbubbles was around 30 μm at low gas velocity. However, as the gas velocity became higher, millibubbles were also produced and the amount of millibubbles increased. When a large bubble (millbubble in this case) flowed upward in the liquid where microbubbles were dispersed, some microbubbles disappeared. From the flow behaviour, it was considered that microbubbles were absorbed into the wake of the large bubble.

Figure 6.2 shows the longitudinal distribution of average diameter of bubble, d_{AVE} for different gas velocities, u_G. The superficial gas velocity was based on the cross-sectional area of the column. The abscissa is the position from the column bottom, z. At the gas velocity of 0.1 mm/s, the average diameter becomes smaller in the direction of gas–liquid flow to the column top. It is considered that microbubbles are generated from the supersaturated liquid as liquid flows upward in the column. At the gas velocities of 0.2 and

0.4 mm/s, on the other hand, the average diameter becomes larger. It was observed that many microbubbles were absorbed as the gas–liquid flows upward at high gas velocity.

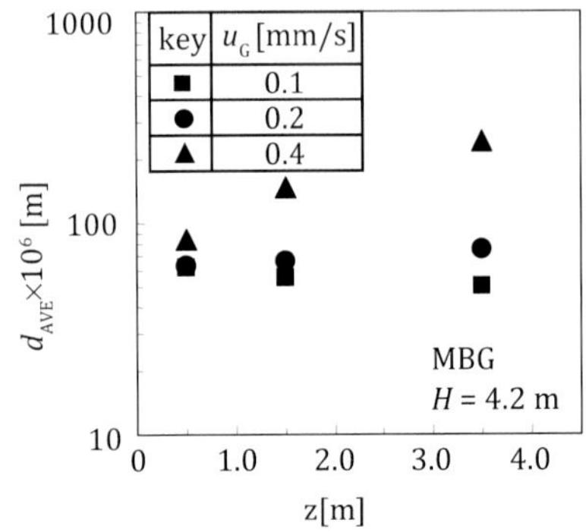

Figure 6.2 Longitudinal distribution of average diameter of bubble for different gas velocities.

Figure 6.3 shows the effect of gas velocity on the volumetric mass transfer coefficient, k_La. The parameter is the column height, H. The volumetric coefficient for SGS increases linearly with the gas velocity and is hardly affected by the column height. The volumetric coefficient for MBG is much higher than that for SGS. This is because the specific interfacial area, a is larger for the former than for the latter. The volumetric coefficient for MBG increases with decreasing column height, and this trend becomes clearer at the higher gas velocity. This is because the fraction of region of microbubble dispersion in the column is larger as the column height decreases. In this figure, the data of Li and Tsuge [9] are also plotted. They have used ozone gas and measured the volumetric coefficient in a column of 0.20 m in diameter and 1.2 m in height. Their volumetric coefficient is higher than that for SGS and similar to that for MBG.

The specific gas–liquid interfacial area was calculated from the average bubble diameter and the overall gas holdup, ε, that is $a = 6\varepsilon/d_{AVE}$. The mass transfer coefficient was obtained by dividing the volumetric mass transfer coefficient by the specific gas–liquid interfacial area. Figure 6.4 shows the plot of mass transfer coefficient against the average diameter of bubbles. The mass transfer coefficient of microbubbles (MBG) is much lower than that of millibubbles (SGS). This is considered because the liquid–phase boundary film is much thicker for microbubbles than for millibubbles. In this figure, shown also are three results calculated from equations reported by Levich [8], Calderbank and Moo-Young [10] and Akita and Yoshida [11]. Both the mass transfer coefficients of micro- and millibubbles are lower than the line calculated from the equation of Levich. The mass transfer coefficient of millibubbles shows a similar tendency to the line of transit region of Calderbank and Moo-Young, through the value is a little higher than the line of the transit region. It is found that most of the mass transfer coefficients of microbubbles exist between the extrapolated lines of equations of Calderbank and Moo-Young, and Akita and Yoshida.

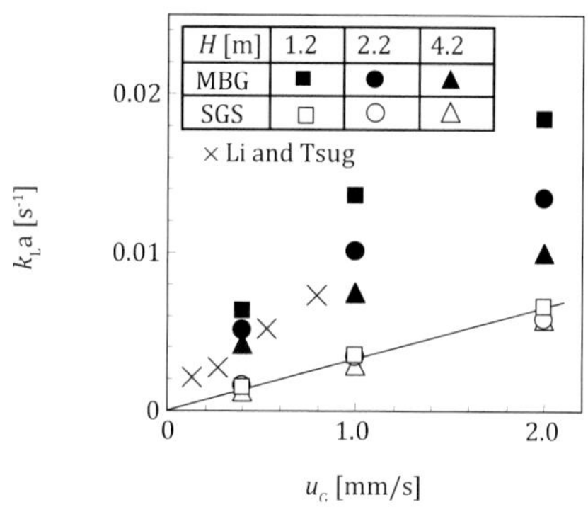

Figure 6.3 Effect of gas velocity on volumetric mass transfer coefficient.

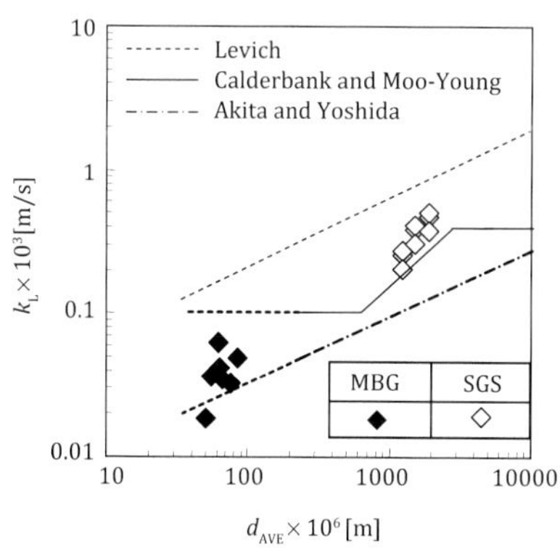

Figure 6.4 Plot of mass transfer coefficient against average diameter of bubble.

6.1.4 Development of Airlift Bubble Column Dispersed with Microbubbles

The draft tube was inserted into the column dispersed with microbubbles, the length and diameter of draft tube were changed and mass transfer and decomposition performance were measured. The design guides of bubble column-type reactor were decided.

The bubble column was made from transparent acrylic resin, and the inner diameter and height were 0.190 and 1.90 m, respectively [12]. The draft tube length was changed from 0.50 to 1.30 m. The distance between the column bottom and the lower end of draft tube was a half of draft tube diameter [13]. The inner diameter of draft tube was changed from 0.070 to 0.160 m. An MBG was set below the bottom of the column. Ozone was supplied from an ozone generator, and the ozone concentration at gas inlet was 60 g/m^3. The superficial gas velocity based on the cross-sectional area of the column was 0.29 mm/s. Liquid height in the column was 1.65 m. For the measurement of flow characteristics, the bubble diameter and volumetric mass transfer coefficient of ozone in water were measured. For the examination of decomposition performance, the methylene-blue aqueous solution (the initial concentration was 0.003 mol/m^3) was used as the liquid phase, and the decomposition conversion was estimated.

For MBG used in this study, microbubbles and millibubbles were coexisted in the column, and the amount of millibubbles increased with increasing gas velocity. In the column without draft tube, the amount of microbubbles decreased at the higher position. In the column with draft tube, on the other hand, microbubbles were separated from millibubbles above the upper end of draft tube and entrained into the annular section by the liquid circulation flow. As a result, the region of microbubble dispersion became larger for the column with draft tube than for the column without draft tube. As the draft tube was longer and larger, microbubbles were observed to flow downwards more smoothly in the annular section, and the average bubble diameter in the annular section was larger.

Figure 6.5 shows the effects of draft tube length, L_D (a) and diameter, D_D (b) on the average bubble diameter in the annular section of bubble column, $d_{ave,A}$. The average bubble diameter in the annular section increases with increasing draft tube length and diameter. This is because larger micro bubbles are entrained into the annular section due to the higher liquid velocity generated by airlift.

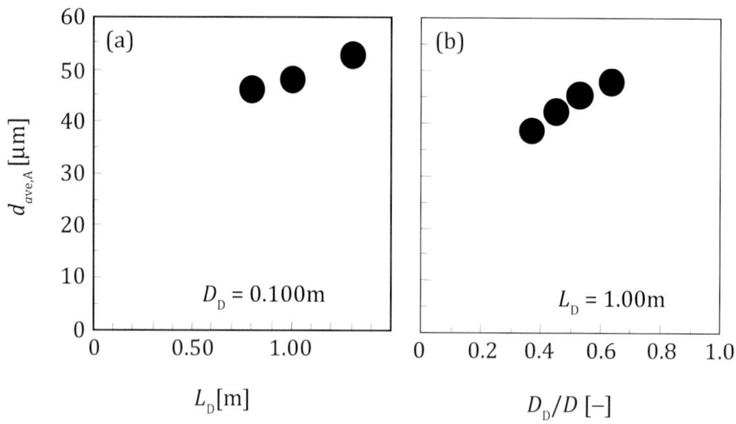

Figure 6.5 Effect of draft tube length (a) and diameter (b) on average bubble diameter in annular section of bubble column.

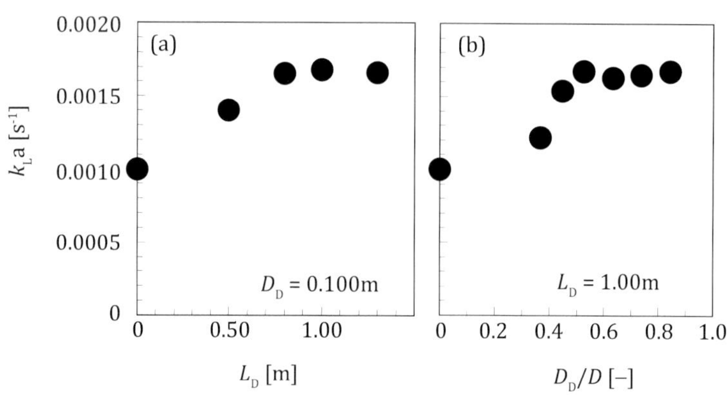

Figure 6.6 Effect of draft tube length (a) and diameter (b) on volumetric mass transfer coefficient of ozone in water.

Figure 6.6 shows the effects of length (a) and diameter (b) of draft tube on the volumetric mass transfer coefficient of ozone in water. A datum at $L_D = 0$ and $D_D = 0$ show the volumetric coefficient for the column without draft tube. The volumetric coefficient for the column with draft tube is much higher than that for the column without draft tube. For the column with draft tube, the volumetric coefficient increases with increasing draft tube length and become nearly constant beyond the length of 0.80 m; and the volumetric coefficient increases with increasing draft tube diameter and become nearly constant beyond the diameter ratio of 0.5. From these data, it is found that the draft tube diameter should be larger than a half of column diameter. The draft tube length is longer than 0.80 m for the column height of 1.90 m.

In Fig. 6.7, the decomposition conversion ratio is plotted against the volumetric mass transfer coefficient ratio. The abscissa is [(ratio of volumetric mass transfer coefficient with draft tube to that without draft tube) – 1], and the ordinate is [(ratio of decomposition conversion with draft tube to that without draft tube) – 1]. The decomposition conversion ratio increases with increasing volumetric mass transfer coefficient ratio, and all data ride on a line regardless of length and diameter of draft tube. From this result, it is clear that the insertion of draft tube enhances the mass transfer and the decomposition performance improves.

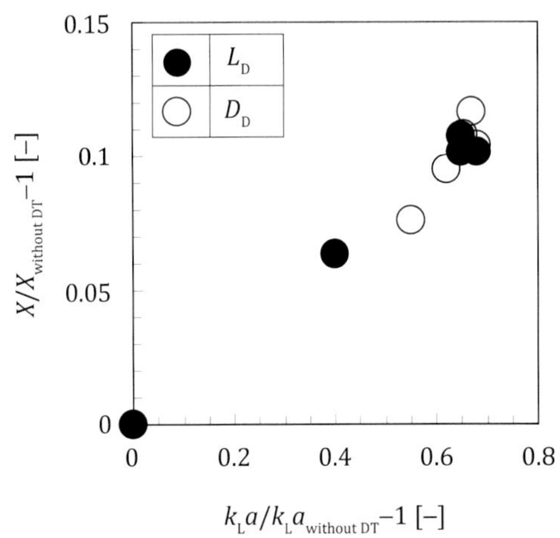

Figure 6.7 Plot of decomposition conversion ratio against volumetric mass transfer coefficient ratio.

References

1. Japan patent publication number 2008-260862 (in Japanese).
2. Japan patent publication number 2006-289992 (in Japanese).
3. Japan patent publication number 2009-057454 (in Japanese).
4. Japan patent publication number 2009-173519 (in Japanese).

5. Daiguji H, Makuta T, Kinoshita H, Oyabu T, Takemura F (2007). Fabrication of hollow melamine-formaldehyde microcapsules from microbubble templates, *J Phys Chem B*, **111**, 8879–8884.
6. Japan patent publication number 2006-281075 (in Japanese).
7. Japan patent publication number 2008-006367 (in Japanese).
8. Bando Y, Yoshimatsu Y, Luo W et al. (2008) Flow characteristics in cocurrent upflow bubble column dispersed with micro-bubbles, *J Chem Eng Japan*, **41**, 562–567.
9. Li P, Tsuge H (2006) Ozone transfer in a new gas-induced contactor with microbubbles, *J Chem Eng Japan*, **39**, 1213–1220.
10. Calderbank PH, Moo-Young MB (1961) The continuous phase heat and mass-transfer properties of dispersions, *Chem Eng Sci*, **16**, 39–54.
11. Akita K, Yoshida F (1974) Bubble size, interfacial area, and liquid-phase mass transfer coefficient in bubble columns, *Ind Eng Chem Process Des Dev*, **13**, 84–91.
12. Yasuda K, Wang Y, Haneda K et al. (2010) Development of airlift bubble column dispersed with micro-bubbles, *Can J Chem Eng*, **88**, 518–522.
13. Bando Y, Hayakawa H, Nishimura M (1998) Effects of equipment dimensions on liquid mixing time of bubble column with draft tube, *J Eng Japan*, **31**, 765–770.

6.2 Development of Antisolvent and Reactive Crystallisation Technique

Kaoru Onoe[a] and Masakazu Matsumoto[b]

[a]*Department of Life and Environmental Sciences, Faculty of Engineering, Chiba Institute of Technology, 2-17-1, Tsudanuma, Narashino, Chiba 275-0016, Japan*
[b]*Department of Basic Science, College of Industrial Technology, Nihon University, 2-11-1 Shinei, Narashino, Chiba 275-8576, Japan*
kaoru.onoe@it-chiba.ac.jp

6.2.1 Introduction

With the development of the microscale bubble generation technique to obtain bubbles whose size is <50 μm, considerable attention has been paid to the utilisation of the so-called "minute-bubbles". Minimising bubble formation in gas–liquid systems helps achieve the following: (i) acceleration of mass transfer and reactive absorption with an increase in the gas–liquid interfacial area, (ii) increase in the average residence time of the bubbles with a decrease in buoyancy, and (iii) occurrence of interactions at the gas–liquid interface [1,2]. Thus, by increasing the residence time of bubbles in the liquid phase, a quasi-homogeneous gas–liquid system can be obtained. Furthermore, when external energy is supplied to an aqueous solution containing minute-bubbles, unique mass-transfer patterns and reaction phenomena are expected to occur [3].

In this chapter, applications of new reaction fields around minute-bubbles to antisolvent and reactive crystallisation are described.

6.2.2 Application to the Miniaturisation of Particle Size during Antisolvent Crystallisation

With regard to improve the production efficiency based on a unit volume of crystalliser, it is necessary to control the crystal size and number using the generation of local supersaturation regions around minute-bubbles. Figure 6.8 shows the antisolvent crystalliser equipped with a minute-bubble generator. Nitrogen (N_2) minute-bubbles are generated in the sodium chloride (NaCl) solution using a self-supporting bubble generator by increasing the impeller shear rate under reduced pressure. Bubbles with an average size (d_{bbl}) of 100–2000 μm are generated using a dispersing-type bubble generator (hole size of dispersion plates: 5–120 μm). The temperature of NaCl solution was kept constant at 298 K, and ethanol (EtOH) was added as an antisolvent to the saturated NaCl solution where the added volume ratio of EtOH (ϕ_{EtOH}) becomes 10 vol%.

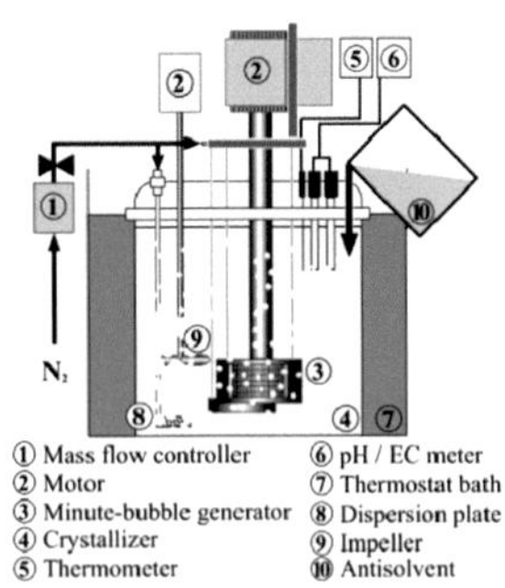

Figure 6.8 The antisolvent crystalliser equipped with a minute-bubble generator.

As a result, the crystal size distributions were shifted to lower zone, and distribution width was narrowed against the miniaturisation of average bubble size. Figure 6.9 shows the effect of average bubble size on rates of nucleation (r_N) and crystal growth (r_G). The nucleation rate increased with a decrease in average bubble size, and the crystal growth rate showed negative correlation with nucleation rate. Owing to the supply of minute-bubbles, the acceleration of nucleation and inhibition of crystal growth may occur during antisolvent crystallisation.

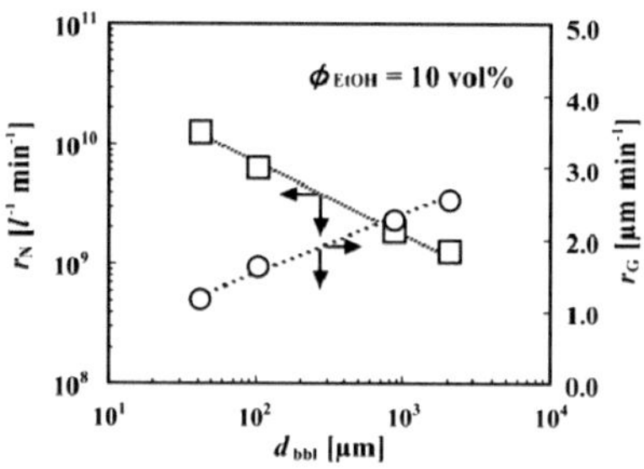

Figure 6.9 Effect of average bubble size on rates of nucleation and crystal growth of NaCl.

6.2.3 Utilisation to Polymorphic Control during Antisolvent Crystallisation

Glycine has three polymorphs—most stable γ-form, metastable α-form and unstable β-form. The temperature of saturated glycine solution was set at 303 K, and methanol as an antisolvent was mixed into the solution (mixture ratio of methanol (ϕ_{MeOH}):10–60 vol%). At the same time as mixing methanol with the saturated glycine solution, N_2 minute-bubbles with an average bubble size of 10 μm were continuously supplied to the mixed solution using a self-supporting bubble generator, and glycine polymorphs were changed [4].

Figure 6.10 shows the time changes in the produced moles of total glycine, α-form, β-form and γ-form based on unit volume (G_i, i = total, α, β, γ) at various ϕ_{MeOH} values when N_2 minute-bubbles were supplied. G_{total} was changed remarkably with ϕ_{MeOH}. The most stable γ-form was not obtained under these experimental conditions. When ϕ_{MeOH} was in the range less than 20 vol%, α-form with high selectivity (almost 100%) was crystallised at all values of crystallisation time t_c. When ϕ_{MeOH} was in the range over 30 vol%, β-form was dominant product in the early stage of crystallisation, and the polymorphic transformation from unstable β-form to metastable α-form was observed with the progress of crystallisation. The rate of polymorphic transformation increased with a decrease in ϕ_{MeOH}. Thus, it was confirmed that the increase in supersaturation in the bulk solution with ϕ_{MeOH} inhibited the polymorphic transformation.

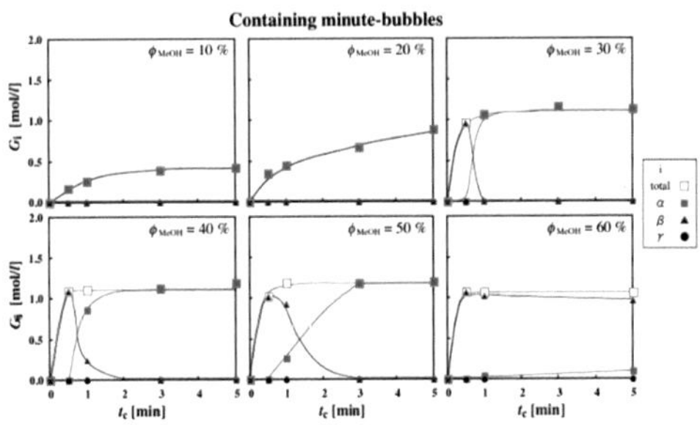

Figure 6.10 Time changes in molar concentration of produced glycine polymorphs (containing minute-bubbles).

When N_2 minute-bubbles were not supplied, the time changes in G_i at various ϕ_{MeOH} values are shown in Fig. 6.11. The α-form was crystallised predominantly and the production of β-form was hardly observed at a ϕ_{MeOH} lesser than 50 vol%. At a ϕ_{MeOH} of 60 vol%, the selectivity of β-form was higher than that of α-form. These tendencies revealed that not only ϕ_{MeOH} increase but also N_2 minute-bubble formation led to the inhibition of polymorphic transformation from unstable β-form to metastable α-form caused by increasing t_c. It is described that the feature of glycine crystallisation using N_2 minute-bubbles is to obtain the unstable β-form glycine with high selectivity at a lower addition ratio of methanol ($\phi_{MeOH} = 30$ vol%), that is, at a lower supersaturation in the bulk solution.

To clarify the effects of N_2 minute-bubble formation on the production of glycine polymorphs, G_α, G_β and G_γ obtained under supplying minute-bubbles were compared with that obtained under no-supplying minute-bubbles. Within 0.5 min crystallisation before the polymorphic transformation, the effects of mixture ratio of methanol on molar concentration of produced glycine polymorphs are shown in Fig. 6.12. When N_2 minute-bubbles were not supplied, in the ϕ_{MeOH} range less than 40 vol%, the α-form production increased remarkably with an increase in ϕ_{MeOH}, and levelled down more than 50 vol%. The production of β-form was observed in the ϕ_{MeOH} range of 50–60 vol%. Meanwhile, when N_2 minute-bubbles were supplied, α-form and β-form were major products at a ϕ_{MeOH} lower than 20 vol% and a ϕ_{MeOH} greater than 30 vol%, respectively. Thus, N_2 minute-bubble formation causes the expansion of the generation region of unstable β-form.

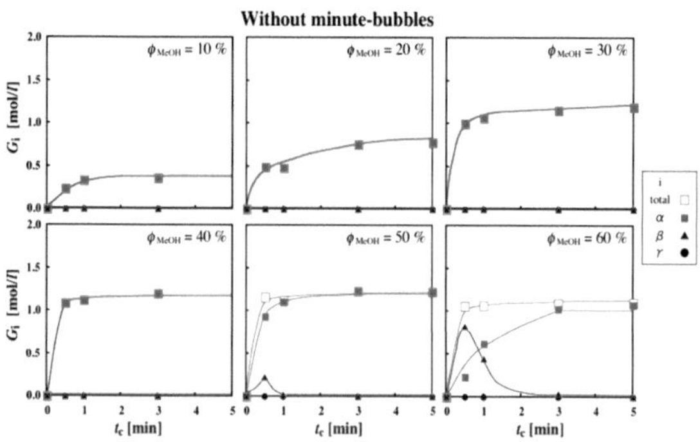

Figure 6.11 Time changes in molar concentration of produced glycine polymorphs (without minute-bubbles).

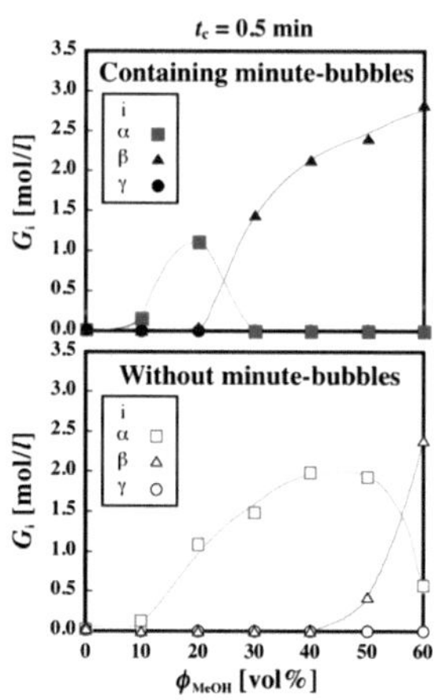

Figure 6.12 Effects of mixture ratio of methanol on molar concentration of produced glycine polymorphs ($t_c = 0.5$ min).

6.2.4 Utilisation to Polymorphic Control during Reactive Crystallisation

Calcium carbonate ($CaCO_3$) has been used as a filler material in a wide variety of industrial fields such as papers, pigments, paints, plastics, rubbers and textiles. Such industrial applications require well-defined $CaCO_3$, which consist of particles with a fine size, narrow size distribution, uniform morphology and crystal structure (polymorph). These physical properties of $CaCO_3$ play a critical role in determining the product properties [5,6].

$CaCO_3$ has three polymorphs, most stable calcite, metastable aragonite and unstable vaterite, which have the crystal system of rhombohedral, orthorhombic and hexagonal, respectively. Under the atmospheric temperature and pressure, the thermodynamic stability of the polymorphs is in order of calcite, aragonite and vaterite. Because the morphology and the physicochemical properties such as solubility and density depend on each polymorph, the polymorphism control of $CaCO_3$ crystals is very important during reactive crystallisation. The polymorph depends mainly on the operational conditions, such as solution temperature, solution pH, feed rate of reactants, solution composition, type and concentration of additives, and mixing condition. These operating parameters are interrelated and interact on each other in a crystallisation process.

In our previous study [6,7], the reactive crystallisation of $CaCO_3$ was progressed using CO_2/NH_3 minute-bubbles with an average bubble size (d_{bbl}) of 40 μm when solution pH was varied in the range of 6.9–12.0, and the effects of solution pH on the polymorphism of $CaCO_3$ were investigated. The results confirmed that calcite and vaterite were major products at a solution pH lower than 9.0 and a solution pH greater than 11.0, and aragonite was crystallised with high selectivity (almost 100%) in the solution pH range of 9.7–10.5. The effects of minimising bubble size and mixing ratio of NH_3 with CO_2 (α_{CO_2/NH_3}) at a constant solution pH are as follows.

Figure 6.13 shows a semi-batch crystallisation apparatus equipped with a gas flow controller, a gas mixer, a pH/EC meter, a self-supporting bubble generator, a crystallisation vessel and a thermostat bath. Minute-bubbles are generated using a self-supporting bubble generator by increasing the impeller shear rate under reduced pressure, with the rotation rate maintained at 1000 min^{-1}, and d_{bbl} is varied in the range of 40–1000 μm by controlling the N_2 flow rate. The physical properties (polymorph, morphology and particle size) of reaction products were measured by using an X-ray diffractometer, a scanning electron microscope and a laser particle size analyser. The selectivity of polymorphs in a product mixture was determined by direct linear analysis [8].

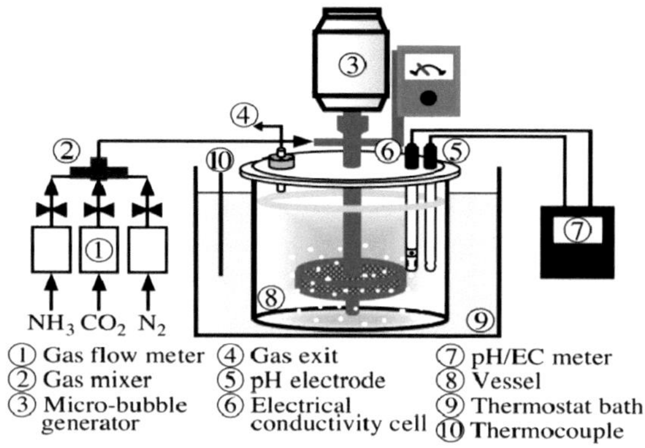

Figure 6.13 Reactive crystallisation apparatus equipped with a minute-bubble generator.

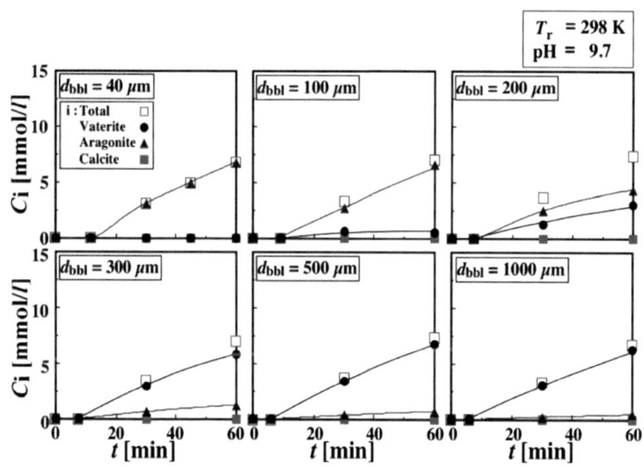

Figure 6.14 Time changes in produced molar concentration of $CaCO_3$ polymorphs ($\alpha_{CO_2/NH_3} = 0.20$).

Figure 6.14 shows the time changes in the produced molar concentration of total $CaCO_3$, vaterite, aragonite and calcite based on unit volume (C_i, i = total, vaterite, aragonite, calcite) when d_{bbl} was varied in the range of 40–1000 μm at a constant α_{CO_2/NH_3} of 0.20. $C_{vaterite}$, $C_{aragonite}$ and $C_{calcite}$ were determined from the total weight of $CaCO_3$ and the selectivity of $CaCO_3$ polymorphs. At a d_{bbl} of 1000 μm, the selectivity of vaterite was higher and the crystallisation of aragonite was not observed. When d_{bbl} decreased to 40 μm, aragonite with high selectivity (almost 100%) was obtained at all values of crystallisation time. The zeta potential of the minute-bubbles with a d_{bbl} of 10–30 μm showed the negative value between −20 and −100 mV, and the negative value increased with an increase in solution pH [9,10]. These findings suggested that the increase in the Ca^{2+} ion concentration in the gas–liquid interfacial area was caused by increasing the electric charge on the bubble surface due to minimisation of bubble size. Therefore, the obtained $CaCO_3$ polymorph differs because of the change in local supersaturation at the gas–liquid interfaces.

Figure 6.15 shows the time changes in C_{total}, $C_{vaterite}$, $C_{aragonite}$ and $C_{calcite}$ at various α_{CO_2/NH_3} values when d_{bbl} was fixed at 40 μm. At α_{CO_2/NH_3} of 1.00, vaterite was crystallised predominantly and $C_{aragonite}$ was relatively smaller than $C_{vaterite}$. With a decrease in α_{CO_2/NH_3} the aragonite crystallisation was progressed gradually and aragonite was major product at α_{CO_2/NH_3} of 0.20. Investigation of the gas absorption of CO_2 and/or NH_3 minute-bubbles cleared that the absorption rate of NH_3 gas was remarkably higher than that of CO_2 gas. Thus, the generation of local pH regions around the gas–liquid interfaces and further minimisation of bubble size were concluded by the superior dissolution of NH_3 due to mixing of NH_3 with CO_2. Furthermore, studies on the kinetics of CO_2 absorption show that the absorption rate of CO_2 increases with an increase in solution pH [11,12]. These findings indicated that the acceleration of CO_2 dissolution was caused by an increase in the local pH in the gas–liquid interfacial area, and then CO_3^{2-} ion concentration increased in the area around the gas–liquid interfaces. Accordingly, mixing of NH_3 gas with CO_2 gas helps the generation of local supersaturation regions with high dissolubility of NH_3 gas. The relation between d_{bbl}, α_{CO_2/NH_3} and the production rates of $CaCO_3$ polymorphs based on unit volume (r_i, i = vaterite, aragonite, calcite) are shown in Fig. 6.16. $r_{vaterite}$, $r_{aragonite}$ and $r_{calcite}$ were determined from the average value of the gradient of the produced moles of each polymorph at t of 20 and 45 min. Under these experimental conditions, no generation of calcite was observed. At all values of α_{CO_2/NH_3}, $r_{aragonite}$ increased with decreasing d_{bbl}, and the aragonite production with minimisation of bubble size was enhanced with a decrease in α_{CO_2/NH_3}. When α_{CO_2/NH_3} value was set at 0.20, $r_{aragonite}$ was higher than $r_{vaterite}$ in the d_{bbl} range less than 200 μm, and the selectivity of aragonite was reached in the range over 95 % at a d_{bbl} of 100 μm. At a d_{bbl} of 40 μm, $r_{aragonite}$ increased remarkably with a decrease in α_{CO_2/NH_3}, and the crystallised polymorph was only aragonite at an α_{CO_2/NH_3} of 0.20. On the contrary, $r_{vaterite}$ was levelled down with decreasing α_{CO_2/NH_3} and then drops to almost nothing at 0.2 in α_{CO_2/NH_3}. These results indicated that d_{bbl} and α_{CO_2/NH_3} are interdependent for aragonite crystallisation, and the simultaneous decrease of d_{bbl} and α_{CO_2/NH_3} is effective to the acceleration of aragonite crystallisation.

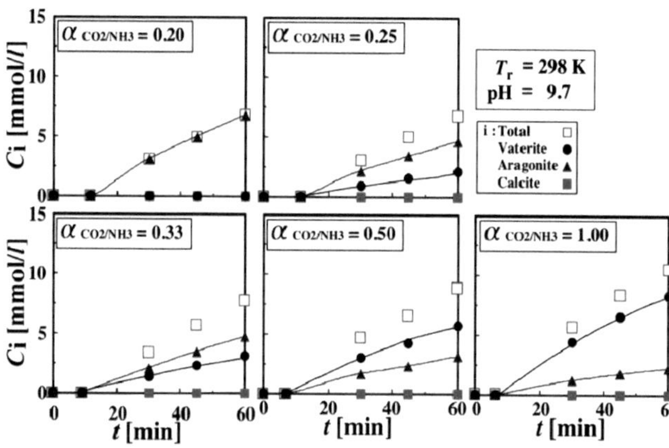

Figure 6.15 Time changes in produced molar concentration of $CaCO_3$ polymorphs ($d_{bbl} = 40$ μm).

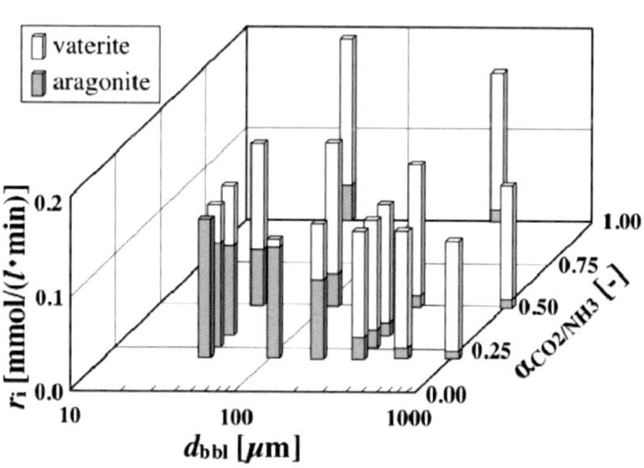

Figure 6.16 Effects of d_{bbl} and α_{CO_2/NH_3} on production rate of aragonite, vaterite.

6.2.5 Application of Microwave Irradiation to Reactive Crystallisation

Recently, the demand for lithium carbonate (Li_2CO_3) has remarkably increased, as it is used in lithium-ion batteries, heat-resistant glassware, and medicines, as well as in the manufacture of other lithium salts [13,14]. To improve the dissolubility of this salt for better Li_2CO_3 utilisation, the following must be achieved: fine particle size, narrow size distribution, uniform morphology and high surface area. Therefore, downsizing and control of the size distribution of Li_2CO_3 particles are very important during Li_2CO_3 crystallisation. Here, microwave irradiation has been recently employed in the solid-phase synthesis of oxides and sulfides and fine metal powders, fabrication of superconductors and semiconductors, pyrolysis of metal-organic precursors for powder synthesis, and crystallisation of inorganic compounds from solution [15–17]. For developing a crystallisation technique that enables the control of the crystal properties of carbonates with high dissolubility by the generation of local supersaturation regions, the reactive crystallisation of Li_2CO_3 using minute gas–liquid interfaces around CO_2 minute-bubbles was activated by microwave irradiation. The solubility of Li_2CO_3 decreases sharply with an increase in temperature. Therefore, when such minute gas–liquid interfaces are considered as new reaction fields, fine crystals with a narrow size distribution can be obtained by inhibiting crystal growth in the liquid phase.

Figure 6.17 shows the schematic of the waveguide-type microwave apparatus comprising a 1.5 kW generator (magnetron), rectangular waveguide (cross-sectional area: 109.2 × 54.6 mm^2), and an irradiated field including a flask-type quartz vessel (internal volume: 500 ml) as a crystalliser. Microwaves with a frequency of 2.45 GHz generated from the magnetron are adjusted by the tuner and made incident on the quartz vessel after passing through a horizontal waveguide.

Figure 6.18 shows the time changes in solution temperature (T_s) under the microwave irradiation when the incident irradiation power (P_W) is varied in the range of 200–400 W. T_s increases linearly with bubble supply time (t_b), and the rate of temperature increase (r_T) is 9.0, 14.5 and 23.0 K/min when P_W is maintained at 200, 300 and 400 W, respectively. This increase in r_T results from the strong vibration of water molecules and ions that selectively absorb the microwave energy.

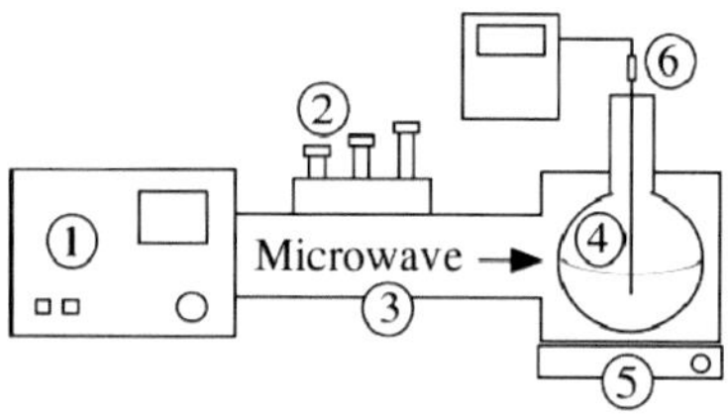

Figure 6.17 Reactive crystalliser under microwave irradiation.

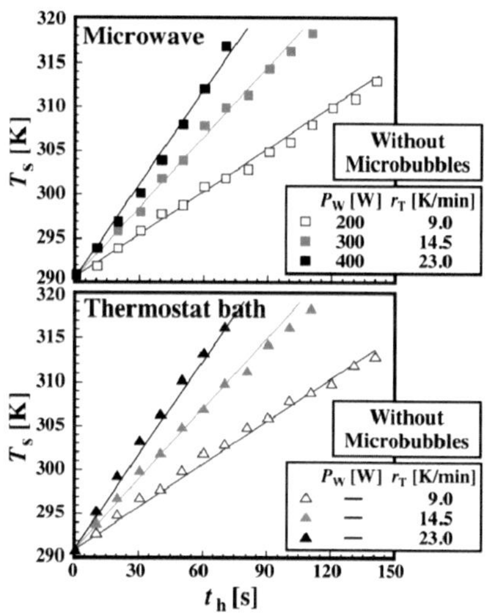

Figure 6.18 Comparison of time changes in T_s.

Figure 6.19 shows the particle size distributions of Li_2CO_3 at various d_{bbl} values when r_T is fixed at 23.0 K/min (P_W = 400 W). Because of the minimisation of bubble formation, the rate of nanoparticle crystallisation was significantly enhanced during microwave irradiation than during thermostat bath heating. Furthermore, the increase in the ratio of the volume of nanosized particles (size less than 1 μm) to the total volume of all the particles ($V_{np}/(V_p)_0$) was more pronounced than that in only the CO_2 minute-bubble formation with a decrease in d_{bbl} during microwave irradiation of the solution with CO_2 minute-bubbles.

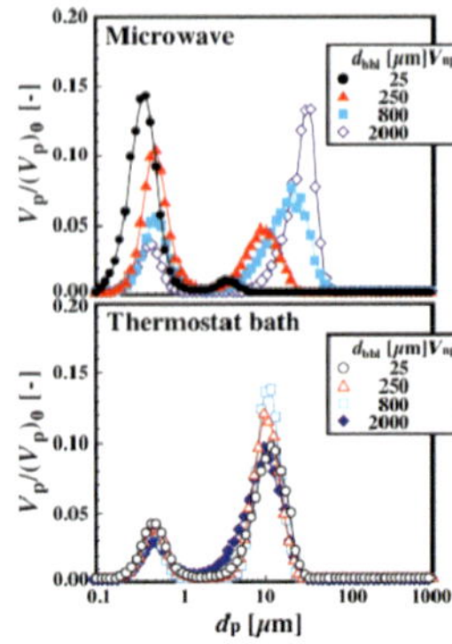

Figure 6.19 Effect of d_{bbl} on particle size distribution of Li_2CO_3 ($r_T = 23.0$ K/min).

From the above results obtained by varying r_T and d_{bbl}, the production rate of Li_2CO_3 nanoparticles with diameters less than 1 μm (r_{np}) and the rate of production of micronsized particles (r_{mp}) are determined. The obtained values are plotted against d_{bbl} and r_T, as shown in Fig. 6.20. r_{np} and r_{mp} were calculated from the total production rate of the Li_2CO_3 particles and the particle size distribution of Li_2CO_3. At all values of r_T, the total production rate increased slightly when bubble formation was minimised. Under microwave irradiation, r_{np} increased remarkably with an increase in r_T and a decrease in d_{bbl}. On the contrary, during thermostat bath heating, the increase in r_{np} was relatively smaller than that under microwave irradiation.

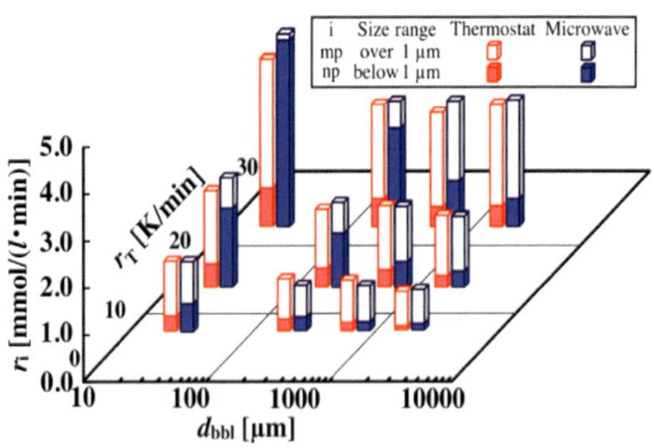

Figure 6.20 Relation between average bubble size and production rate of micro/nano particles.

References

1. Li P, Tsuge H (2006) Ozone transfer in a new gas-induced contactor with microbubbles, *J Chem Eng Japan*, **39**, 1213–1220.
2. Onoe K, Matsumoto M, Akiya T (2002) Reactive crystallization of calcium carbonate using CO_2 micro bubbles, *Bull Soc Sea Water Sci Jpn*, **56**, 357–361 (in Japanese).
3. Matsumoto M, Isago, M, Onoe K (2004) The application of micro bubbles for dissolution and crystallization of calcium carbonate in gas-liquid-solid system, *Bull Soc Sea Water Sci Jpn*, **58**, 475–485.
4. Matsumoto M, Wada Y, Suzuki M, Yoshida M, Onoe K (2011) Prediction of local supersaturation at minute gas-liquid interfaces—Utilization of polymorphic change during antisolovent crystallization, *Bull Soc Sea Water Sci Jpn*, **65**, 301–309.
5. Onoe K, Matsumoto M (2008) Synthesis of carbonate fine particles by reactive crystallization using CO_2 micro-bubbles, *Chem Ind*, **59**, 539–545 (in Japanese).
6. Matsumoto M, Fukunaga T, Suzuki M, Onoe K (2010) Reactive crystallization of $CaCO_3$ using CO_2/NH_3 bubbles—Effect of bubble size and molar ratio of CO_2/NH_3 on polymorphism, *Bull Soc Sea Water Sci Jpn*, **64**, 11–18.
7. Matsumoto M, Fukunaga T, Onoe K (2008) Polymorph control of calcium carbonate by reactive crystallization using microbubble technique, *Chem Eng Res, Des*, **88**, 1624–1630.
8. Klug HP, Alexander LE (1974) *X-ray Diffraction Procedures*, 2nd ed, Wiley, New York.
9. Matsumoto M, Isago M, Onoe K (2004) Effects of feed condition of CO_2/NH_3 micro bubbles on reactive crystallization of $CaCO_3$, *Proceedings 10th APCChE*, No. 762.
10. Matsumoto M, Fukunaga T, Onoe K (2006) Effects of $CO_2/NH_3/N_2$ bubble size on reactive crystallization of $CaCO_3$ fine particles, *Proceedings 11th Asian Pacific Confederation of Chemical Engineering Congress*, No. 99.
11. Sherwood TK, Pigford RJ, Wike CR (1975) *Mass Transfer*, pp. 336–343. McGrow-Hill, New York.
12. Bulter JM (1982) *Carbon Dioxide Equilibria and Their Application*, pp. 75–78, Addision-wesley, London.
13. Tsuge H, Kuze I, Fujii K (2004) The effective utilization of beneficial resources dissolved in seawater—Crystallization characteristics of

lithium carbonate, *Bull Soc Sea Water Sci Jpn,* **58**, pp. 579–584 (in Japanese).

14. Matsumoto M, Morita Y, Yoshinaga M, Hirose S, Onoe K (2009) Reactive crystallization of lithium carbonate nanoparticles by microwave irradiation of aqueous solution containing CO_2 microbubbles, *J Chem Eng, Japan,* **42**, Suppl. 1, pp. s242–s248.
15. Wada Y, Kuramoto H, Kitamura J, Sakata T, Mori H, Yanagida S (2001) Microwave-assisted size control of CdS nanocrystallites, *J Mater Chem,* **11**, pp. 1936–1940.
16. Guo L, Luo H, Gao J, Guo L, Yang J (2006) Microwave hydrothermal synthesis of barium titanate powders, *Mater Lett,* **60**, pp. 3011–3014.
17. Heuser JA, Spendel WU, Pisarenko AN, Pechan MJ, Pacey GE (2007) Formation of surface magnetite nanoparticles from iron-exchanged zeolite using microwave radiation, *J Mater Sci,* **42**, 9057–9062.

6.3 Preparation of Hollow Microcapsules

Toshionori Makuta,[a] Hirofumi Daiguji,[b] and Fumio Takemura[c]

[a]*Graduate School of Science and Engineering, Yamagata University, Yonezawa, Yamagata 992-8510, Japan*

[b]*Graduate School of Frontier Sciences, University of Tokyo, Kashiwa, Chiba 277-8563, Japan*

[c]*Institute for Energy Utilization, National Institute of Advanced Industrial Science and Technology (AIST), Tsukuba, Ibaraki 305-8565, Japan*

makuta@yz.yamagata-u.ac.jp

6.3.1 Introduction

Hollow microcapsules are gas-filled spherical particles with diameters between 1 and 1000 μm. They have good potential because of their advantageous properties such as low effective density and high specific surface area. They are therefore widely used for applications, such as the reduction of material weight, the encapsulation and immobilisation of bioactive and catalytically active substances, the modification of the impact strength of compounds, and the improvement of thermal and acoustical insulation. In the medical and pharmaceutical fields, ultrasound contrast agents, drug-delivery systems and the synthesis of artificial cell structures using hollow microcapsules of diameter less than 10 μm are of increasing interest [1].

Recently, the fabrication of tailor-made hollow sphere structures using processes such as spray drying and dripping, emulsion [2], fluidised bed [3] and suspension [4] techniques have been gaining increasing interest. In these processes, template techniques, that is, the deposition of solid layers on dispersed templates, such as liquid droplets or solid particles, are usually used. To obtain a hollow sphere, a liquid or solid core inside the layer is removed by dissolution, evaporation or thermolysis in the final stage of the fabrication process. However, the capsule shell is often damaged in the core removal process. We therefore developed a method of fabricating hollow microcapsules using a microbubble template technique. Figure 6.21 shows a simplified schematic diagram of this fabrication method. If microbubbles can be used as templates instead of liquid droplets or solid particles, damage during the core removal process is completely avoided. In addition, the microbubble template technique can be used for a variety of wall component materials such as polymers and inorganic materials. However, the microbubbles do not remain in the liquid for a period longer than the period for which the dispersed liquid or solid cores do. Some microbubbles merge into large bubbles and float into the air, and others disappear as a result of dissolution of the gas into the liquid. Thus, in the microbubble template technique, the encapsulation rate should be faster than the dissolution rate of the microbubbles.

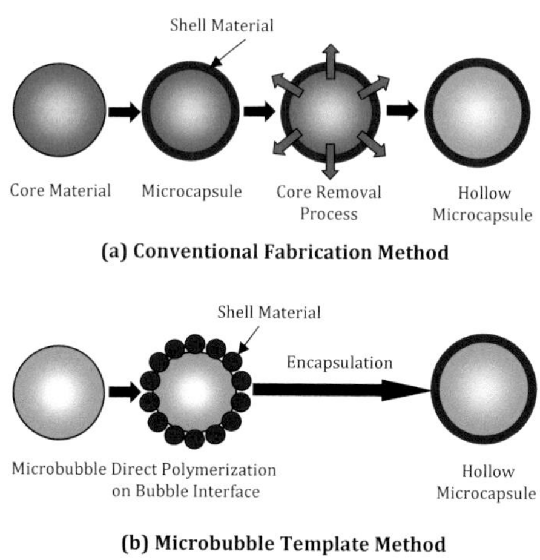

Figure 6.21 Schematic diagram of fabrication methods for hollow microcapsule: (a) conventional method, (b) microbubble template method.

Here, we introduce two fabrication methods for hollow microcapsules using microbubbles as templates. One is the fabrication of melamine–formaldehyde hollow microcapsules using polymerisation at the gas–liquid interfaces of microbubbles [5], the other is the fabrication of poly(lactic acid) (PLA) hollow microcapsules using solvent drying around microbubbles [6].

6.3.2 Hollow Microcapsules Fabricated by in-situ Polymerisation Method with Microbubble Template

In this section, we give an example of hollow microcapsules produced using melamine–formaldehyde resin as the shell component by an in-situ polymerisation method. Melamine–formaldehyde microcapsules have been widely used in carbonless copy paper and other pressure-sensitive products, such as carriers for fragrant oils and pest repellents. In-situ polymerisation methods are one of the conventional fabrication processes for microcapsules. In this method, dispersed liquid or solid cores are encapsulated in a continuous phase, and polymerisation occurs exclusively in the continuous phase. The polymerisation rate of melamine–formaldehyde resin is faster than that of other polymers, and bubbles in the continuous phase are available as core materials in the in-situ polymerisation method if the bubbles are stable in the liquid. We therefore demonstrate the fabrication of melamine–formaldehyde hollow microcapsules from microbubble templates.

6.3.2.1 Materials and methods

The continuous phase was a 4% poly(vinyl alcohol) (PVA) aqueous solution produced from anionic PVA (Gohsenal T-350, Nippon Gohsei Co., Ltd., Japan) and aniline (Wako Pure Chemical Industries, Ltd., Japan). The shell material was prepared from melamine, formaldehyde (37% w/w aqueous solution) and gelatin (all from Wako Pure Chemical Industries, Ltd., Japan). The pH was adjusted by adding 10% w/w Na_2CO_3 aqueous solution obtained from anhydrous sodium carbonate and glycine (Wako Pure Chemical Industries, Ltd., Japan). All chemicals were reagent grade. Distilled water (Direct-Q3UV, Millipore Co., Ltd., USA) was used for all experiments.

First, 100 g of 4% PVA aqueous solution with 1.3 g of aniline and 0.08 g of gelatin was poured into a 300 mL pressure vessel; the air was compressed up to 4 atm, and the vessel valve was closed. The vessel was maintained at 4 atm and 300 K, and some of the air dissolved into the solution. After 15 min, the valve was opened and the pressure decreased to atmospheric pressure. During this period, bubbles were generated in the solution. Large bubbles rose into the air, but microbubbles remained in the solution. The solution containing microbubbles was poured into a 300 mL beaker and was heated up to 338 K. Finally, 0.33 g of glycine was added to the solution as an anionisation agent for the melamine–formaldehyde polymer.

For the melamine–formaldehyde pre-polymer preparation, 20 g of 37% formaldehyde aqueous solution, 6.6 g of melamine and 26.6 g of distilled water were poured into a 300 mL beaker. The pH was adjusted to 10.0 using 10% Na_2CO_3 aqueous solution, and the mixture was stirred on a hot stirrer at 338 K for 15 min. During stirring, the methylolmelamine formation reaction occurred [1,3]:

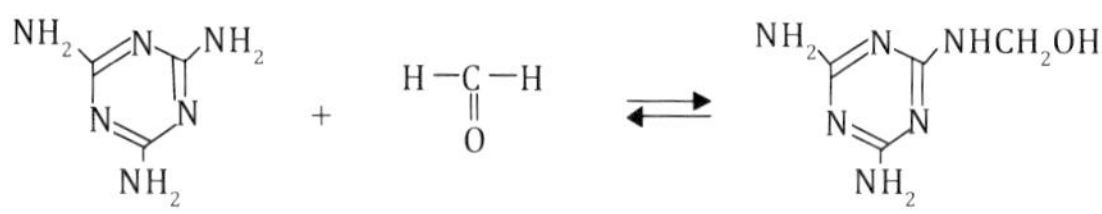

(6.1)

The melamine–formaldehyde pre-polymer was added to the PVA solution containing the microbubbles, and encapsulation of the microbubbles by in-situ polymerisation was executed by stirring this mixture at 338 K for 15 min. During stirring, melamine–formaldehyde oligomeric derivatives were produced by the formation of two different types of bridges between triazine rings, namely, a methylene-ether bridge and a methylene bridge [1][3]:

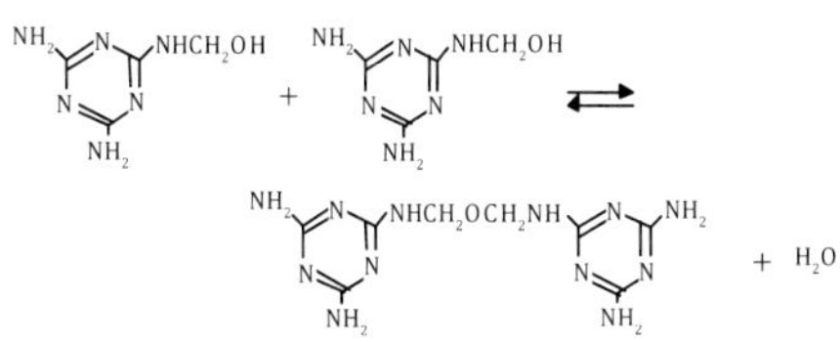

(6.2)

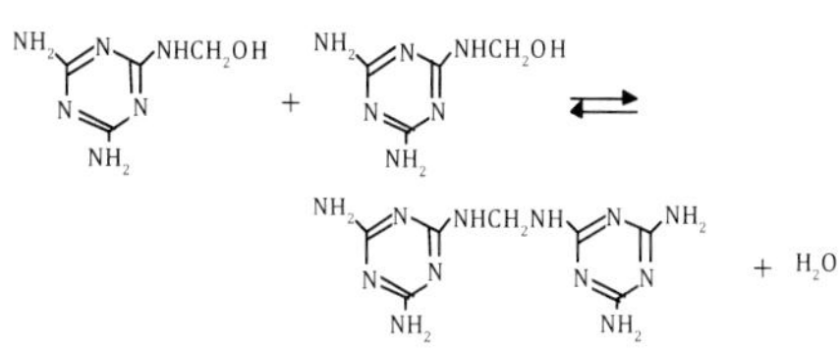

(6.3)

A sequentially crosslinked network was formed by a polycondensation reaction. After completion of the encapsulation reaction, the reaction was quenched by adjusting the pH of the solution to 9.5 by adding 10% Na_2CO_3 aqueous solution. After these processes, the fabricated hollow microcapsules floated on the surface of the PVA solution.

6.3.2.2 Experimental results

Figure 6.22 shows a scanning electron microscope image of melamine–formaldehyde hollow microcapsules located on the surface. Figure 6.23 shows the optical microscope image of freeze-dried hollow microcapsules. Figure 6.22 reveals that microcapsules ranging in size from 40 μm to 100 μm were fabricated by the encapsulation process described above. The microcapsules shown in Fig. 6.23 have cracks on their shells, indicating that the microcapsules are hollow; the shell thickness is of the order of 100 nm. Further evidence that the produced spheres are not solid particles but hollow capsules is that the spheres float and accumulate on the liquid surface. If the spheres are solid particles, they will sink because the density of melamine–formaldehyde resin is larger than that of 4% PVA aqueous solution. Most of the solution sediment is spherical polymer particles, as shown in Fig. 6.22. In this fabrication technique, hollow capsules are easily fabricated and purified because they accumulate on the liquid surface. This method could be useful for mass fabrication of hollow microcapsules. However, as Fig. 6.23 shows, the stiffness of the microcapsules is low, and cracks and breakages tend to occur easily because the shells are extremely thin.

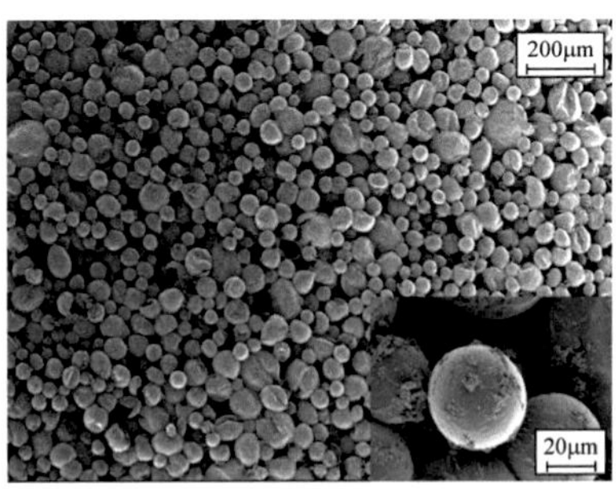

Figure 6.22 Scanning electron microscope image of melamine–formaldehyde hollow microcapsules fabricated by in-situ polymerisation.

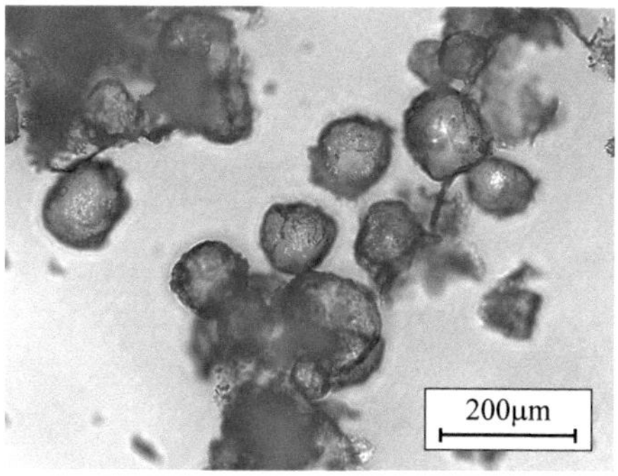

Figure 6.23 Optical microscope image of freeze-dried melamine–formaldehyde hollow microcapsules.

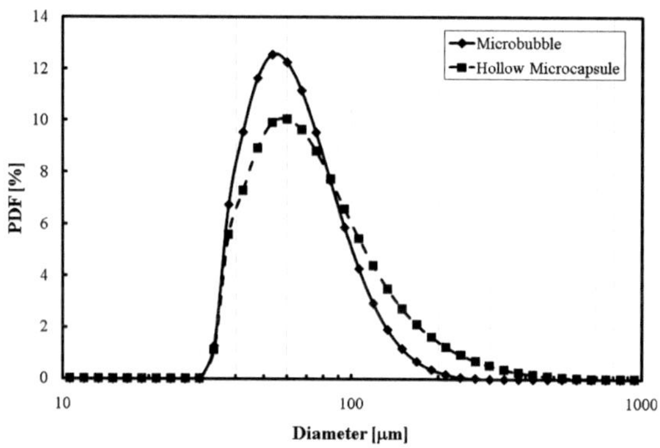

Figure 6.24 Diameter distributions of microbubbles and fabricated hollow microcapsules.

In this fabrication process, microbubbles in the PVA solution are encapsulated by the melamine–formaldehyde resin; therefore, we measured the diameter distributions of the microbubbles and hollow microcapsules, that is, the reaction field of encapsulation. Figure 6.24 shows the diameter distributions of microbubbles and hollow microcapsules determined using a laser diffraction particle analyser (Mastersizer 2000, Malvern Instruments Co., Ltd., UK) and calculated using the Beer–Lambert law [7]. Figure 6.24 shows that the distribution of hollow microcapsules is approximately the same as that of microbubbles, especially at the lower detection limit of 30 μm and the peak at around 50 μm. These facts indicate that microbubbles of diameter less than 30 μm dissolve completely, and the bubbles of diameter more than 500 μm float and disappear quickly, before the melamine–formaldehyde polymerisation. When microbubbles of a low-solubility gas (sulfur hexafluoride) are used as the microcapsule template, smaller hollow microcapsules are fabricated, and the diameter distribution of the microcapsules shifts to the smaller side [8]. Experimental results under parameter-changed conditions revealed that the surface morphology of the microcapsules could be controlled by adjusting the reaction time, glycine concentration, or pre-polymer concentration [5]. In addition, the capsule shell thickness could be controlled by adjusting the aniline concentration.

The melamine–formaldehyde hollow microcapsules fabricated by this process consist mainly of gas because of the negligible shell thickness relative to the capsule size, and they are classified easily by flotation or sieving. Therefore, these microcapsules are expected to be useful as new industrial functional materials for thermal and sound insulation, and weight saving.

6.3.3 Hollow Microcapsules Fabricated by Solvent Evaporation Method with Microbubble Template

In this section, we give an example of hollow microcapsules prepared by the solvent evaporation method using PLA resin as the shell component. PLA is a biodegradable polymer that can be decomposed by microbes such as bacteria. Because of its slow degradation rate, PLA is often used for suture threads inside the human body [9]. Biodegradable polymers are synthesised from crude materials via a complex multistep fabrication that includes evaporation, distillation and polymerisation [10]. Covering microbubbles with a biodegradable polymer shell using a direct polycondensation reaction is therefore difficult because, as a result of dissolution and bursting, the microbubbles do not stay in the liquid during the polycondensation reaction time. We therefore developed a simple method of fabricating hollow microspheres physically covered with biodegradable polymer, using microbubbles as templates.

6.3.3.1 Materials and methods

A solvent evaporation method consisting of the following four steps was used to fabricate PLA hollow microspheres.

Step 1: A 2 g/L methylene chloride solution of PLA (MW 6000–16,000, Polyscience, USA) and a 2% w/w PVA aqueous solution were prepared, and then droplets of the methylene chloride (Wako Pure Chemical Industries, Ltd, Japan) solution of PLA, about 1 mm in diameter, were dispersed in the PVA aqueous solution using a microsyringe. Because of the surface active properties of PVA, the droplets did not coalesce in the aqueous solution.

Step 2: The droplets and solution were then enclosed in a vessel, which was then pressurised using air to over 300 kPa. The air was then fully dissolved into both the droplets and the solution.

Step 3: The pressure of the vessel was then decreased to atmospheric pressure, thus yielding microbubbles in the droplets as well as in the solution as a result of degassing. The microbubbles in the droplets have a uniform diameter and remained stable, as shown in Fig. 6.25, whereas those in the solution soon disappeared by flotation. Step 2 and Step 3 are performed to increase the number of microbubbles in the droplets, but Daiguji reported that these processes were not necessarily needed to generate hollow PLA microcapsules [11].

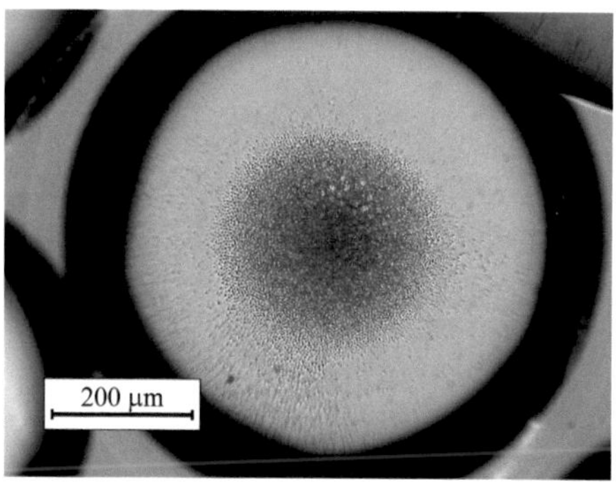

Figure 6.25 Representative photograph of a cloud of bubbles with uniform diameter of 2 μm formed inside a methylene chloride droplet.

Step 4: Hollow PLA microspheres with inner diameters, the same as those of the microbubbles, were released spontaneously from the droplets into the PVA aqueous solution. Methylene chloride has extremely low solubility in water, but is not insoluble; therefore, the droplets shrank as a result of the dissolution of methylene chloride into the PVA aqueous solution. In contrast, PLA is completely insoluble in water; thus, as the droplets shrank, the PLA concentration in the methylene chloride droplets gradually increased and the PLA finally solidified. During the PLA solidification phase, PLA hollow microcapsules are released from the droplet surfaces.

6.3.3.2 Experimental results

Figure 6.25 shows a methylene chloride droplet containing microbubbles in the PVA aqueous solution a few minutes after depressurising (Step 3) and Fig. 6.26 shows representative photographs of the release of PLA-covered bubbles from the droplet during the dissolution of methylene chloride (Step 4). Both images were obtained using an optical microscope (VB-7000, Keyence Co., Ltd., Japan).

Figure 6.25 shows that a cloud of bubbles with uniform diameters of 2 μm was generated inside the methylene chloride droplets as a result of degassing. The surface tension acting on the interface between the PVA solution and the methylene chloride droplets prevented the bubbles from leaving the droplet. According to the theory derived by Ward et al. [12], bubbles in a closed volume of a liquid–gas solution can be thermodynamically stable when the number of bubble nuclei is sufficiently large. The observed system was not completely closed because methylene chloride was transferred to the surrounding PVA aqueous solution. However, the dissolution rate of methylene chloride into the PVA aqueous solution was much smaller than the relaxation rate needed to achieve liquid–vapor equilibrium. Therefore, the observed system could be regarded as a closed system. In addition, the surface active properties of PLA prevented the bubbles from coalescing. Consequently, microbubbles remained inside the droplets for several hours.

Figure 6.26 shows that 3 μm diameter spheres moved away spontaneously from the droplets into the PVA aqueous solution in the final shrinking phase of the methylene chloride droplets as a result of the dissolution of methylene chloride into the PVA aqueous solution. These 3 μm diameter spheres were released from the droplets after degassing, but they are rarely released from the droplet just after degassing. After 4 h, spheres were being released continuously from the droplets, and their release rate was increasing; the droplets completely solidified over 5 h. Because the transfer rates of both PLA and air into the PVA aqueous solution were much smaller than that of methylene chloride, the concentration of PLA and the number density of microbubbles in the droplets increased gradually, and the release rate of hollow PLA microspheres also increased gradually.

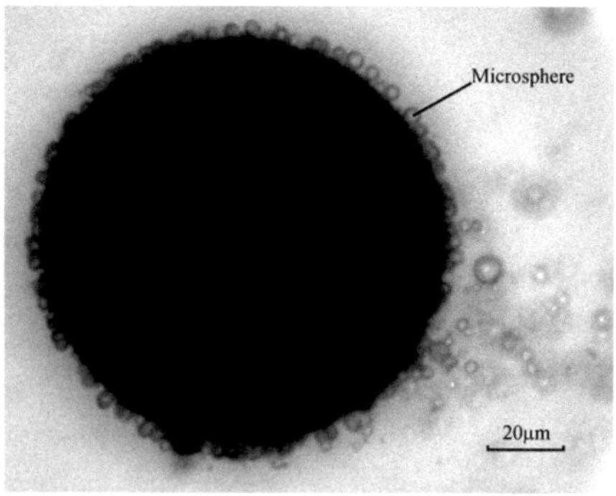

Figure 6.26 Representative photograph of PLA-covered bubbles rising to the surface of a methylene chloride droplet, and of bubbles leaving the droplet during the dissolution of methylene chloride.

Figure 6.27 shows the fluorescence image of PLA microspheres fabricated by this method. For the images in Fig. 6.27, PLA stained with Nile Red was visualised using a confocal laser microscope (TCS-SL, Leica) focused at the horizontal section near the particle center. The excitation wavelength was 488 nm. Figure 6.27a shows low-fluorescence-intensity rings with a dark area inside the spheres, clearly indicating that the fabricated microspheres were hollow, with a single internal void.

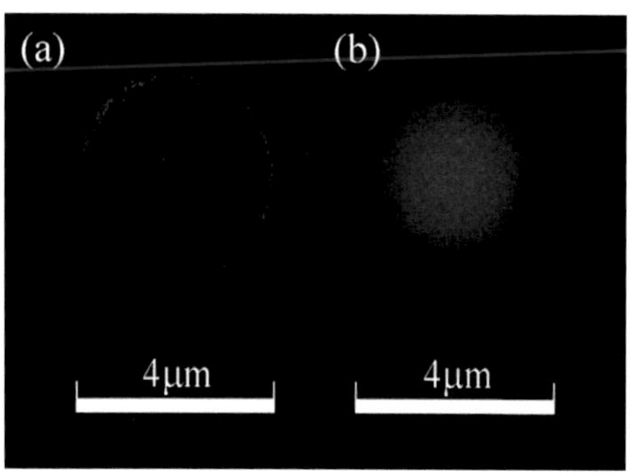

Figure 6.27 Fluorescence images of (a) a hollow PLA microsphere and (b) a PLA solid particle, which does not enclose a microbubble.

The precise mechanism of microbubble encapsulation has not yet been clarified, but the following two important results were obtained from this experiment. (1) The inner diameters of the hollow PLA microspheres were almost the same as those of the template microbubbles. (2) The hollow PLA microspheres were released from the methylene chloride droplets spontaneously, and their release rate increased as the droplets shrank. These results suggest that PLA-adsorbed microbubbles approached the interface between the droplet and the PVA aqueous solution, and the PLA solidified around the microbubbles because of drying of the methylene chloride, and then, hollow PLA microspheres were released from the droplet into the PVA aqueous solution.

These hollow PLA microcapsules are smaller than blood capillaries and are composed of a biocompatible polymer, and they have harmonic ultrasound imaging abilities [12]. Cornejo et al. reported that the size and uniformity of hollow PLA microcapsules can be controlled by the initial PLA concentration, addition of PVA to the aqueous medium and the PLA molecular weight [13]. Therefore, hollow PLA microcapsules could be used as ultrasound contrast agents and are expected to be generated at low cost and to be controllable in size.

6.3.4 Conclusions

In this chapter, we introduced two types of methods for fabricating hollow microcapsules using microbubbles as a template. One is fabrication of melamine–formaldehyde hollow microcapsules by chemical polymerisation at the gas–liquid interface of the microbubbles; the other is fabrication of hollow PLA microcapsules by physical precipitation on the microbubbles. Other examples of hollow microcapsules produced using microbubbles as a template are calcium carbonate hollow microcapsules [14] and cyanoacrylate hollow microcapsules [15]. Most of these studies of hollow microcapsules are at a basic level, and improvements in fabrication yield, safety and size controllability are necessary before they can be used in practical applications. However, these methods make hollow microcapsule fabrication easy because they eliminate the core removal process for generating the hollow structure in the microcapsules. In the medical field, these microcapsules are expected to be used as new ultrasound contrast agents with polymer shells, since the biodegradable hollow microcapsule has an ultrasound frequency response close to that of a microbubble and the microcapsule structure is similar to a single-void structure and allows for fluid circulation in a blood vessel.

References

1. Bertling J, Blomer J, Kummel R (2004) Hollow microspheres, *Chem Eng Technol*, **27**, 829–837.
2. Omi, S, Katami K, Yamamoto A, Iso M (1994) Synthesis of polymeric microspheres employing SPG emulsification technique, *J Appl Polym Sci*, **51**, 1–11.
3. Strohm H, Sgraja M, Bertling J, Lobmann P (2003) Preparation of TiO_2-polymer hybrid microcapsules, *J Mater Sci*, **38**, 1605–1609.
4. Caruso F, Caruso RA, Mohwald H (1999) Production of hollow microspheres from nanostructured composite particles, *Chem Mater*, **11**, 3309–3314.
5. Daiguji H, Makuta T, Kinoshita H, Oyabu T, Takemura F (2007) Fabrication of hollow melamine-formaldehyde microcapsules from microbubble templates, *J Phys Chem B*, **11**, 8879–8884.
6. Makuta T, Takada S, Daiguji H, Takemura F (2009) Simple fabrication of hollow poly-lactic acid microspheres using uniform microbubbles as templates, *Mater Lett*, **63**, 703–705.
7. Couto HJB, Nunes DG, Neumann R, Franca SCA (2009) Micro-bubble size distribution measurements by laser diffraction technique, *Miner Eng*, **22**, 330–335.
8. Tamakawa Y, Makuta T (2012) Yield improvement and durability enhancement of hollow microcapsules fabricated by bubble template method, *J Jpn Soc Exp Mech*, **12**, 17–22.
9. Heino A, Naukkarinen A, Kulju T, Tormala P, Pohjonen T, Makela EA (1996) Characteristics of poly(L-)lactic acid suture applied to fascial closure in rats, *J Biomed Mater Res*, **30**, 187–192.
10. Gruber PR, Hall ES, Kolstad JJ, Iwen ML, Benson RD, Borchardt RL (1992) Continuous process for manufacture of lactide polymers with controlled optical purity, U.S. Patent No. 5142023.
11. Daiguji H, Takada S, Cornejo JJM, Takemura F (2009) Fabrication of hollow poly(lactic acid) microcapsules from microbubble templates, *J Phys Chem B*, **113**, 15002–15009.

12. Ward CA, Tikuisis P, Venter RD (1982) Stability of bubbles in a closed volume of liquid-sas solution, *J Appl Phy*, **53**, 6076–6084.

13. Cornejo JJM, Daiguji H, Takemura F (2011) Factors affecting the size and uniformity of hollow poly(lactic acid) microcapsules fabricated from microbubble templates, *J Phys Chem B*, **115**, 13828–13834.

14. Hadiko G, Han YS, Fuji M, Takahashi M (2005) Synthesis of hollow calcium carbonate particles by the bubble templating method, *Mater Lett*, **59**, 2519–2522.

15. Makuta T, Tamakawa Y, Endo J (2011) Hollow microspheres fabricated from instant adhesive, *Mater Lett*, **65**, 3415–3517.

Chapter 7

Food Industry

Xu Qingyi, Nobutaka Nakamura, and Takeo Shiina
National Food Research Institute,
National Agriculture and Food Research Organization,
Ibaraki, 305-8642, Japan
shiina@affrc.go.jp

7.1 Introduction

Bubble technology is largely concerned with the food processing or waste water disposal methods of food factories and thus influences the quality, production efficiency or cost of food. In addition, the food developed using microbubble (MB) or micro-nano bubble (MNB) technology is expected to have new characteristics because the quantity and size of the food bubbles are important factors that contribute to the appearance, physical properties and texture of the food. Furthermore, control over the growing and pre-postharvest physiological properties of fresh produce, sterilisation of food or equipment and treatment of public water and wastewater are proposed as applications of MNB technology in the food industry.

Recently, considerable studies have been performed regarding the utilisation of unique MNB characteristics; however, the basic

Micro- and Nanobubbles: Fundamentals and Applications
Edited by Hideki Tsuge

ISBN 978-981-4463-10-2 (Hardcover), 978-981-4463-11-9 (eBook)
www.panstanford.com

characteristics and efficient application methods of MNB remain unclear.

In this chapter, we describe our research concerning the generation of an MNB method for the food industry and the trends of MNB applications in food-related areas.

7.2 Characteristics of MNBs Generated with Different Conditions

7.2.1 Effects of the Generated Method on MNB Properties

MB technology has been gradually accepted as cost- and environmentally effective techniques with great potential for applications to almost every field in the food industry. The selection of a suitable generation method is significant to generate MBs with the desired properties. Mechanical agitation and sonication are two commonly used methods for MB generation. However, systematic information regarding their generation efficiency and effects on the properties of MBs is not available. A comparative study was conducted in this paper.

As attention towards the potential applications of MBs increases, the information concerning the efficiency of the generation methods and their effects on the properties and stabilities of MBs are crucial for selecting an appropriate method. Xu et al. (2008) [26] evaluated the generation efficiencies of two commonly used methods, mechanical agitation and sonication, in two surfactant systems.

The generation efficiency of MBs was evaluated in terms of gas stability, bubble number, interfacial area and bubble size. The results showed that sonication is more effective in the generation of MBs than mechanical agitation. MBs prepared by sonication exhibit a larger stability, smaller diameter, higher bubble number, larger interfacial area and better uniformity. The behaviour of MBs depends on the generation method employed and surfactant used. The existence of a critical diameter for MBs to shrink was found. MBs prepared by sonication have a smaller critical diameter than those prepared by mechanical agitation. The results indicate that the critical diameter and rate of shrinkage are also affected by the surfactant type. The surfactant may affect not only the generation efficiency but also the properties of MBs. From the results of this paper, the size effect of MBs may be greatly affected by the generation method; thus, the use of a suitable generation method is of particular important for the applications of MBs.

7.2.2 Effect of Substance on MNB Properties

As interest in the applications of MBs grows, an understanding of the factors that affect the formation and property of MB is significant for the effective generation of MBs. The factors that contribute to the reduction of surface tension and the structural enhancement of interfacial films are among the most important parameters for generating and stabilising bubbles. The effects of surfactant concentration and the addition of electrolyte on bubble size distribution and stability were investigated [24]. An anionic surfactant, sodium dodecyl sulfate (SDS), was used as the surfactant. The bubble diameters were found to decrease with increasing SDS concentration, reaching a minimum diameter at SDS concentrations above the critical micelle concentration (CMC) (Fig. 7.1).

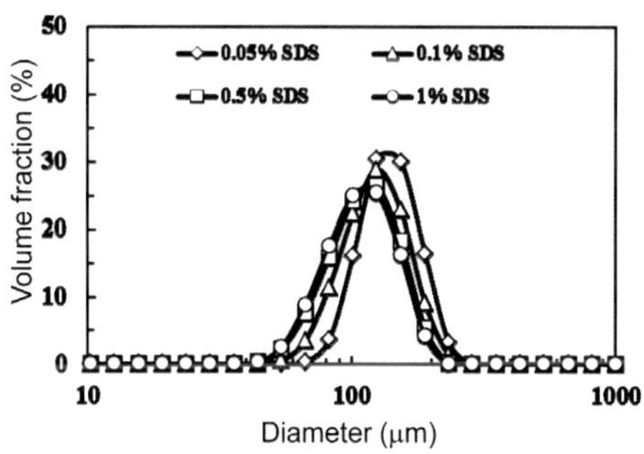

Figure 7.1 Effect of SDS concentration on the size distribution of MBs.

The stability of the bubbles increased with increasing SDS concentration, and a maximum stability was obtained with SDS concentrations above the CMC. The effect of the addition of electrolyte was investigated with sodium chloride (NaCl) at an SDS concentration below the CMC. The addition of NaCl decreased the bubble size and increased the stability to a certain extent (Fig. 7.2). The influence mechanism of NaCl was examined in terms of the changes in SDS micelle surface tension and zeta potential. The addition of low salt concentration did not affect the surface tension. The surface tension was reduced with increasing salt concentration and reached a constant value with NaCl concentrations above 0.25%. The presence of NaCl resulted in a significant decrease in zeta potential, implying a reduction in the surface charge of SDS micelles. The result suggests that the presence of NaCl may improve the generation and stability of bubbles through enhancing the adsorption monolayer and interfacial film structures.

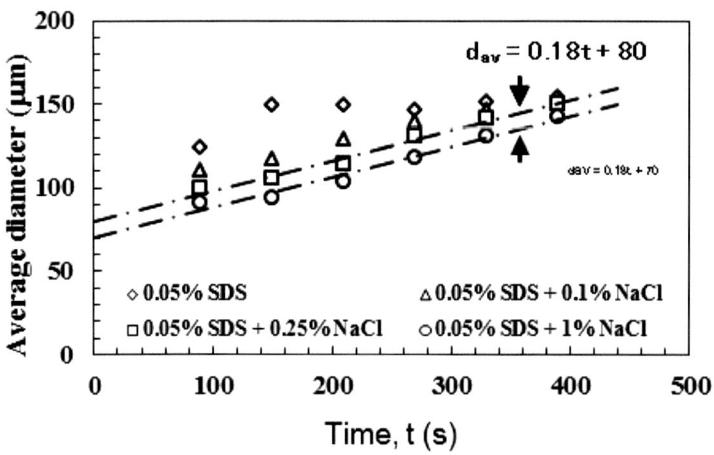

Figure 7.2 Effect of NaCl concentration on MB stability under the SDS concentration below CMC.

7.2.3 Effect of Biosurfactant on MNB Properties

Xu et al. (2010) [25] characterised the physico-chemical properties of the soybean oil-derived polysoap surfactant, Palozengs R-004 (R-004). The surface activity of R-004 is comparable with the reported activities of the biosurfactants produced by microorganisms and higher than some of the conventional synthetic surfactants. R-004 exhibits unique aggregation behaviour. Specifically, small aggregates (pre-micelles) form at very low concentrations. The zeta potential measurements showed that the micelles of R-004 are negatively charged due to the presence of carboxylic groups. The ability of

R-004 to form and stabilise MBs was evaluated and was found to be greatly affected by filtration while remaining independent of the R-004 concentration over the concentration range studied (Fig. 7.3). These results suggest that a very low level of surfactant can be used to produce MBs without affecting their properties. These results also indicate that soybean oil-based polymers are promising surfactants for food, pharmaceutical and cosmetic applications.

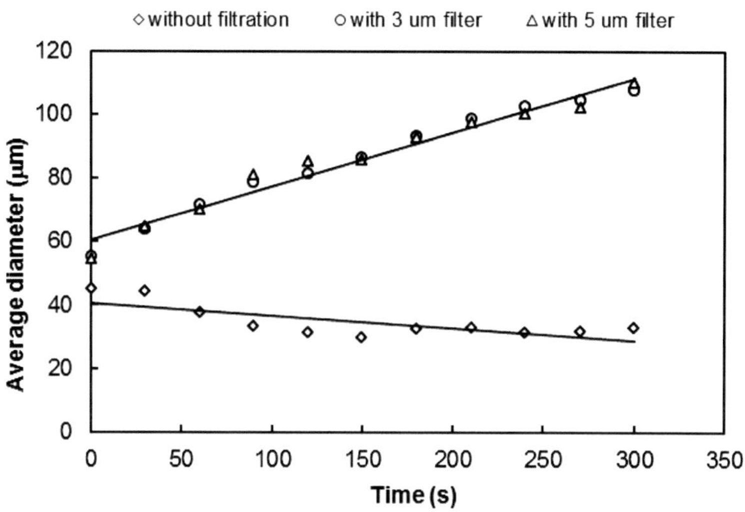

Figure 7.3 Effect of the filtration of surfactant used on the changes of MB size after generation.

7.2.4 Potential of MNB in the Food Industry

Newly developed or quality-improved food/food materials are expected to be produced by microstructure control. As an example regarding the process of cream production, Kukizaki (2009) [6] reported that the formability, water liberation and over run, which are main factors of cream physicality, can be controlled using MNB technology. Controlling the surface characteristics of MNB is considered to contribute to improving the stability [23] and supplementing novel MNB functionality. Moreover, MNB is expected to be used as a delivery system of functional or aroma components in foods and used as a carrier of drug or antimicrobial agents to maintain food hygiene or safety.

7.3 Trends of MNB Applications in the Food Industry

In this section, recent examples of MNB applications in food industries and the novel findings of MNB, which are expected to be applied to food-related areas, are introduced.

7.3.1 Control of Dissolved Oxygen Using MB Treatment

MB treatment is being applied in fisheries in Japan, as described in Chapter 8.2. In this chapter, therefore, examples of dissolved oxygen (DO) control applications using MB treatment (with the exception of fisheries) are presented.

Weber and Agblevor (2005) [22] reported that fermentations with MB treatment showed five-fold higher volumetric oxygen transfer coefficient (kLa) values and nearly two-fold higher cell mass productivities of *Trichoderma reesei* (cellulose-forming bacterium). Sakai et al. (2003) [10] reported that the quality change of Japanese Sake during storage was slowed after lowering the DO using nitrogen gas MB treatment while bottling. In the case of cultivating tomatoes, the number of days from the first flowering to the harvest fruit is reported to be shortened by aerating the culture liquid in a hydroponic system using MB [9], whereas the more puffy fruits were created by longer aeration durations [5]. Although the DO control using MB treatment can be applied to various areas, the merits and drawbacks of MB treatment should be examined before being employed.

7.3.2 Separation or Degradation Using MB Treatment

Recently, the regulation of effluents has become increasingly strict worldwide. In Japan, water quality regulations were enforced under the Water Pollution Control Law and Sewerage Raw in connection with those changes. Thus, effective effluent treatment methods are required. MB treatment is one of the suitable treatment methods for effluents from factories due to its effective gas solubility and flotation separation of suspended solids.

The components included in wastewater from food factories vary significantly. Thus, selecting the suitable treatment method for certain sewerage is significant. Goto et al. [3] showed that MB removed oil from oil contaminated soil due to its effective flotation separation. Terasaka et al. (2007) [20] investigated the removal of small particles (iron oxide fine particles) suspended in water using MB generated with surfactant. The authors also showed that the small particles in the water were effectively removed using MB with opposing electrical charge and that the zeta potential of MB can be controlled with the pH of the solution from intended particles. In addition, the separation rate of the small particles from water using MB could be described using a model of particle separation and re-entrainment [21].

Nakamura et al. (2008) [8] verified that ozone MB treatment rapidly decreased the pesticide concentration and total organic carbon in water when compared with ozone milli-bubble treatment. Takagi et al. (2007) [17] examined a pH neutraliser system for alkali waste water from food factories using MB treatment with CO_2 gas. By using effluent CO_2 gas from a food factory, the establishment of wastewater treatment systems with less environmental burden is expected. In addition, ozone MB treatment is applied to degrade sludge, downsize wastewater treatment tanks [2] and reduce processing time and ozone consumption when compared with normal ozone bubble treatment (Bando et al., 2008) [1].

7.3.3 Sterilisation Using MB Treatment

Heating and the use of chemicals are common sterilisation methods for the food industry. However, excess heating causes quality deterioration (appearance, texture, taste or the amount of nutrients); the chemical residues can give rise to safety and/or quality problems. Therefore, a non-thermal/chemical-free sterilisation method with fewer environmental loads and fewer health risks is required, particularly for fresh food because of its poor tolerance against heat and chemicals.

When MB treatment is applied for sterilisation, some biocidal gas is often incorporated. With the combined effect of ozone bubbling and ozone nanobubble treatments, inoculated Feline calicivirus (a surrogate of norovirus) in oyster bodies was inactivated more than 99% after 12 hours of treatment [18]. Tamaki et al. (2007) [19] demonstrated that ozone MB treatment was effective for the suppression of *Fusarium* spp. in the hydroponic system when compared with ozone bubbling treatment. Shirai et al. (2008)[14] and Takagi et al. (2009)[16] reported that ozonated water (10–12 ppm, nano-pico ozone water™) decreased foot and mouth disease virus, swine vesicular disease virus, *Salmonella* Enteritidis [14], and the disinfecting effect continued for 60 min after generation [16]. Shimoda et al. (2001) [13] showed that the addition of CO_2 MB to pressurised water can significantly decrease the bacteria in the water.

As the application for washing, air MBs are often used. Soli et al. (2009) [15] reported that slightly acidic hypochlorous water treatment in combination of pre-treatment with air MBs (with a sucrose fatty acid ester surfactant) was effective for the decontamination of fresh produce. Sharma et al. investigated the influence of fluid shear and MB on bacterial detachment from surfaces and reported that the detachment of the tenaciously adhering bacteria, which were not detached by flow alone, is allowed by adding MBs to the flow [11]. However, MB treatment became ineffective for larger bacterial entities, such as adhering coaggregates [11], and the detachment properties varied depending on the bacteria type [12].

The effect of MB treatment on the reduction of surfactant use is also reported: 1/4 of the CMC (cloth washing) [4], 1–2% of CMC (flotation separation of iron oxide fine particles in solution) [21] and 10–100 mg per use (lavatory basin cleaning) [7].

7.4 Conclusions

MNBs represent a novel technology for new prospective applications in food industries. However, the safety of MNB should be clarified. For example, ozone is allowed as an existing food additive in Japan because it typically degrades rapidly. If ozone gas remains in MNB without degradation for long periods of time, revising the safety standards of ozone as a food additive may be necessary. In Japan, on March 31, 2009, the Labour Standard Bureau, Ministry of Health, Labour and Welfare released the "Notification on Precautionary Measures for Prevention of Exposure etc. to Nanomaterials (Notification No.0331013)". However, this notification does not address the issues of MNB. Accumulating scientific knowledge of

MNB, including safety issues, is critical for encouraging the utilisation of MNB for the food industry.

References

1. Bando Y, Hoshimatsu T, Wang Y, Yasuda K, Sugie T, Asai T (2008) Influence of micro-bubble on ozone-decomposition of excess sludge, *Progress in Multiphase Flow Research*, **3**, 51–57. (in Japanese)
2. Gong X, Takagi S, Huang H, Matsumoto Y (2007) A numerical study of mass transfer of ozone dissolution in bubble plumes with an Euler-Legrange method, *Chem Eng Sci*, **62**, 1081–1093.
3. Goto Y, Serizawa A, Eguchi T, Tanaka H, Izumi M (2006) Oil separation from oil polluted soil by micro bubble injection and separation mechanisms, *Japanese J Multiphase Flow*, **20**(1), 39–49. (in Japanese)
4. Himuro S (2007) A new washing method using microbubbles, *Progress in Multiphase Flow Research*, **3**, 39–45. (in Japanese)
5. Kageyama Y (1981) Studies on two trusses tomato culture in hydroponics: IV. Effect of oxygen concentration in nutrient solution on the yield of fruit and uptake of nutrient and water, *Sci Rep Fac Agr Okayama Univ*, **58**(1), 23–29. (in Japanese)
6. Kukizaki M (2009) Micro-nano bubble generation from porous glass membrane in *Food Nano Technology* (Nakajima M, Sugiyama S, eds) pp 158–166. (in Japanese)
7. Nakamura T, Matsunaga H, Maeda Y, Kawamoto S (2007) Bubble generation mechanism for automatic toilet bowl cleaning, *National Technical Report*, **55**(4), 99–103. (in Japanese)
8. Nakamura Y (2008) Development of technology for removing pesticide residue of greengrocery using ozone micro- and nano-bubble, *Urakami Foundation Memoirs*, **16**, 160–164. (in Japanese)
9. Namiki T, Hukushima M, Nishi S, Takashima S (1976) Studies on production of vegetable crops in water culture XIV: Effects of different levels of dissolved oxygen supply at different stages of growth on growth and yield of the tomato, *Sci Rep Kyoto Pref Univ Agr*, **28**, 31–39. (in Japanese)
10. Sakai Y, Nakama H, Shiinoki M, Kashiwagi T, Ohnari H (2003) Development of sake storage technique using micro bubble generation technology, *Regional Science Promotion Program 2003, JST*, 131–135. (in Japanese)

11. Sharma PK, Cibcus MJ, van der Mei HC, Busscher HJ (2005a) Influence of fluid shear and microbubbles on bacterial detachment from a surface, *Appl Environ Microb*, **71**(7), 3668–3673.
12. Sharma PK, Gibcus MJ, van der Mei HC, Busshcer HC (2005b) Microbubble-induced detachment of coadhering oral bacteria from salivary pellicles, *Eur J Oral Sci*, **113**(4), 326–332.
13. Shimoda M, Osazima Y (2001) Inactivation of Microorganisms with micro-bubbles of supercritical CO2, *Journal of Antibacterial and Antifungal Agents*, **29**(2), 99–105. (in Japanese)
14. Shirai J, Matsumura E, Hagiwara N (2008) Experimental study of disinfection for animal disease viruses using nano pico ozone water, *J Jpn Vet Med Assoc*, **61**, 233–239. (in Japanese)
15. Soli KW, Yoshizumi A, Motomatsu A, et al. (2010) Decontamination of fresh produce by the use of slightly acidic hypochlorous water following pretreatment with sucrose fatty acid ester under microbubble generation, *Food Control*, **21**(9), 1240–1244.
16. Takagi M, Iwata T, Akiba M, Matsumura E, Hagiwara N, Kamio T (2009) Bactericidal effect of superfine and high density ozone water on Salmonella enteritidis, *J Jpn Soc Poult Dis*, **44**, 150–157. (in Japanese)
17. Takagi S (2007) Development of energy-saving system for wastewater treatment using microbubbles, Technical Report of Industrial Technology Research Grant Program in FY2006. (in Japanese)
18. Takahashi M (2004) Potential of micro-nano bubbles in food industry, *Shokuhin Kogyo*, **47**(16), 43–49. (in Japanese)
19. Tamaki M (2007) The Trends and the potential of microbubble application of sterilization and purification of hydroponic solutions, in *Latest Technology on Microbubbles and Nanobubbles* (Tsuge H, ed), CMC Publishing, Tokyo, pp 204–214. (in Japanese)
20. Terasaka K, Shinpo Y (2007) Separation of fine particles suspended in water using microbubble flotation, *Japanese Journal of Multiphase Flow*, **21**(1), 77–83. (in Japanese)
21. Terasaka K, Aoki S, Kobayashi D (2008) Removal of iron oxide fine particles from waste water using microbubble flotation, *Progress in Multiphase Flow Research*, **3**, 43–50. (in Japanese)
22. Weber J, Agblebor FA (2005) Microbubble fermentation of Trichodermareesei for cellulase production, *Process Biochem*, **40**, 669–676.
23. Xu QY, Nakajima M (2007) Production of shelled microbubble and its apparatus. Japanese Patent, Code P09A015190.

24. Xu QY, Nakajima M, Ichikawa S, et al. (2009) Effects of surfactant and electrolyte concentrations on bubble formation and stabilization, *J Colloid Interf Sci*, **332**, 208–214.

25. Xu QY, Liu ZS, Nakajima M, et al. (2010) Characterization of a soybean oil-based biosurfactant and evaluation of its ability to form microbubbles, *Bioresource Technol*, **101**, 3711–3717.

26. Xu QY, Nakajima M, Ichikawa S, Nakamura N, Shiina T (2008) A comparative study of microbubble generation by mechanical agitation and sonication, *Innov Food Sci Emerg*, **9**, 489–494.

Chapter 8

Agricultural, Marine and Stock-Raising Fields

8.1 Sterilisation of Hydroponics Solution and Removal of Pesticides in Vegetables

Masahiko Tamaki
School of Agriculture, Meiji University, Kanagawa 215-0035, Japan
mtamaki@meiji.ac.jp

8.1.1 Introduction

In present agriculture, hydroponic culture increases year by year because it offers the prospect of high productivity in limited cultivation space. Nutrient solutions used in hydroponic culture are commonly disposed after use. However, from the point of view of environmental protection, reusing nutrient solutions may be needed in the near future. However, because cyclic use of nutrient solutions causes rapid spread of disease by infection via roots throughout an entire cultivation facility, sterilisation is indispensable.

Micro- and Nanobubbles: Fundamentals and Applications
Edited by Hideki Tsuge

ISBN 978-981-4463-10-2 (Hardcover), 978-981-4463-11-9 (eBook)
www.panstanford.com

Up to now, Kobayashi et al. [1] reported that ozone microbubbles (O_3MB) have a much greater disinfectant effect on *Fusarium oxysporum* f. sp. *melonis* and *Pectobacterium carotovorum* subsp. *carotovorum* than O_3 millibubbles (O_3MM).

In addition, present agriculture has enabled mass and stable production by using agricultural pesticides. However, agricultural pesticides can have an adverse effect on the environment in addition to being harmful to humans, animals and marine life. Health hazards to farmers as well as the residue in crops is also a global problem. Recently, the safety of crops including contamination with agricultural pesticides is a major concern to both the producer and consumer, and the development of a method to remove pesticides before marketing has been eagerly awaited.

Moreover, although there have been several studies on the use of O_3MM for removing residual pesticides from vegetables and fruits, few studies have reported on the use of O_3MB to remove them. Therefore, because microbubbled gas is highly soluble in water and ozone (O_3) is a powerful decomposer of organic matter, O_3MB were expected to remove residual pesticides efficiently from vegetables and fruits.

O_3 is generated by the passage of air or oxygen gas through a high voltage electrical discharge or by ultraviolet light irradiation [2]; it then has a strong oxidative power and is used for sterilisation, virus inactivation, deodorisation, bleaching (decolouration), decomposition of organic matter, mycotoxin degradation and so on [3–6].

Microbubbles (MB) have special properties such as generation of free radicals, self-pressurisation and negative charge, rising slowly in water and the interior gas is completely dissolved in water [6], and their use in the field of food science and agriculture is attracting attention [7,8].

In this study, the disinfectant effect of O_3MB on *Fusarium oxysporum* f. sp. *melonis* and *Pectobacterium carotovorum* subsp. *carotovorum* as well as the fertiliser composition of nutrient solution treated with O_3MB and its effect on the growth of cultivated plants and the effects of O_3MB on the removal of pesticide (fenitrothion, FT) infiltrated into vegetables have been mentioned earlier.

8.1.2 Experiment 1: Ozone Microbubbles as a Disinfection in Nutrient Solution and Their Effects on the Composition of Fertiliser and the Growth of Cultivated Plants

8.1.2.1 Materials and methods

8.1.2.1.1 Deep-flow technique hydroponics systems

Figure 8.1 shows a diagram of the equipment used in the experiment. The system consists of a cultivation bed (width 120 cm × length 565 cm × water depth 7.5 cm), a plastic cylindrical container (diameter 28 cm × height 48 cm), an MB generator, an O_3 generator and two pumps. Another system consisted of a cultivation container (width 40 cm × length 60 cm × water depth 32 cm), a plastic cylindrical container (diameter 28 cm × height 48 cm), an MB generator, an O_3 generator and two pumps. Nutrient solution in these systems is circulated only when O_3MB is generated. Tap water (20°C) was placed into a cultivation bed, and after letting it naturally de-chlorinate for 24 h, the removal of the chlorine in the tap water was confirmed by a chlorine comparator. Then, hydroponics fertilisers (Farm Ace, Nos. 1 and 2 at 1.5 and 1 g L^{-1}, respectively) were added.

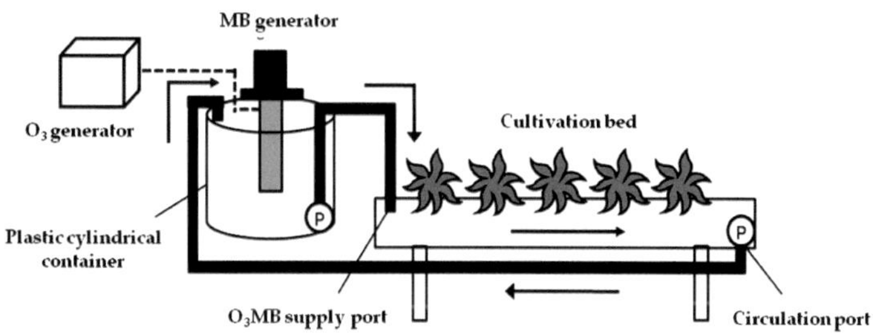

Figure 8.1 Diagram of the cultivation equipment.

8.1.2.1.2 Disinfectant effect of O_3MB against F. oxysporum *f. sp. melonis and* P. carotovorum *subsp. carotovorum*

The cultivated *F. oxysporum* f. sp. *melonis* (5.0 × 10^3 spores mL^{-1}) and *P. carotovorum* subsp. *carotovorum* (5.0 × 10^3 cells mL^{-1}) were individually placed into the hydroponics system, and the nutrient solution was circulated for 60 min so that they reached all parts of the system. Then, O_3MB treatment was performed for 120 min. Sequential samples of nutrient solution in which a pathogenic fungus and bacterium were suspended were taken and the number of surviving fungus and bacterium was measured. Samples of nutrient solution were acquired from two locations: the region where the nutrient solution containing O_3MB entered the cultivation bed (O_3MB supply port) and the lower part of the cultivation bed (circulation port), where the submerged pump circulating the nutrient solution was located.

8.1.2.1.3 *Preparation of lettuce plants*

Lettuce plants (*Lactuca sativa* L., cv: Melbourne MT) that had grown for 5 days after seeding were used in the experiments.

8.1.2.1.4 *Relationship between the number of O_3MB treatments and the fungicidal effect against* F. oxysporum *f. sp. melonis*

Hydroponics fertiliser was put into 60 L of dechlorinated water (20°C) held in cultivation containers. Into this nutrient solution, *F. oxysporum* f. sp. *melonis* was added at a concentration of $1.0 \times 10^3 - 1.0 \times 10^4$ spores mL^{-1}. Each container was covered with styrofoam boards. Eight lettuces were planted in each container, and O_3MB treatment was performed for 5 min according to the above generating conditions. The dO_3 within the containers at this time was approximately 0.2 ppm. Four treatments were conducted as follows: 1. No O_3MB treatment, 2. O_3MB treatment once per week (Monday), 3. O_3MB treatment twice per week (Monday and Wednesday) and 4. O_3MB treatment three times per week (Monday, Wednesday and Friday). The lettuces in each container were harvested on the 28th day after planting and presence of disease was examined visually.

8.1.2.1.5 *Influence of O_3MB treatment on the fungicidal effect against* F. oxysporum *f. sp.* melonis *and the growth of lettuce*

The cultivated *F. oxysporum* f. sp. *melonis* (5.0×10^3 spores mL^{-1}) was placed into the hydroponics system, and the nutrient solution was circulated for 60 min. Four treatments were conducted as follows:

1. No O_3MB generated and no *F. oxysporum* f. sp. *melonis* (control), 2. No O_3MB generated, but *F. oxysporum* f. sp. *melonis* added (*F. oxysporum* f. sp. *melonis*), 3. O_3MB generated, but no *F. oxysporum* f. sp. *melonis* added (O_3MB) and 4. *F. oxysporum* f. sp. *melonis* added and O_3MB generated (*F. oxysporum* f. sp. *melonis* + O_3MB). O_3MB treatment was performed by generating O_3 for 30 min twice a week according to the above generating conditions. A total of 64 lettuces were planted in all treatments. They were harvested 28 days after planting and were visually examined for the presence of disease. The fresh and dry weights of the above-ground parts of the lettuce plants were measured as well as the length of the longest roots. Dry weight was measured after 2 days of open-air drying at 80°C.

8.1.2.2 Results and discussion

Twenty minutes after O_3MB treatment, numbers of *F. oxysporum* f. sp. *melonis* and *P. carotovorum* subsp. *carotovorum* at the O_3MB supply port were reduced to levels below the detection limit. At the circulation port as well, numbers of both fungus and bacterium were below the detection limit 60 min after O_3MB treatment (Fig. 8.2). The dO_3 was 0.15 ppm at the O_3MB supply port after 20 min of O_3MB treatment and was 0.2 ppm at the circulation port after 60 min of O_3MB treatment. Therefore, at extremely low dO_3 concentrations of approximately 0.15–0.2 ppm, *F. oxysporum* f. sp. *melonis* and *P. carotovorum* subsp. *carotovorum* were both inactivated. O_3MB treatment requires substantial time to kill fungus and bacterium in locations far away from where the treatment to sterilise nutrient solutions is performed. When O_3MB in concentrations of 0.15 ppm or more circulates throughout the hydroponics system, *F. oxysporum* f. sp. *melonis* and *P. carotovorum* subsp. *carotovorum* throughout the entire system can be killed.

The relationship between the number of times O_3MB is generated and its fungicidal effect on *F. oxysporum* f. sp. *melonis* is shown in Table 8.1. Because it is more difficult for O_3MB to kill *F. oxysporum* f. sp. *melonis* than *P. carotovorum* subsp. *carotovorum* (Fig. 8.2), the following experiments were performed using only *F. oxysporum* f. sp. *melonis*. While the rate of Fusarium disease was very high at 62.5% in the control, the rate was only 8.3% with O_3MB treatment once per week. With O_3MB treatment twice and three times per week, no disease caused by *F. oxysporum* f. sp. *melonis* was observed. Because dO_3 within the containers used in this experiment was approximately

0.2 ppm, it is possible to kill *F. oxysporum f. sp. melonis* in the nutrient solution at 20°C using O_3MB once or twice per week at a low-level concentration of 0.2 ppm.

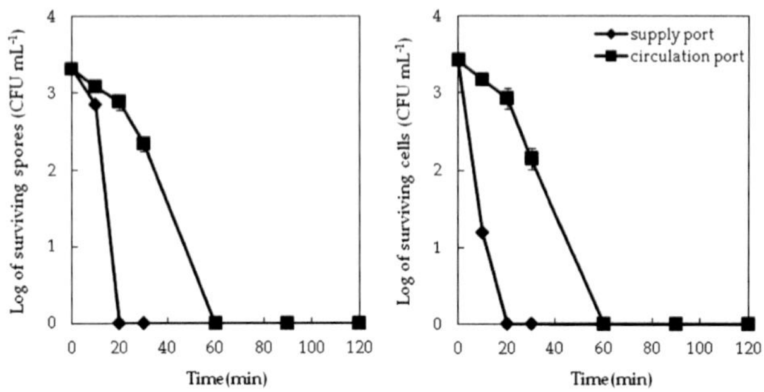

Figure 8.2 Sequential changes in disinfectant effect of O_3MB against *F. oxysporum* f. sp. *melonis* (left) and *P. carotovorum* subsp. *carotovorum* (right).

Table 8.1 The relationship between the number of times O_3MB treatment and the rate of Fusarium disease outbreak

Number of times O_3MB (number of times/week)	Rate of fusarium disease (%)
0	62.5
1	8.3
2	0
3	0

Next, in the *F. oxysporum* f. sp. *melonis* treatment, the rate of disease outbreak was 48.4%, and approximately half the plants suffered from disease outbreak and poor growth. But in the *F. oxysporum* f. sp. *melonis* + O_3MB treatment, the rate of disease outbreak was extremely low at 4.7%. Also, compared to the control, the plant growth with O_3MB treatment was unaffected (Table 8.2). With *F. oxysporum* f. sp. *melonis* + O_3MB treatment, plants that suffered disease outbreak were near the circulation port of the cultivation bed. After 30 min of O_3MB treatment, dO_3 at the circulation port was 0.05 ppm; so, it is conceivable that *F. oxysporum* f. sp. *melonis* was not completely killed because of the low dO_3 level.

Table 8.2 Fungicidal effect of O_3MB treatment against *F. oxysporum* f. sp. *melonis* and its influence on the growth of lettuce

Treatments	Infection rate (%)	Fresh weight (g)	Dry weight (g)	Longest root (cm)
Control	0	59.6 ± 9.8a*	2.5 ± 0.7a	46.4 ± 6.3a
F. oxysporum f. sp. *melonis*	48.4	30.6 ± 17.9b	1.4 ± 0.7b	30.2 ± 11.3b
O_3MB	0	56.4 ± 7.9a	2.3 ± 0.4a	43.7 ± 6.6a
F. oxysporum f. sp. *melonis* + O_3MB	4.7	55.9 ± 9.7a	2.3 ± 0.6a	45.6 ± 10.3a

* Different letters indicate a difference significant at the 5% level by Turkey–Kramer test between treatments ($n = 3$).

These results indicate that O_3MB has an extremely strong disinfectant effect against plant pathogenic fungus and bacterium in the nutrient solution.

8.1.3 Experiment 2: Removal of Residual Pesticides in Vegetables Using Ozone Microbubbles

8.1.3.1 Materials and methods

8.1.3.1.1 Treatment with agricultural pesticides

FT (Sumithion emulsion, 50% of methylnitrophos) was 1000 fold diluted in tap water, and three drops of a spreading agent (Haiten power, Hokko Sangyo, Co., Ltd. Tokyo, Japan) were added. The concentration of FT solution was 500 ppm. Lettuce leaves and fruits of cherry tomatoes and strawberries were immersed in this solution (60 L) for 1 min and left in a cool dark room for 24 h to infiltrate FT into vegetables. Thereafter, they were washed in tap water for 1 min and treated with O_3.

8.1.3.1.2 O_3 treatment

Forty litres of tap water was pooled in a cylindrical vessel (55 cm × 32 cm i.d.) and kept at 20°C in a room to remove chlorine in tap water for 24 h.

O_3MM were generated in the dechlorinated water with an O_3 generator at a flow rate of 2.5 L/min. The maximum amount of O_3 that could be dissolved under these conditions was 0.2 ppm. Then the vegetables were immersed in this solution for 0, 5 or 10 min. O_3MB were generated in the dechlorinated water using a gas-water circulation type MB generator (FS101-L1, Fuki Co. Ltd., Saitama, Japan) combined with an O_3 generator at a flow rate of 2.5 L/min. Ozonated water solutions were produced containing 0.5, 1.0, or 2.0 ppm dissolved O_3. Then, the vegetables were immersed in these solutions for 0, 5, or 10 min. A further treatment was set up where MBs were continuously generated during vegetable immersion in the ozonated solutions (bubbling O_3MB). In these treatments, O_3 microbubbling was continued to maintain the concentration of dissolved O_3 at 2.0 ppm for 0, 5 or 10 min vegetable immersions. A control treatment was also conducted where the vegetables were immersed in dechlorinated water. The solution temperature was maintained at 20°C during all treatments and the concentrations of dissolved O_3 were measured using an OZ-21P O_3 analyzer with a DO_3 electrode. All analyses were performed in triplicate.

8.1.3.1.3 Residual pesticide analysis

The effects of FT on vegetables were analyzed using analytical methods of pesticide residues as described by Ueji et al. [9]. Analyses were run in triplicate.

8.1.3.2 Results and discussion

8.1.3.2.1 Removal of residual pesticides in vegetables using O_3MB dissolved by different concentrations

Figure 8.3a shows the reduction in residual FT in lettuce treated with the O_3MB and O_3MM solutions. Before treatment with O_3MM or O_3MB solutions, but after washing with tap water, the concentration of residual FT in lettuce was 212.2 ppm (data not shown). The residual FT in lettuce was reduced to 67%, 55% and 45% after 5 min of treatment with the 1.0 ppm O_3MB, 2.0 ppm O_3MB and 2.0 ppm bubbling O_3MB solutions, respectively. After 10 min of treatment, the respective amounts of residual FT had been further reduced to 49%, 45% and 42%. The similarly high reductions in residual FT achieved with the 1.0 ppm O_3MB, 2.0 ppm O_3MB and 2.0 ppm bubbling O_3MB, indicates that immersion of lettuce in an O_3MB solution containing 1.0 ppm or more dissolved ozone may be sufficient to effectively remove residual FT from the lettuce, possibly because lettuce has thin leaves. In contrast, after 10 min of treatment with the O_3MM and 0.5 ppm O_3MB solutions, the residual FT had only been reduced to 87% and 78%, respectively.

The dissolved ozone in the O_3MB treatment solutions generates hydroxyl radicals that are highly effective at decomposing organic molecules like the residual FT [7,8]. Hydroxyl radicals are generated by the collapse of O_3MB in solution, and so, the 1.0 ppm O_3MB, 2.0 ppm O_3MB and bubbling O_3MB solutions would have had a high enough concentration of dissolved ozone to produce a large amount of hydroxyl radicals. However, the O_3MB treatment was not nearly so effective because the concentration of dissolved ozone was much lower and so far fewer hydroxyl radicals would have been generated.

Figure 8.3b shows the reduction of residual FT in cherry tomatoes for each treatment. The starting concentration of residual FT in the cherry tomatoes was 3.0 ppm (data not shown) prior to ozone solution treatments. Removal of residual FT by the various treatment solutions was much less in the cherry tomatoes than in the lettuce. After 10 min of treatment, residual FT had been reduced to 65% in 2.0 ppm bubbling O_3MB solution, but remained at >90% for all other treatments. The most likely explanation for the lower reduction of residual FT in the cherry tomatoes is that the dissolved ozone and hydroxyl radicals could not penetrate the thick pericarp of the tomatoes and to reach the sarcocarp and were inactivated by contact with the pericarp. The greater effectiveness of the bubbling O_3MB solution was probably because the concentration of dissolved ozone remained high, and so, hydroxyl radicals continued to be generated throughout the treatment.

Figure 8.3c shows the reductions in residual FT in strawberries for each treatment. The starting concentration of residual FT in strawberries was 37.8 ppm (data not shown). After 10 min of treatment, the greatest reduction in residual FT was in the 2.0 ppm bubbling O_3MB treatment where 75% residual FT remained. The other treatments ranged from 85% residual FT remaining with the 2.0 O_3MB treatment to 91% with the O_3MM treatment. The amount of FT that was removed with the 2.0 ppm bubbling O_3MB solution was lower than that in the cherry tomatoes. Strawberries have a rougher surface and larger surface area than cherry tomatoes, and this may cause ozone and hydroxyl radicals to lose their specific activity upon contact with the surface of strawberries, preventing them from removing FT in the sarcocarp.

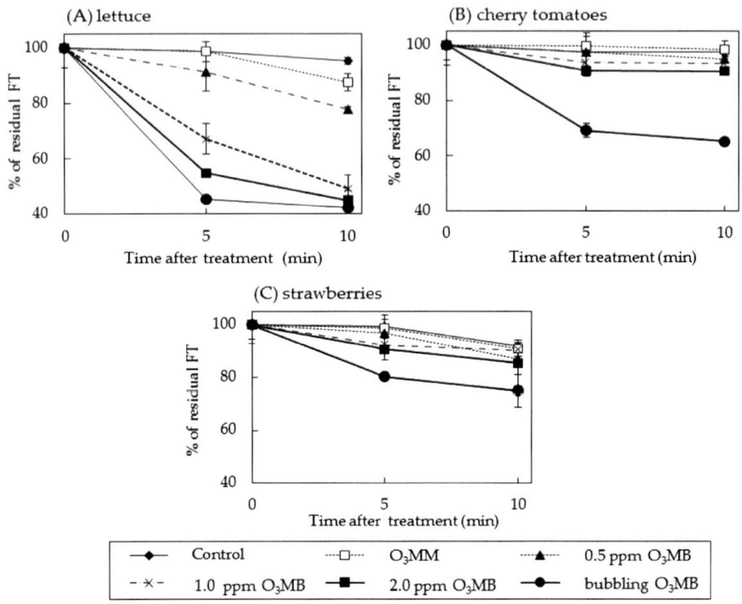

Figure 8.3 Change in the percentage of residual FT in lettuce (a), cherry tomatoes (b), strawberries (c) treated with each O_3MB and O_3MM.
Vertical bars show standard errors.

Clearly, residual pesticides in leafy vegetables can be removed by immersion in ozonated solutions, but the concentrations of residual pesticide in the earlier studies were low. In the present study, high concentrations of residual FT found in lettuce (>200 ppm) could be reduced to less than 100 ppm in 5–10 min by treatment with 1.0–2.0 ppm O_3MB solution. Such a large reduction may be possible because the chemical structure of FT is similar to diazinon, which can be easily decomposed by hydroxyl radicals [10], and so, the oxidative powers of ozone and hydroxyl radicals may act in concert to effectively degrade FT. This effective joint action was only possible in the MB-generated solutions because the millibubble generated solutions could not achieve high enough dissolved ozone concentrations and not generate hydroxyl radicals.

These results showed that O_3MB can remove high concentrations of residual pesticides within a short time from not only leafy vegetables but also fruity vegetables. Thus, O_3MB could be useful for removing residual pesticides from a wide range of vegetables.

References

1. Kobayashi F, Ikeura H, Ohsato S, Goto T, Tamaki M (2011) Disinfection using ozone microbubbles to inactivate *Fusarium oxysporum* f. sp. *melonis* and *Pectobacterium carotovorum* subsp. *carotovorum*, *Crop Prot*, **30**, 1514–1518.
2. Mahapatra AK, Muthukumarappan K, Julson JL (2005) Applications of ozone, bacteriocins, and irradiation in food processing: a review. *Critical Rev. Food Sci. Nut.,* **45**, 447–461.
3. Cataldo F (2008) Ozone decomposition of patulin—A micotoxin and food contaminant, *Ozone Sci Eng*, **30**, 197–201.
4. Karaca H, Velioglu YS (2009) Effects of some metals and chelating agents on patulin degradation by ozone, *Ozone Sci. Eng.*, **31**, 224–231.
5. Karaca H, Velioglu YS, Nas S (2010) Mycotoxins: contamination of dried fruits and degradation by ozone, *Toxin Rev*, **29**, 51–59.
6. Takahashi M, Chiba K, Li P (2007a) Free-radical generation from collapsing microbubbles in the absence of a dynamic stimulus, *J Phys Chem B*, **111**, 1343–1347.
7. Sumikura M, Hidaka M, Murakami H, Nobutomo Y, Murakami T (2007) Ozone micro-bubble disinfection method for wastewater reuse system, *Water Sci Technol*, **56**, 53–61.
8. Takahashi M, Chiba K, Li P (2007b) Formation of hydroxyl radicals by collapsing ozone microbubbles under strongly acidic conditions. *J Phys Chem B*, **111**, 11443–11446.
9. Ueji M, Kobayashi Y, Nakamura K (2002) *Analytical Methods of Pesticide Residues*, Softscience Co., Japan, 238–239.
10. Kouloumbos VN, Tsipi DF, Hiskia AE, Nicolic D, Van Breeman RB (2003) Identification of photocatalytic degradation products of diazinon in TiO2 aqueous suspensions using GC/MS/MS and LC/MS with quadrupole time-of-flight mass spectrometry, *J Am Soc Mass Spectrom*, **14**, 803–817.

8.2 Application of a Microbubble Generator to Aquaculture

Hiroaki Tsutsumi,[a] Sarawood Srithongouthai,[b] Daigo Hama,[b] Ichiro Takase,[c] and Tetsuo Nishi[c]

[a]*Faculty of Environmental and Symbiotic Sciences, Prefectural University of Kumamoto, Kumamoto 862-8502, Japan*
[b]*Keiten Co., Ltd., Amakusa, Kumamoto 863-0044, Japan*
[c]*Taikohgiken, Ltd., Kumamoto 860-0047, Japan*
hiro@pu-kumamoto.ac.jp

Application of microbubble generators to aquaculture was studied in the 1980s because various farms including shellfish culture and fish culture suffered from shortage of oxygen supply due to high-density culturing. Prof Hirofumi Ohnari, Tokuyama College of Technology in Japan, invented a microbubble generator and used it to control DO levels of the water in an oyster farm in Hiroshima Bay in Seto Inland Sea [1,2]. His original microbubble generator adopts a nozzle that creates a "Rotational Flow of the water" in the nozzle, using a water pump and produces microbubbles, breaking the air leading to the centre of the rotational flow of the water. He demonstrated that the addition of microbubbles is very effective as a way to add oxygen to the water of the oyster farm, and this technology drew the attention of many researchers, and various different types of microbubble generators have been invented in the last decade in Japan [3–8].

We have studied ways of controlling DO levels at a fish farm of red seabream, *Pagrus major*, farms in Amakusa, Kyushu, western Japan, using a microbubble generator that created the "Rotational Flow" of the water in the nozzle [9–11]. In this study, we used a compressor to increase the air flow rate to the nozzle to increase the production of microbubbles. This system had a great advantage to treat a large amount of the water in the fish farms and had a big effect on controlling DO levels in water and production of fish, although the whole microcrobubble generating system was compact, and consumed much less energy compared to the conventional air bubbling devices.

In this chapter, we would like to emphasise two key points on the use of a microbubble generator in aquaculture, show the results of DO control experiments in net pens with a microbubble generating system at a fish farm to demonstrate its great potential for controlling the DO levels of the water and introduce the latest developments in the evolution of the microbubble generator that created the "Rotational Flow" of the water in the nozzle.

8.2.1 Key Points on the Use of a Microbubble Generator in Aquaculture

8.2.1.1 When should the microbubble generator be used?

In demonstration and preliminary studies of the microbubble generator, bubbling with this device was often performed in the field of aquaculture during daytime. However, additional oxygen supply is needed at aquaculture farms at night because the photosynthetic activities by phytoplankton have a potential for keeping DO around saturated levels in the water of the farms during the daytime, but it consumes oxygen for respiration at night. The rearing aquatic organisms in the farms consume oxygen throughout the day at a rate that mainly depends on the water temperature. Therefore, the DO of the water in the farms tends to decrease to a critical level for reared aquatic organisms during the night in warm seasons [9,12,13].

We have conducted experiments to prevent a decrease in the DO levels of the water of the fish farm at nights by using the microbubble generator [9]. Figure 8.4 compares the fluctuation of DO of the water in a net pen of the red seabream farm over the course of a day with and without the addition of microbubbles. Without the addition of microbubbles, the DO of the water decreased at night and fell to its lowest levels early in the morning (6.06 mg/L at 8:00 on July 27 and 6.13 mg/L on July 28, 2005). With microbubbles from a microbubble generator, which was set in the net pen and worked for 14 h a day between 18:00 and 8:00, the DO of the water in the net pen was maintained to around saturated levels in the early morning (6.6 and 7.1 mg/L in the early morning of July 24 and 25, 2005).

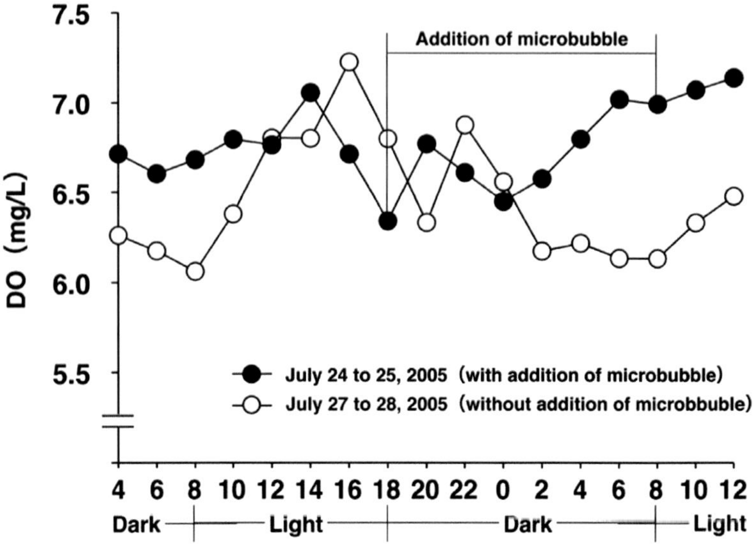

Figure 8.4 Fluctuation of DO of the water in a net pen of the red seabream farm in a day in Kusuura Bay, Amakusa, Kyushu, Japan, between the conditions with and without addition of microbubble (unpublished).

8.2.1.2 How is the power supply prepared?

If aquaculture farms are located far from the shore, a power supply must be prepared independently. One of the easiest ways is to get power from the engine of a fishing boat. However, it is difficult to use it for a long period. A solar battery is not useful for the operation of the microbubble generator at nights, and the cost is high. We adopted a microbubble generating system manufactured by Tashizen Technoworks Co. Ltd. for the experiments. It was made of a small diesel engine generator, a compressor, a switchboard and air control board and maximum six sets of the microbubble generator. We set it on a small raft and moored it to the net pens in the fish farm (Fig. 8.5). If the aquaculture farm is established in an enclosed bay to protect from strong wind and wave, this system will work safely.

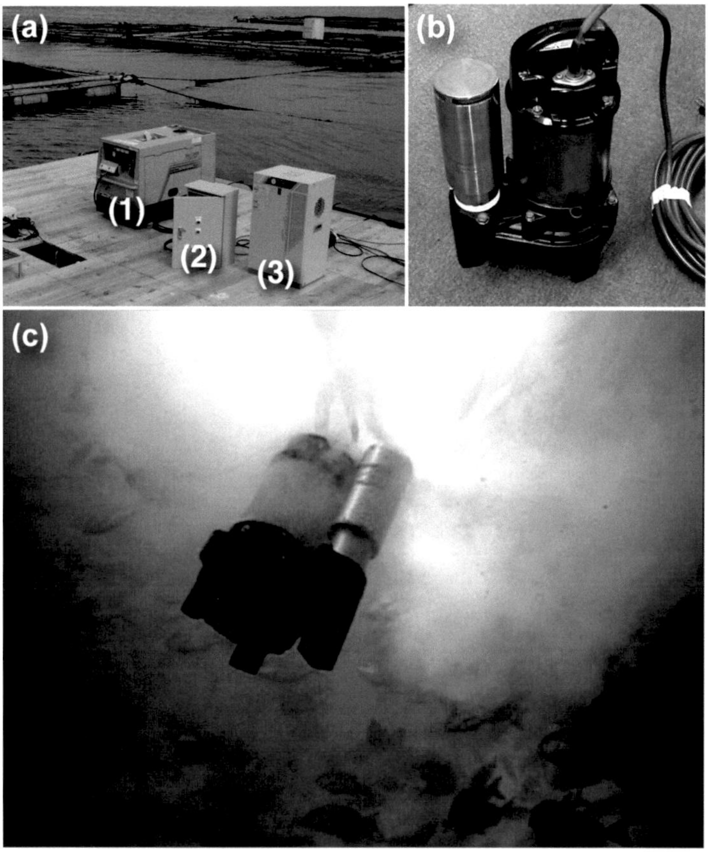

Figure 8.5 Microbubble generating system and microbubble generator, which were manufactured by Tashizen Technoworks Co. Ltd. (a) This system includes (1) a small-engine generator (output 7.5 kVA, 220 V), (2) a switch board, (3) a compressor (0.75 kVA, capacity: 78 L/min) (from the left side) and maximum of six sets of (b) microbubble generators (400 VA, 22 cm in length, 36 cm in height). When the generator and compressor are set on the raft beside the fish farm, they are covered by stainless housings. (c) The microbubble generator is releasing microbubbles in the net pen rearing red seabream.

8.2.2 The Results of DO Control Experiments in Fish Farm Net Pens with a Microbubble Generating System

We conducted DO control experiments in the fish farm of red seabream with the microbubble generating system in Kusuura Bay, Amakusa, Kumamoto Prefecture, Kyushu, western Japan, between 2003 and 2007 (Fig. 8.6). The water depth of the bay is between 16 and 20 m at the centre of the bay. The annual temperature of the surface water ranged from 19.7°C to 29.3°C between May and October 2007 (unpublished data). This study is financially supported by a Research and Development Program for New Bio-industry Initiatives of the Bio-oriented Technology Research Advancement Institution (BRAIN) of Japan.

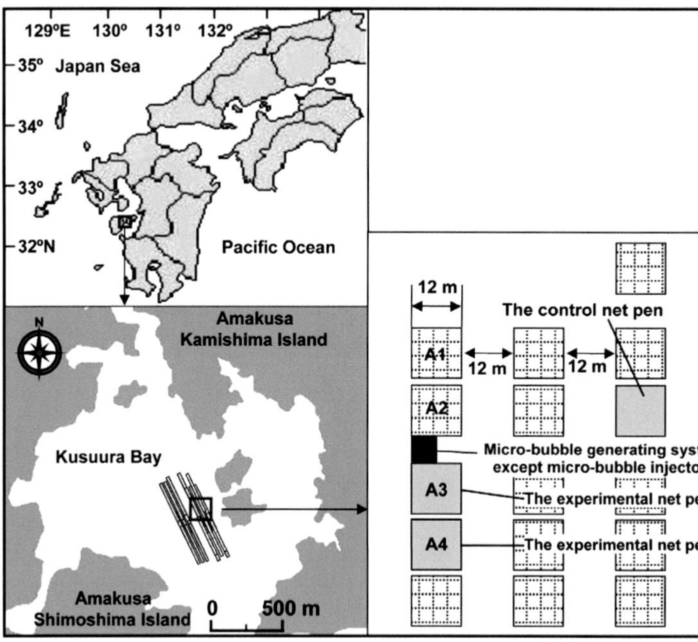

Figure 8.6 Study area, Kusuura Bay, Amakusa, Kumamoto Prefecture, Kyushu, Japan.

8.2.2.1 Experimental setting

We set the microbubble generating system (except microbubble generators) on a small raft (5 m × 5 m), which was fixed at the middle of the four net pens (A1–A4) arranged in a row along the tidal current of the bay, at the centre of the fish farm (Fig. 8.6). The DO of the net pens tends to decrease most seriously at a depth of 5–6 m, due to respiration of reared fish [9]. The microbubbles released from the generator can increase DO not only in the upper layers from the generator but also in the lower layers due to the extremely small buoyancy of microbubbles and the downward water flow caused by the difference in current velocity between the inside of the net pen and the intermediate layer below the net pen [10]. The two net pens (A3 and A4) are referred to as "the experimental net pen E1" and "the experimental net pen E2", respectively. Because the release of microbubbles from a single microbubble generator set in a net pen could contribute to increase the DO of the water within a radius of approximately 10 m [10], we set "the control net pen" approximately 50 m apart from the experimental net pens, which are referred to as "the control net pen".

In the experimental net pens, microbubble generators released microbubbles (Fig. 8.5c) for 14 h per day between the evening and the next morning (17:00–7:00) between June and October 2007. We set up a multi-parameter probe (6600 EDS, YSI/Nanotech) to monitor the water conditions at a depth of 5 m in the experimental net pen E1, which was located approximately 4 m away from the microbubble generator, and in the control net pen. The measurement data of the water conditions were stored in the memory of the probes. We downloaded the data to a personal computer and cleaned the probes bi-weekly.

We reared approximately 10,000 individuals of 3 year old red seabream in the same feeding conditions in all three net pens (experimental net pens E1 and E2 and the control net pen) for 50 days between June 18 and August 6, 2007. In the experimental net pen E1 and the control net pen, the rearing experiments were continued until October 3, 2007 (for 108 days in total). Thirty individual varieties of fish were captured from each of these three net pens with a handy net; their size and weight were measured to monitor the growth of the fish, and they were released to the original net pens at the beginning (June 18, 2007), the middle (August 6, 2007) and the end of the experiments (October 3, 2007).

8.2.3 Results of DO Control Experiments

8.2.3.1 Control of DO levels in the net pens at night

Figure 8.7 compares the daily DO fluctuation patterns of the water at a depth of 5 m between the experimental net pen E1 and the control net pen at four different periods in the experiment. From June 21 to 23 2007, the DO of the water in both of the net pens was kept above 5 mg/L. The water temperature ranged between 21.9°C and 23.8°C. Due to the relatively low temperature at the beginning of the warm seasons, the respiration rates of the fish in the net pens and the phytoplankton in the water passing through the net pens were not yet markedly elevated. At nights (18:00–6:00), the mean DO level in the experimental net pen E1 (6.47 mg/L) was 5.1% higher than that in the control net pen (6.16 mg/L).

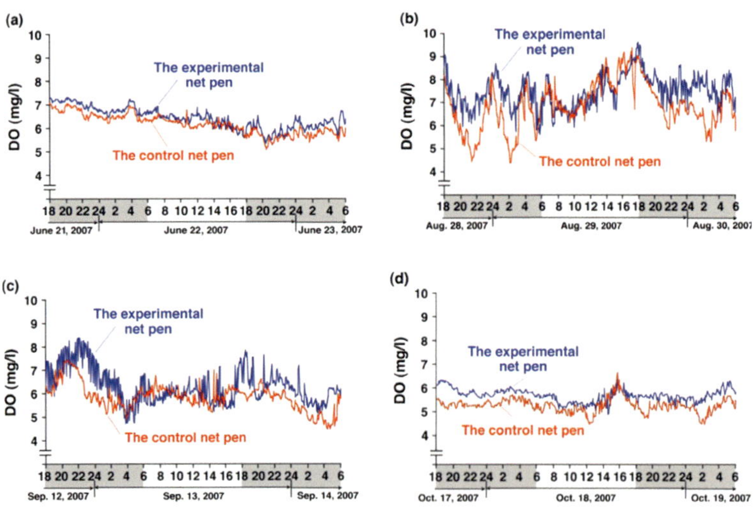

Figure 8.7 Daily DO fluctuation patterns of the water at a depth of 5 m in depth between the experimental and control net pens at four different periods in the experiment between June 18 and October 3, 2007 ((a) June 21–23, 2007, (b) August 28–30, 2007, (c) September 12–14, 2007, (d) October 17–19, 2007).

From August 28 to 30, 2007, when the respiration of the fish and the phytoplankton became most active due to high water temperatures (27.8–30.0°C), DO in the water decreased to levels that imposed physiological stress on the fish (less than 5 mg/L) (the lowest DO was 4.37 mg/L, 67.4% of the saturation level) in the control net pen at nights, even if the DO levels became over-saturated over 9 mg/L (approximately 140% of the saturated levels) during the daytime. In the experimental net pen E1, DO of the water remained around the DO saturated level (approximately 6.5 mg/L) or even in over-saturated conditions during most of the night. The mean DO level in the experimental net pen in dark conditions was 7.52 mg/L, which was 15.7% higher than that in the control net pen (6.50 mg/L).

From September 12 to 14, 2007, the water temperature was relatively high still (between 27.1°C and 28.2°C), and DO of the water in the control net pen decreased markedly during the night, although the DO of the water in both the control and experimental net pens fluctuated over the same range in the day (6:00–18:00) (6.01 mg/L and 5.94 mg/L in mean, respectively). The mean DO level in the experimental net pen E1 in dark conditions (6.47 mg/L) was 11.2% higher than that of the control net pen (5.82 mg/L).

From October 17 to 19, 2007, the water temperature decreased to between 24.6°C and 25.5°C. However, the DO of the water in both the net pens fluctuated in the lowest range (mostly at less than 6 mg/L) in the four periods because vertical mixing of the water occurred in the whole water column including the water in the bottom layer with low DO conditions (at an approximate depth of 16 to 18 m) due to the cooling effect of autumn air on the surface water and/or strong wind and waves. During dark conditions, the DO of the water in the control net pen fluctuated between 4.42 mg/L and 5.70 mg/L, while the DO range rose from 5.22 mg/L to 6.31 mg/L in the experimental net pen E1. The mean DO level of the experimental net pen was 9.7% higher than that of the control net pen.

Thus, we can see that the microbubble generator has ample potential to prevent the DO of the water in the net pen from decreasing to unhealthy low levels of less than 5 mg/L for the reared fish in the warm seasons, even though it is only a small electric device (400 VA) (Fig. 8.5).

8.2.3.2 Enhanced growth of fish by controlling DO levels

The results of the experiments indicate that the control of DO levels of water in the net pen using the microbubble generator improved the growth rate of the fish markedly (Fig. 8.8). In the control net pen, the body weight of the fish was 1.25 ± 0.17 kg in wet weight (mean ± SD, $n = 30$) at the beginning of the experiment on June 18, 2007, and they gained 1.39 ± 0.25 kg and 1.70 ± 0.26 kg by August 8 (50 days later) and October 3, 2007 (108 days later), respectively. In the experimental net pen E1, the body weight of the fish was 1.19 ± 0.24 kg in wet weight (mean ± SD, $n = 30$), slightly lighter than that of the control net pen at the beginning of the experiment. The difference in the mean body weight between the experimental net pen E1 and the control net pen was not statistically significant ($p = 0.15$, Mann–Whitney's U test, $n = 30$). The fish grew faster than that in the control net pen. The body weight reached 1.83 ± 0.21 kg in wet weight at the end of the experiment (108 days later), which was 0.13 kg heavier than that in the control net pen. The difference in body weight between these two net pens was statistically significant ($p = 0.02$, Mann–Whitney's U test, $n = 30$). The increase in the body weight of the fish in the experimental net pen E1 during the period of the experiment for 108 days was 0.64 kg in wet weight, which was 42% more than that in the control net pen (0.45 kg).

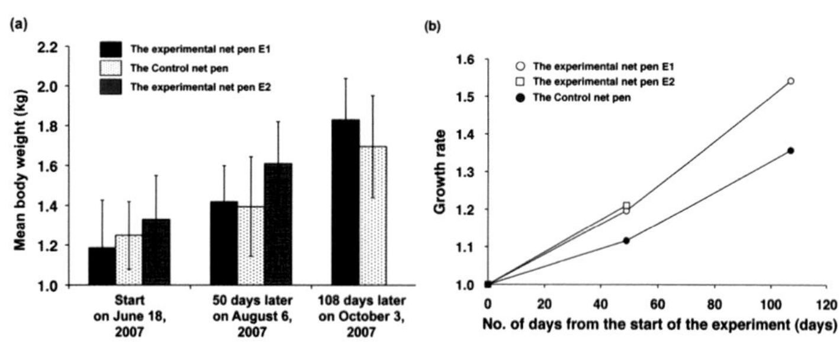

Figure 8.8 Changes of (a) the mean body weight and (b) the growth rate of 3 year old red sea bream in the experimental and the control net pens in the experiment. Error bars indicate the standard deviation ($n = 30$) (unpublished).

In the experimental net pen E2, the body weight of the fish was 1.33 ± 0.22 kg in wet weight (mean ± SD, $n = 30$), slightly heavier than that of the control net pen at the beginning of the experiment. The difference in the mean body weight between the experimental net pen E2 and the control net pen was not statistically significant ($p = 0.15$, Mann–Whitney's U test, $n = 30$). The fish grew faster with almost the same growth rate as the fish reared in the experimental net pen E1 for 50 days (The growth rates of fish in the experiment net pen E1 and E2 were 1.19 and 1.21, respectively, while that in the control net pen was 1.11, Fig. 8.8b). The body weight of the fish reached 1.61 ± 0.21 kg in wet weight 50 days later from the start, which was 0.22 kg heavier than that in the control net pen. The difference of the mean body weight between these two net pens already became statistically significant ($p < 0.01$, Mann–Whitney's U test, $n = 30$).

Thus, DO control of the water in the net pen using the microbubble generator can bring a remarkable acceleration of growth of the fish.

8.2.4 Further Development of the Microbubble Generator to Improve the Potential for Oxygen Supply to the Water

In aquaculture farms, we need to treat a large amount of water to control DO levels suitable for farming in a relatively short time. For example, the popular size of a net pen for fish farming in coastal waters is 10–15 m in diameter or square and 5–20 m in depth, and the water in the net pens is exchanged due to tidal current. In the case of shrimp farms, each pond is more than 10,000 m^2 in size and 1–2 m in depth usually, although the water exchange is much restricted. To control DO levels of the aquaculture farms more effectively, we need to modify the microbubble generator even more to maximise its DO addition potential.

We have modified the inner structure of the nozzle of the microbubble generator to obtain even greater production of microbubbles. The basic idea for the modification is to find a way to obtain faster rotation of the water is created inside the nozzle. To realise this idea, we designed a new nozzle with "Double-chamber/Dual-vortex High-speed Rotation System: DDHRS" (Fig. 8.9) [14,15]. It has a double chamber with a wider entrance of water at the bottom of the outer chamber to draw a larger amount of water into the top of the inner chamber and creates a double vortex of the water in it. The inner vortex of the water can rotate much faster than the outer one, and it enables the production of much larger amounts of microbubbles, reacting with the air introduced to the inner chamber.

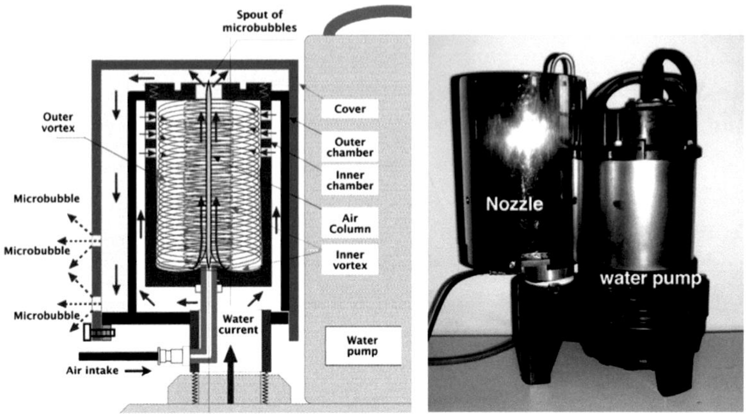

Figure 8.9 The inner structure of a new nozzle for production of microbubble with "Double-chamber/Dual-vortex High-speed Rotation System: DDHRS" [14,15]. This nozzle is fitted to a water pump for releasing microbubble.

Oxygen gas dissolution potential by bubbling the water with air was compared with the potential of a conventional diffuser, a microbubble generator with a nozzle of SSRS (Single-chamber/Single-vortex Rotation System) and the microbubble generator with a nozzle of DDHRS, named as "eco-Bubble®" (Fig. 8.10). The microbubble generator with a nozzle of SSRS is the same device that was used in the DO control experiments at the red seabream farm (Fig. 8.5). It took 200 min to increase DO of 1.4 m^3 of sea water in a tank from 20% to 95% saturation by bubbling with the conventional diffuser at an air flow rate of 2 L/min. The microbubble generator with a nozzle of SSRS could intake air into the nozzle at a flow rate of 0.7 L/min to release microbubbles, but took just over half that time, 114 min, to achieve the same elevation of DO saturation level in the water. Bubbling with the microbubble generator demonstrated approximately 4.5 times higher DO dissolution potential for the water compared to that using conventional diffuser. In the case of bubbling the water by the microbubble generator with a nozzle of DDHRS, it can take the air to the nozzle at a flow rate of 2.0 L/min.

This indicates that it can produce an approximately three times larger negative air pressure for intake of air to the nozzle than the microbubble generator with a nozzle of SSRS, presumably due to the faster rotational flow of the water in the nozzle. Therefore, we have great expectations for this new microbubble generator with DDHRS for controlling DO of the aquaculture farms. Now, we are conducting DO control experiments at a shrimp farm using the microbubble generator. The results of the experiments will be reported elsewhere.

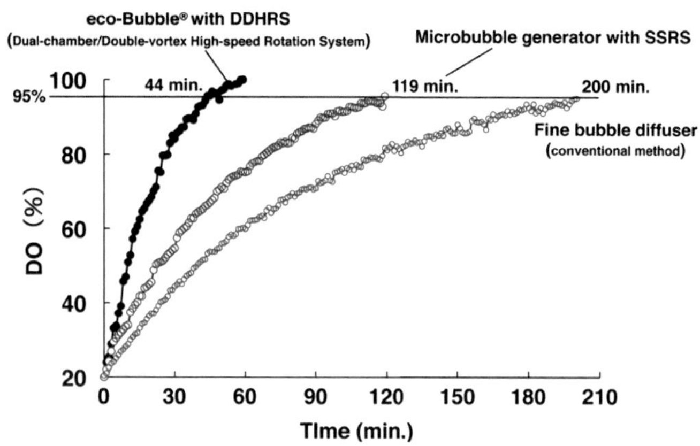

Figure 8.10 Comparison of oxygen gas dissolution potential by bubbling the water among a conventional diffuser (the flow rate of air: 2 L/min), a microbubble generator with SSRS (Single-chamber/Single-vortex Rotation System) (the flow rate of air: 0.7 L/min) and a microbubble generator with DDHRS (eco-Bubble®) [15] (the flow rate of air: 2 L/min) (unpublished). These two microbubble generators used the same water pump in the experiments. Experimental conditions: 1.4 m^3 of seawater with 33.0 in salinity, 23°C in water temperature.

Acknowledgements

We would like to express our thanks to Mr. Richard Lavin for his critical reading of the manuscript. This work was supported by the Research and Development Program for New Bio-industry Initiatives of the Bio-oriented Technology Research Advancement Institution (BRAIN) of Japan.

References

1. Onari H (2001) Fisheries experiments of cultivated shells using micro-bubbles techniques, *J Heat Transfer Soc Jpn,* **40**, 2–7. (in Japanese)
2. Onari H, Maeda K, Matsuo K, Yamahara Y, Watanabe K, Ishikawa N (2002) Effect of micro-bubble technique on oyster cultivation, *Annu J Hydraulic Eng,* **46**, 1163. (in Japanese)
3. Matsuo K, Nakayama T, Onari H, Shimose T (2001) Study on scallop cultivation by using micro bubble technique. *Proc Annu Conf Jpn Soc Civil Eng,* **2**, 384–385. (in Japanese)
4. Nobui B, Onari H, Onari H, Shimose T, Maeda K (2002) Study on pearl cultivation by using micro bubble technique, *Proc Annu Conf Jpn Soc Civil Eng,* **7**, 499–500. (in Japanese)
5. Takahashi M, Kawamura T, Yamamoto Y, Ohnari H, Himuro S, Shakutsui H (2003) Effect of shrinking microbubble on gas hydrate formation, *J Phys Chem B,* **107**, 2171–2173.
6. Sadatomi M, Kawahara A, Kano K, Ohtomo A (2005) Performance of a new micro-bubble generator with a spherical body in a flowing water tube, *Exp Therm Fluid Sci,* **29**, 615–623.
7. Tsuge H (2007) Production methods, in *The Latest Technology on Microbubbles and Nanobubbles* (Tsuge H ed), CMC Books, Tokyo, 15–30. (in Japanese)
8. Tsuge H (2010) Production and analysis of microbubble, in *The Latest Technology on Microbubbles and Nanobubbles II* (Tsuge H ed), CMC Books, Tokyo, 6–10. (in Japanese)
9. Srithongouthai S, Endo A, Inoue A, Kinoshita K, Yoshioka M, Sato A, Iwasaki T, Teshiba I, Nashiki H, Hama D, Tsutsumi H (2006) Control of dissolved oxygen levels of the water in the net pens for fish farming with a microscopic bubble generating system, *Fish Sci,* **72**, 485–493.
10. Endo A, Srithongouthai S, Nashiki H, Teshiba I, Iwasaki T, Hama D, Tsutsumi H (2008) DO-increasing effects of a microscopic bubble generating system in a fish farm, *Mar Pollut Bull,* **57**, 78–85.
11. Tsutsumi H, Kinoshita T, Tamaki S, Kunihiro T, Ohwada K, Wada M, Zhang D, Nishimura M, Kogure K, Endo A, Yoshioka M, Takase I, Nashiki H, Srithongouthai S, Hama D (2008) Bioremediation and microscopic-bubble techniques for environmental management of fish farms and enclosed bays. Concept and outline of the research project, *Océanis,* **34**, 183–198.
12. Hirata H, Kadowaki S (1990) DO management in coastal fish farms, in *Coastal Fish Farms and the Environment* (Watanabe T ed), Kouseisha-Kouseikaku Tokyo, 28–38. (in Japanese)

13. Hirata H, Kadowaki S, Ishida S (1994) Evaluation of water quality by observation of dissolved oxygen content in mariculture farms, *Bull Nat Res Inst Aquac Suppl*, **1**, 61–65. (in Japanese)
14. Tsutsumi H (2011) Control of oxygen concentration of the water in the bays with a compact micro-bubble generator, *J Jpn Inst Mar Eng*, **46**, 86–89. (in Japanese)
15. Taikohgiken Ltd. (2012) Home page, "Welcome to the world of eco-Bubble®," http://eco-bubble.jp/ (in Japanese)

8.3 Following in the Footsteps of Food Valley—Cutting-Edge Micro- and Nanobubble Technology for Better Agriculture and Aquaculture

Hisatsune Nashiki
Tashizen Technoworks, Higashi-ku, Kumamoto-shi, Kumamoto 861-8046, Japan
fvgw7410@mb.infoweb.ne.jp

8.3.1 Introduction

Recently, micro- and nanobubble technologies have come into practical use in many fields for both biological and non-biological systems.

The Company started its engineering development of micro- and nanobubble technologies in 1997 for the purpose of revitalizing aqueous environments, including small streams/rivers and closed water areas, because I had been involved in firefly-related activities for a long time.

We worked with the goal of improving the water quality, and our efforts resulted in unexpectedly high improvements in firefly habitat conservation and increased aquatic plant and animal activity. Consequently, we thought about using such new technologies in the agriculture and fishing industries and have developed small but robust equipment. We started locally but have gradually increased our scope to include various types of plants and fish to improve crop quality, which contributes, in turn, to increases in revenue for farmers we work with.

The use of micro- and nanobubble system in the farming industry for hydroponically cultivated cutting lettuce (*Lactuca sativa* var *angustana*) shows that it has vitalized plants' roots and increased crop yields, according to the comparative results through measurement of dehydrogenase activity.

For soil-cultured *Juncus decipiens*, the height of the plant is 10 cm taller than usual, and they have been sold at higher prices as quality materials for tatami matting, resulting in less work in the field because work in a muddy field is not required, resulting in positive economic effects.

Regarding the cultivation of flowers and ornamental plants using micro- and nanobubbles, the diameter of gerbera daisy stem is more than double for the same breed, and its flower diameter is also almost twice as large as usual. Such gerbera daisies have been shipped as high quality cut flowers because they keep fresh longer.

Thanks to micro- and nanobubbles, dissolved oxygen at the bottom of the sea increases, living organisms that improve water quality such as *Capitella* and others increase, farmed red sea bream grows faster, and 3-year-old fish just before shipment have lower fat and contain more amino acids, which are known as components of *umami*. Furthermore, by applying this technology to pig farming, pigs get substantially healthier and grow faster; the meat quality of these pigs is evaluated as extremely delicious by professional cooks in taste tests, as is the case with red sea bream.

We realized, indeed, that "water is the source of life," considering increased bioactivity in the biological systems (agriculture, livestock, fisheries, environment and health) in the process of our activities.

8.3.2 Our Technologies and Technical Characteristics

8.3.2.1 Technical characteristics

- Portable type available
- Can work with low-pressure pumps
- Small floating substances (frequent cause of failure) cause no problem for our products
- Equipment with anti-cavitation measures, preventing damage due to high energy generation

8.3.2.2 Our technologies

- Swirl, two-phase flow
- Robust
- From small size to large size (10–204 l/min)
- Portable type for agricultural use: nanobubble DBON(R), 50 l/min for small size/green house; for water culture/soil culture
- Fixed large equipment (multi-flow, direct coupling type) = for agricultural site (240 l/min), for mid-size/large-size; for green house; for water culture/soil culture
- Drinking water for livestock, etc.

8.3.3 Current Situation of Agricultural Water and Nanobubble Water

In many cases, deep ground water for lowland farming is oxygen deficient while water in dams and rivers has insufficient percolation.

8.3.3.1 Features of treated water

i. Dissolved oxygen is hard to attenuate because nanobubbles remain for a longer time.
ii. With higher percolation of water, water retention in the soil is improved.
iii. The surface of microscopic bubble is negatively charged.
iv. Negative ions are generated in large quantity.
v. Ultrasonic waves are generated due to crush behavior.

8.3.3.2 Analysis of influence of micro- and nanobubble water on crop growing

i. Influence of micro- and nanobubble water on root morphology and physiological activity
   Small roots/rootlets of plants get bigger, resulting in higher dehydrogenase activity (which shows state of roots' activity and respiration). Such roots are pure white in color because ferric oxide does not occur around such roots due to hydrogen generated by dehydrogenase.

ii. High-temperature resistance
Vital energy of such roots appears, among others, at high temperature. Mustard spinach (hydroponic culture), for which the upper limit of water temperature is 26°C, will neither suffer from oxygen starvation nor cause root rot or icterus even if the water temperature temporarily increases to around 30°C.

iii. Growth promotion
Growth of agricultural crops is promoted because of an increase in oxygen supply and nutrient absorption.

iv. Quality improvement
Since nitrate nitrogen does not accumulate within the crop plant, its quality as a low-sulfated crop is improved. In addition, the physiological balance within the plant is maintained normal all the time, contributing to improvements in disease resistance.

v. Increase in agricultural revenue
Cultivation with micro- and nanobubble water is a major contributing factor to achieve high-quality and high-yield production due to growth-promotion as well as high-quality products due to good physiological balances. Such crops are in good demand in the market, leading to increases in revenue.

vi. Lower environmental load
Since disease and pestilence tolerance increases due to healthy growing of crops, the frequency of pest control can be decreased to be more environmentally friendly. In addition, more oxygen is supplied into water systems, such as paddy fields; therefore, biological activities in the water/soil and purification are promoted further.

vii. Prevention of damages caused by nitrate acid ions
Under the condition of high concentration of nitric acid ions, nitric acid ions tend to combine with proteins, resulting in nitrosamine, which causes cancer, cyanosis (extrication of purple fry), grass tetany of livestock (unexpected death of cattle, etc.), and others.

Supply of micro- and nanobubble water enables us to grow vegetables with low concentrations of nitric acid ions for hydroponic cultivation as well as for soil cultivation. We believe this is because amino acid can be generated faster in such environments.

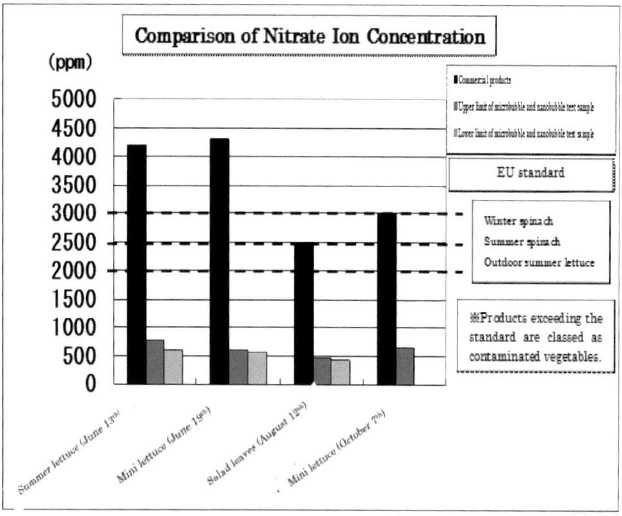

8.3.4 Application of Micro- and Nanobubbles in Livestock Raising

For livestock raising, we have conducted a joint research of pig urine treatment.

We have also conducted experiments to improve pork quality while we have developed a dedicated equipment of drinking water for them.

1. It is reported that the micro- and nanobubbles have the following effects on biological body:
   i. Assisting blood flow
   ii. Growth hormone secretion
   iii. Neurotransmitter secretion
   iv. Relief of stress
2. Result of the experiment on drinking water for pigs (a before and after comparison):
   i. Rate of accident (death) is reduced
      The number of accidental death caused by mutual biting and others is reduced because of stress reduction effects. Infectious diseases and death due to diseases are reduced by improved blood circulation.

8.3.5 Technologies for High Quality and Stable Production

It is expected that food security or insufficiency of food resources will be focused among countries, as global warming, increase of population, and rapid economic development of developing countries make progress in the 21st century. Therefore, it will be important to develop innovative technologies in agriculture, forestry and fisheries for stability of the nation. Our technologies are considered as useful and helpful to improve self-sufficiency in food for the future. Furthermore, high-quality crops will create export markets to achieve a new world for agriculture and fishing industries.

8.3.6 Examples of Application: Nanobubble DBON; Multi-Flow; Direct Coupling Type and Others

8.3.6.1 Example of application

Agriculture: sprout, spinach, mustard spinach, garland chrysanthemum, arugula, lettuce, small green onion, sweet herbs, cucumber, green pepper, bitter gourd, strawberry, grape tomato, tomato, eggplant, melon, water melon, shiitake mushroom (*Lentinus edodes*), *J. decipiens*, rice plant, lotus root, gerbera daisy, chrysanthemum, pot flower, rose, *Phalaenopsis* orchid, orchid, carnation, and *Eustoma russellianum*.
Fruit tree: Dekopon (*Citrus reticulata* Siranui).
Landscaping: flowers; as measures for old trees.

Fish culture industry

Fresh water: eel, young sweetfish, edible carp, and freshwater laver.
Salt water: prawn, tiger puffer, flounder, abalone, ear shell (*Sulculus diversicolor supertexta*), *Seriola quinqueradiata*, greater amberjack, red sea bream, and sea grapes.

Dried laver seaweed: seed for aquaculture; for laver processing.
Drinkable water for livestock: pig, beef cattle, and chicken.
Processed food product: bread (wheat flour/rice flour), tofu (soybean curd), miso (bean paste), soy sauce, precut vegetable, bean sprout, cooked rice, and noodles.

8.3.6.2 Collaborative research

Agriculture: Professor Manabu Katano, School of Agriculture, Tokai University
Livestock raising: Assistant Professor Yasunori Kawagoe, Faculty of Engineering, Kumamoto University (swine manure treatment)
Noriyuki Tsuchiyama, general manager of Hotel Kumamoto Terrsa (ex. executive chief of Akasaka Prince Hotel; Japanese food) (guidance, eating-quality testing, etc.)

8.3.7 Future Prospects

8.3.7.1 Agricultural production

Food-supply crisis is getting more serious in the 21st century. Since self-sufficiency rate for food in Japan is low, more high production efficiency is required for agriculture. Among others, considering the production of staple diet, the self-sufficiency rate of rice, wheat, corn, soybeans, and others is significant.

The key function is soil improvement and water quality. Micro- and nanobubble water contains plentiful air in it and seeps into the earth deeply, supplying both water and air, altogether owing to its excellent seepage force and oxygenation, which leads to the creation of a better environment for plants' roots. Agricultural crops grow faster and healthier. Accordingly, production efficiency will be improved remarkably.

Especially, considering the increase in global warming, the agricultural industry will face hotter conditions in the future. Therefore, it is believed that the use of nanobubble water that "helps to improve heat resistance" will lead directly to better productivity. In addition, it can be available and useful as an advanced technology of Japan for industry exchange with other countries.

8.3.7.2 High-quality, heavy cropping

For horticultural crops such as vegetables, fruit trees, and so on, it is expected that the use of micro- and nanobubble water will increase yields and improve quality at the same time, which will lead to better economic efficiency. Fresh and healthy vegetables will make people who eat it healthier. Among others, the better the disease resistance of the crops, the lesser the use of agricultural chemicals, which leads to significantly reduced environmental load.

8.3.7.3 Livestock raising

The use of micro- and nanobubble water for pig and beef cattle will help to make them healthier, improve feeding efficiency, and reduce stress, resulting in their faster growth. In addition, it is possible to supply more healthy and delicious meat, resulting in better management of livestock farming with supply of such quality meat as well as higher economic effects. Healthy meat will make people who eat it healthier.

8.3.7.4 Foods

Food processing with micro- and nanobubble water has positive effects on the high-quality of bread (wheat/rice flour), cooked rice, noodles, tofu (bean curd), bean sprout, and others.

8.3.7.5 Environment

Micro- and nanobubble water is essential to recover ecological systems and promote environmental purification because it increases oxygen supply in the water, contributing much to sludge biodegradation in the water. In near future, it is likely to be available for environmental problems, with active international technical exchanges.

8.3.7.6 Conclusion

Establishment of "Research and Development Partnership concerning Micro- and Nanobubble for Agriculture, Forestry and Fisheries/Food Industry" had been anticipated very much. Since "the Act on Research and Development Partnership" was revised in April 2009, the first research association was authorized by the Minister of Agriculture, Forestry and Fisheries in February 2010, and the Company joined it as a member. It is expected that the association will conduct relevant research and development with a goal of practical realization and commercialization with a view to global market, and we place great expectations in it.

In addition, one of the largest themes on the earth for the 21st century is "water issues." We would like to make a contribution to creating a "Food Valley" where "Agriculture," "Foods," "Environment," and "Health" are circulating around "Water" from a conceptual view.

Chapter 9

Medical Field

9.1 Medical Applications of Microbubbles

Shin-ichiro Umemura
Graduate School of Biomedical Engineering, Tohoku University,
Aoba-ku, Sendai, Miyagi 980–8579, Japan
sumemura@ecei.tohoku.ac.jp

9.1.1 Introduction: Features of Microbubbles for Medical Applications

The ultrasonic properties of biological tissues can be reasonably well approximated by those of water. The most basic acoustic parameters of air, which can represent the gas species enclosed in a microbubble, are compared with water in Table 9.1. The acoustic impedance of air is smaller than that of water by more than three orders of magnitude because the density and compressibility of air are three orders of magnitude smaller and four orders of magnitude higher than those of water, respectively. Therefore, acoustic waves are reflected well by a gas body in either water or biological tissues because they are reflected by acoustic impedance mismatches. Strain is two orders of magnitude greater in air than in water at the same ultrasonic intensity. Therefore, nonlinear phenomena can be induced more easily in air than in water in the same ultrasonic intensity range.

Micro- and Nanobubbles: Fundamentals and Applications
Edited by Hideki Tsuge

ISBN 978-981-4463-10-2 (Hardcover), 978-981-4463-11-9 (eBook)
www.panstanford.com

Table 9.1 Comparison of basic acoustic parameters of air and water

	Density ρ (kg/m^3)	Compressibility K (Pa = N/m^2)	Sound Speed $c = \sqrt{K/\rho}$ (m/s = μm/μs)	Acoustic Impedance $\rho c = \sqrt{\rho K}$ (kg/m^2/s)
Air	1.3	1.5×10^5	340	440
Water	1000	2.2×10^9	1500	1.5×10^6

Even more distinctive phenomena occur when such gas is enclosed in water in the form of a bubble. The volumetric oscillation of a bubble consists of the inertia of water, with relatively high density, and the restoring force of air, with relatively high compressibility. Therefore, the ratio between the size and the oscillation period of a bubble can be estimated in the order of 10 μm/μs, which is one to two orders of magnitude smaller than the sound speeds in air and water. In other words, the size of a resonant bubble should be one to two orders of magnitude smaller than the wavelengths in air and water at the same frequency. This is a coincidence, convenient for the medical application of microbubbles, resulting in microbubbles capable of passing through microcapillaries can be resonant to ultrasound in the MHz range, providing both penetration and resolution proper for medical applications. This characteristic of microbubbles has been used for a decade as contrast agents in medical diagnosis in the form of stabilised microbubbles. Use of such ultra lightweight contrast agents makes a good contrast with X ray diagnosis, in which heavyweight contrast agents with high X ray absorbance due to high electron density such as iodine are used.

9.1.2 Ultrasonic Response of a Microbubble and Its Medical Applications

A typical response of a microbubble to ultrasonic pressure is shown in Fig. 9.1. A pulsed waveform at 3 MHz typical for diagnostic ultrasound has been chosen. The static diameter of the microbubble has been chosen to be 1 μm so as to be resonant at 3 MHz. Because of the resonance, the volume of the microbubble continues to oscillate for 0.5–1 μs even after the ultrasonic driving pressure reduced to zero. The scattered acoustic pressure, emitted by the microbubble, has an extremely high peak at the moment when the volume is minimum, resulting in plenty of harmonic content which the driving pressure does not originally have. These harmonics are utilised in contrast imaging as the echo signals specific to microbubbles.

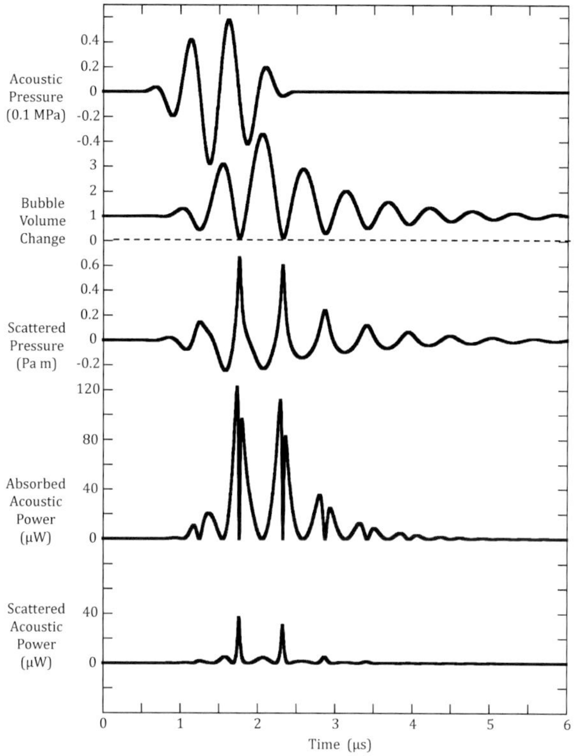

Figure 9.1 Typical response of a microbubble to pulsed ultrasound. It was numerically calculated for a microbubble 1 μm in static diameter in water at a central ultrasonic frequency of 3 MHz.

Figure 9.1 also shows that the absorbed acoustic power, converted to heat, has peaks immediately before and after the moment of the minimum volume. The temperature of the gas enclosed in the microbubble tends to increase rapidly at the moment of the minimum volume by being rapidly compressed. These high temperature and pressure can induce sonochemical reactions. In summary, the acoustic energy can be converted at and around the moment of the minimum volume to other modes of energy such as thermal and chemical energy, which can induce biological effects. Therapeutic applications of such biological effects are under development. Other than the administration of stabilised microbubbles, microbubbles can be generated in living tissues by acoustic cavitation induced by high-intensity ultrasound.

9.1.3 Effect of Gas Species Enclosed in Microbubble

In the analyses to calculate the behaviour of microbubbles such as shown in Fig. 9.1, a Rayleigh–Plesset equation

$$\rho\left[R\frac{d^2R}{dt^2}+\frac{3}{2}\left(\frac{dR}{dt}\right)^2\right]=\left(P_0+\frac{2\sigma}{R}\right)\left[\left(\frac{R_0}{R}\right)^{3\kappa}-1\right]-\frac{4\mu}{R}\frac{dR}{dt}-P_A(t)$$

or its modified versions ([1] for example) are widely used. The left hand side denotes the inertial force of the liquid, and the first term on the right hand side denotes the restoring force due to the compressibility of the gas. By solving this equation, the temporal change of the radius, R, of a microbubble subjected to acoustic pressure, $P_A(t)$, can be obtained. Here, P_0 is the static pressure, σ is the surface tension of the microbubble, ρ is the density of the liquid and μ is the viscosity of the liquid. This viscosity converts the acoustic energy into heat. κ is the polytropic coefficient of the gas enclosed in the microbubble. κ equals to one in isothermal change. In adiabatic change, κ equals to the specific heat ratio of the gas, which is 5/3 for monatomic gas such as argon, 1.4 for diatomic gas such as oxygen and nitrogen and close to one for polyatomic gas such as perfluorocarbons.

Figure 9.2 shows the numerically calculated acoustic power scattered and absorbed by a microbubble, 1 μm in static diameter in water, plotted against the ultrasonic frequency with the polytropic coefficient, κ, as a parameter. The ultrasonic intensity was chosen to be 1 mW/cm^2, at which nonlinear effect can be ignored. The surface tension, σ, was set to the half of water, considering the effect of the agent stabilising the microbubble. The resonant frequency is the lowest; in other words, the microbubble is the softest for isotropic change ($\kappa = 1$). The resonant frequency increases; in other words, the microbubble becomes harder as κ increases. The resonant frequency at $\kappa = 5/3$ is 30% higher than at $\kappa = 1$.

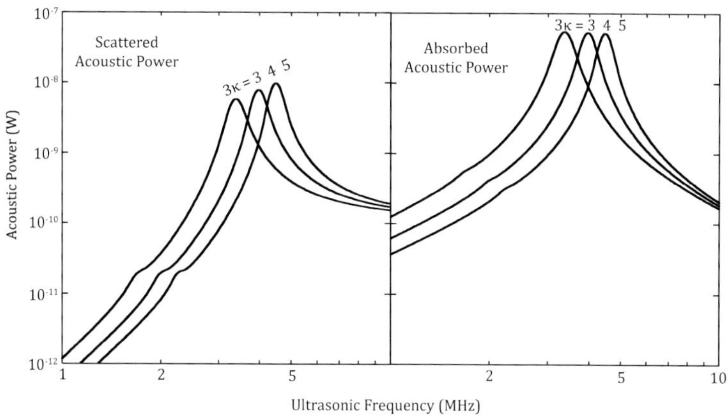

Figure 9.2 Influence of enclosed gas on frequency response of a microbubble. It was numerically calculated for a microbubble 1 μm in static diameter in water for isothermal change ($3\kappa = 3$) and the adiabatic change of monatomic gas species ($3\kappa = 5$).

The gas species, such as perfluorocarbons, enclosed in a microbubble of the second generation contrast agents is polyatomic. Therefore, its specific heat ratio is close to one, and κ can be reasonably approximated to be one. On the other hand, a microbubble of the first generation contrast agents contains air, whose main components are diatomic with a specific heat ratio of 1.4. A microbubble resonant at medical ultrasonic frequency is small enough to be dominated by heat conduction except for those containing gas with low heat conductance, such as xenon, krypton and chlorine. Therefore, assuming κ = 1–1.4 is reasonable for microbubbles in medical applications.

9.1.4 Nonlinear Ultrasonic Response of Microbubble

Figure 9.3 shows the numerically calculated acoustic power scattered and absorbed by a microbubble, 1 μm in static diameter in water, plotted against the ultrasonic frequency with the ultrasonic intensity as a parameter. The ultrasonic intensity is varied by octaves from 1 mW/cm^2, at which nonlinear effects can be mostly ignored, to 1 W/cm^2, at which nonlinear responses are apparent. At 1 mW/cm^2, a nonlinear phenomenon is only observed as a slight hump at 1.7 MHz, which is half the fundamental resonant frequency. At intensities above 30 mW/cm^2, nonlinear humps in the frequency response start to appear not only at the half but also the third of the fundamental frequency. At an intensity of 1 W/cm^2, a nonlinear hump appears even at the two thirds of the fundamental frequency. The fundamental resonance frequency itself shifts towards lower frequency as the ultrasonic intensity increases, especially above 0.1 W/cm^2. It shifts from around 3.4 MHz at 1 mW/cm^2 to around 2.3 MHz at 1 W/cm^2.

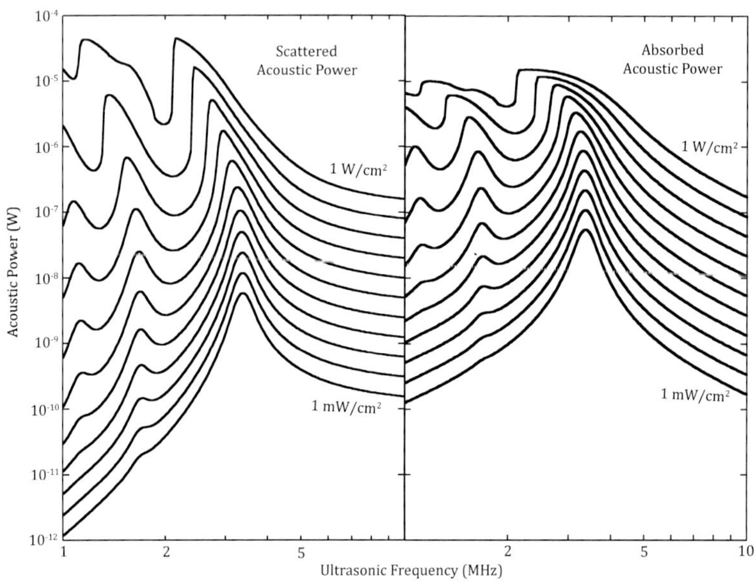

Figure 9.3 Acoustic power scattered and absorbed by a microbubble. They were numerically calculated for a microbubble 1 μm in static diameter in water for ultrasonic intensities varied by octaves from 1 mW/cm^2 to 1 W/cm^2.

Both absorbed and scattered acoustic power increase nonlinearly as the ultrasonic intensity increases. The absorbed power increases at a lower rate and the scattered power increases at a higher rate than that of the intensity. The former is an order of magnitude higher than the latter at 1 mW/cm^2, but it is rather lower than the latter at 1 W/cm^2.

These nonlinear behaviours of microbubbles may be roughly categorised into two classes: 1) those arising from the limited space for inertia-bearing liquid enclosing a microbubble when the microbubble volume is close to the minimum and 2) those arising from the strong volume expanding tendency at acoustic pressure above the static pressure of 0.1 MPa (corresponding to an ultrasonic intensity of 0.3 W/cm^2). The behaviour of the scattered acoustic power increasing at a higher rate than that of intensity is categorised in class 1. The behaviour of the resonance shifting towards lower frequency as the ultrasonic intensity increases is categorised in class 2. The frequency of the scattered echoes shifts lower than the transmit frequency at which the microbubbles are driven. This nonlinear behaviour is so particular for microbubbles that nonlinear propagation of waves cannot cause a phenomenon even similar to this.

9.1.5 Extracting Nonlinear Echo Component by Pulse Inversion

In order to utilise the nonlinear scattering behaviour of microbubbles for contrast imaging, it is necessary to extract the harmonic components from the echo signals of microbubbles received by the ultrasonic probe. A bandpass filter can be used for this purpose [2], but its output signal tends to be elongated in time, deteriorating the range resolution. A pulse inversion method [3] was introduced to solve this problem. A number of modified versions of the method have been adapted to diagnostic ultrasound imaging instruments and used clinically.

In the pulse inversion sequence, transmission pulses with carrier phases of 0° and 180°, as shown in the left side of Fig. 9.4, are used. Each of the two pulses is transmitted, and the corresponding echoes are received. When the pulses are scattered by a microbubble, the two series of received echoes are obtained as shown in the right side of the figure. By taking the sum of the two, the linear components of them are cancelled as shown, and the nonlinear components, which are specific to microbubbles, are extracted. This method reduces the frame rate by twice, but it does not deteriorate the range resolution as a bandpass filter does.

The echoes from biological tissues contain nonlinear components even without microbubbles because a transmit pulse accumulates them while it nonlinearly propagates in tissue. These nonlinear echo components are also extracted by a pulse inversion method and used for diagnostic purposes because imaging, with a higher acoustic signal to noise ratio than conventional linear imaging, can be achieved.

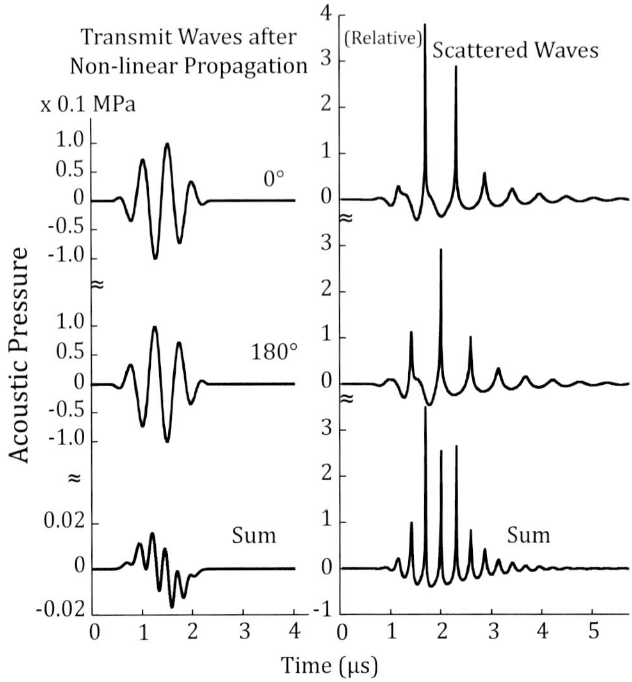

Figure 9.4 Extraction of nonlinear echo component by pulse inversion. It was numerically calculated at a central ultrasonic frequency of 3 MHz assuming isothermal change of a microbubble 1 μm in static diameter in water.

9.1.6 Acceleration of Ultrasonic Heating by Microbubbles

Figure 9.3 implies that not only microbubbles in the size of the fundamental resonance but also those half or a third in size can convert acoustic energy to heat through their nonlinear oscillation at relatively high ultrasonic intensities. Having multiple resonance is beneficial for using actual microbubbles to enhance ultrasonic heating, considering that the stabilised microbubble agents, developed for diagnostic purposes, have a relatively large size distribution up to several times.

Figure 9.5 shows the temperature rise in murine kidney, measured with an extremely thin thermocouple inserted in the tissue, induced by focused ultrasound exposure at 3.2 MHz. The temperature rise is compared before and immediately after the intravenous administration of Optison®, which is a second-generation stabilised microbubble agent, at a dose of 0.2 mL/kg, which is within the clinically approved range. The ultrasonically induced temperature rise is enhanced four to five times by the microbubbles [4]. This effect can be utilised to achieve an ideal focused ultrasound treatment if one can make the microbubbble concentration in the tissue to be treated significantly higher than in the surrounding normal tissues.

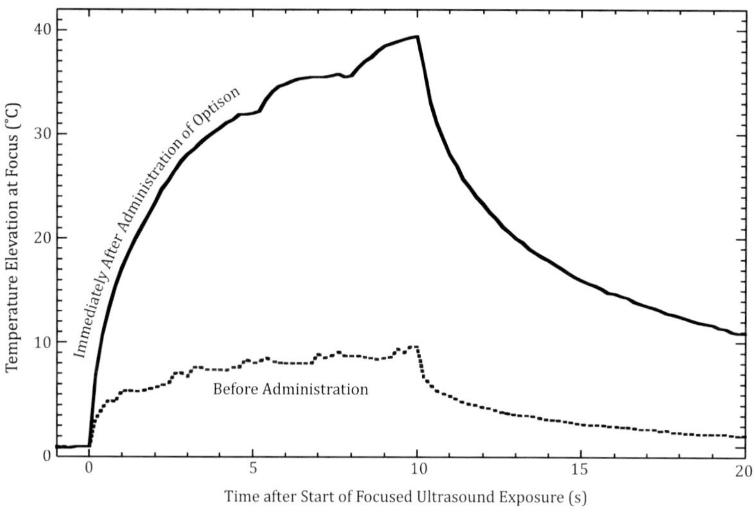

Figure 9.5 Acceleration of in vivo ultrasonic heating by microbubbles. The temperature rise in murine kidney tissue induced by focused ultrasound at 290 W/cm^2 at 3.2 MHz for 10s was measured before and after intravenous administration of a stabilised microbubble agent, Optison, at a dose of 0.2 mL/kg.

9.1.7 Summary

Regarding the medical application of microbubbles, their basic characters were overviewed, and then, their nonlinear scattering and absorption of ultrasound were looked at in detail. Followed was a brief explanation of pulse inversion method to extract nonlinear components, specific to microbubbles, from the echoes for ultrasonic diagnosis. Further followed was a brief explanation of the acceleration of ultrasonic heating by microbubbles for focused ultrasound treatment.

References

1. Keller JB, Miksis MJ (1980) Bubble oscillations of large amplitude, *J Acoust Soc Am*, **68**, 628–633.
2. Schrope BA, Newhouse VL (1993) Second harmonic ultrasonic blood perfusion measurement, *Ultrasound Med Biol*, **19**, 567–579.
3. Burns PN, Wilson SR, Simpson DH (2000) Pulse inversion imaging of liver blood flow: improved method for characterizing focal masses with microbubble contrast, *Invest Radiol*, **35**, 58–71.
4. Umemura S, Kawabata K, Sasaki K (2005) Acceleration of ultrasonic tissue heating by microbubble agent, *IEEE Trans Ultrason Ferroelectr Freq Control*, **52**, 1690–1698.

9.2 Contrast-Enhanced Ultrasound Using Microbubble Contrast Agent

Fuminori Moriyasu
Department of Gastroenterology & Hepatology, Tokyo Medical University, Shinjuku-ku, Tokyo 160–0023, Japan
moriyasu@tokyo-med.ac.jp

9.2.1 Introduction

Ultrasonography has been applied clinically since the 1970s. While non-enhanced B-mode ultrasonography was used for a long time initially, Doppler mode ultrasonography using the Doppler effect became available thereafter. Two-dimensional colour Doppler ultrasonography, in which Doppler signals are superimposed on the B-mode images, has been applied clinically to visualise blood flow. This method has been useful for differential diagnosis between benign and malignant tumours by allowing visualisation of tumour vessels and diagnosis of vascular lesions.

In the colour Doppler method, Doppler shift frequencies of ultrasound scattered by moving scatterers, namely, blood cells, particularly red blood cells, are converted to signals. However, when the Doppler shift frequency is set at 3–4 MHz, which is used in abdominal ultrasound, only blood flow velocities of at least 5 mm/s can be visualised. Therefore, visualisation of the microcirculation from arterioles to capillaries is not possible.

In living bodies, mainly composed of water, the greatest difference in acoustic impedance is seen between fluids and gases. Therefore, to intensify signals from blood flow, ultrasound contrast agents composed of microbubbles that easily pass through capillaries have been developed. Furthermore, the harmonic technology has been developed, which allows visualisation of non-linear signals, making it possible to visualise blood flow from the great vessels to the capillaries.

In addition, some microbubble contrast agents with a diameter of ≤4 μm are phagocytosed by macrophages in the vascular lumen and do not leak out from the blood vessels into the interstitium. Visualisation of macrophages, i.e. cell-targeted imaging, became possible by visualising microbubbles that are phagocytosed and retained by the cells. Kupffer cells in the liver are representative macrophages in the endothelium. Therefore, these microbubble contrast agents that are phagocytosed have two contrast effects as blood pool agents, that is, as conventional sensitisers for intravascular blood flow and as Kupffer cell agents, i.e. cell-targeted contrast agents. Accordingly, use of these contrast agents allows much useful information to be obtained in liver ultrasonography.

This chapter reviews the basic and clinical aspects of ultrasound contrast imaging.

9.2.2 Contrast Agents

9.2.2.1 Types of ultrasound contrast agents

Many contrast agents for contrast-enhanced ultrasonography have been developed and tested in clinical trials. Some of them are already on the market in Japan as well as in foreign countries, as shown in Table 9.2. These agents have been developed mainly by Western pharmaceutical companies. Albunex was launched in the US and Japanese markets in the 1990s, but is no longer used at present. Levovist has been used in Europe and many countries around the world; however, in recent years, this agent has been used almost exclusively in Japan. In Europe, Optison, Definity and SonoVue have been in clinical use. In the US, Optison, Definity and Imagent are used for cardiac imaging. In China, SonoVue has been on the market since 2004 and is used for liver and cardiac imaging.

Table 9.2 Ultrasound contrast agents

Name	Company	Size	Shell material	Inner Gas
Albunex	MBI Mallinckrodt	4.3 mm	Albumin	Air
Levovist	Schering	2–4 mm	No (palmitic acid)	Air
Optison	MBI Amersham Health	3.0–4.5 mm	Albumin	C3F8 + air
Definity	Bristol-Mayers	1.1–3.3 mm	Lipid + surfactant	C3F8 + air
Imagnet	Alliance	5 mm	Surfactant	Perfluoro-carbon
SonoVue	Bracco	2.5 mm	Lipid	SF6
Sonazoid	GE Health Care	3 mm	Phospholipid	Perfluoro-carbon
Quantison	Quadrant Healthcare	3.2 mm	Albumin	Air
Myomap	Quadrant Healthcare		Albumin	Air
Echogen	Sonus, Abbott	3–5 mm	Surfactant	C5F10
Cardipsphere	Point Biomedical		Polymer + albumin	Air
AI700	Acusphere	–2.2 mm	Polymer (PLGA)	Perfluoro-carbon

The development of Sonazoid was started by Nycomed in Norway in the 1980s. It was tested in clinical trials for the diagnosis of neoplastic liver diseases; however, no application for approval of the agent was made in Western countries.

In Japan, phase I, phase II and phase III clinical trials of Sonazoid were started in 1998, 1999 and 2001, respectively, and an exploratory test was conducted in 2003 to examine the effectiveness of this agent in the follow-up evaluation after radiofrequency treatment of liver cancer. During this period, Nycomed was taken over by Amersham Health, and in turn, Amersham Health was taken over by GE Health Care (GEHC). Therefore, GEHC owns the rights to develop Sonazoid in countries other than Japan at present.

In Japan, Sonazoid was developed by Daiichi Pharmaceutical Co., Ltd. (the present Daiichi Sankyo Co., Ltd.). This drug was approved and launched on the market on January 10, 2007 as the first second-generation contrast agent in Japan. The launch of Sonazoid in the market in Japan, ahead of other countries, was an epoch-making event.

In clinical trials of Sonazoid, a comparative test of diagnosability was conducted by blind reviewers, using contrast-enhanced helical

CT as control. The diagnostic ability of Sonazoid contrast-enhanced ultrasonography was found to be equivalent to or better than that of CT, and use of this agent was approved by the Ministry of Health, Labour and Welfare. In particular, it was shown that the ability of Kupffer cell images obtained over 10 min after treatment with Sonazoid for the diagnosis of neoplastic liver lesions was higher than that of CT. This facilitated the approval.

The indication for the use of Sonazoid at the time of the approval was diagnosis of liver tumours, and diagnosis of tumours in other organs is still not covered by health insurance. This is largely because developmental clinical trials of Sonazoid involved only patients with neoplastic liver diseases. In addition, another reason is that Sonazoid was designed so that it can be phagocytosed by the Kupffer cells of the liver as described above.

However, as a blood pool agent, Sonazoid does have a strong ability to allow diagnosis of tumours in other organs, such as pancreatic, breast and renal cancers. Therefore, it is highly desirable that the indications for the use of Sonazoid be expanded in the future. Daiichi Sankyo Co. Ltd., which developed and sells Sonazoid, is planning clinical trials in the future for expanding the indications.

9.2.2.2 Pharmacokinetics of Sonazoid (Fig. 9.6)

When injected intravenously, Sonazoid passes through the right side of the heart and reaches the left side of the heart via pulmonary circulation. During this period, it is considered that only a small amount of the bubbles disappear due to pressure changes. This is because the blood concentration of Sonazoid decreases more slowly as compared with that of other contrast agents even after repeated circulation between the right and left sides of the heart. If the recommended dose (0.015 mL/kg) of Sonazoid is injected, its contrast effect lasts for more than 10 min in blood vessels such as the portal vein.

Sonazoid is trapped by reticuloendothelial organs such as the liver and spleen, which greatly contributes to the decrease in its blood concentration. If the recommended dose is injected intravenously, approximately 25% of the injected bubbles are considered to accumulate in the liver.

During the first cycle of circulation, the blood concentration of Sonazoid is highest in the hepatic artery and portal vein, and a large amount of the contrast agent flows into the hepatic vein. During the second and later circulatory cycles, the signal intensity is much lower in the hepatic vein than that in the hepatic artery and portal vein (Fig. 9.6). This means that the Sonazoid bubbles are trapped at a high frequency by the hepatic sinusoids during the first circulation cycle (Fig. 9.7).

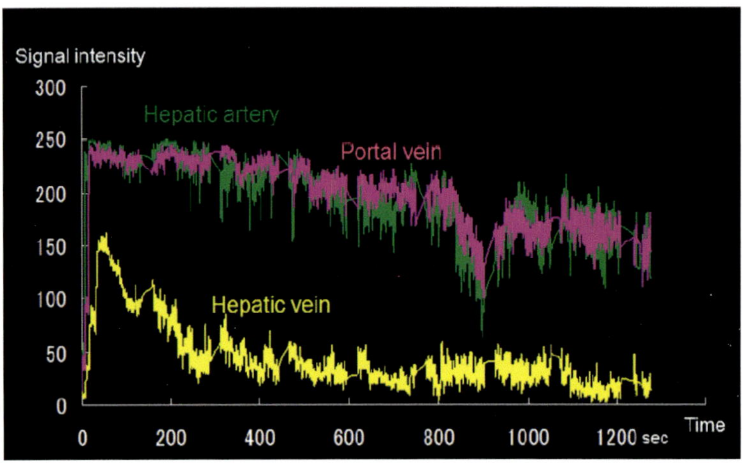

Figure 9.6 Time-signal intensity curves in the hepatic artery, portal vein and hepatic vein after intravenous injection of Sonazoid in healthy individuals. There are no differences in the signal intensity between the hepatic artery and portal vein, indicating that the bubbles were not trapped in the capillaries of organs such as the intestine. There are large differences in the signal intensity among the hepatic artery, portal vein and hepatic vein, indicating that many bubbles were trapped by the sinusoids during a single cycle of circulation.

On the other hand, there are no differences in the contrast agent concentrations between the superior mesenteric artery and vein or between the renal artery and vein, indicating that the bubbles are not trapped in the capillaries of these organs.

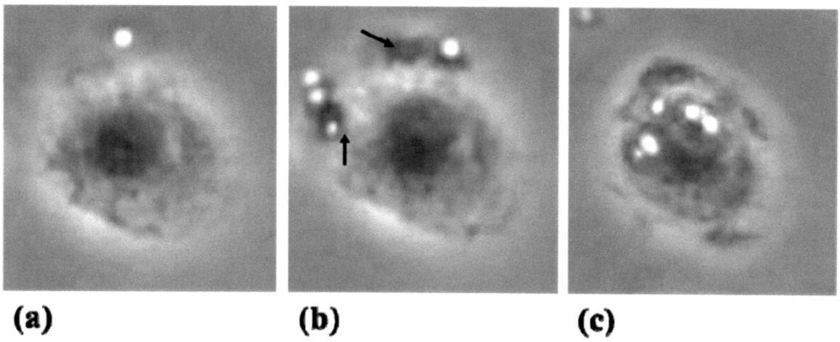

Figure 9.7 Microscopic photograph of phagocytosis of Sonazoid bubbles by cultured Kupffer cells. When the Sonazoid solution was poured into the culture medium of isolated Kupffer cells, bubbles were actively phagocytosed by the Kupffer cells. The Kupffer cells extended tentacles (arrows) to capture and phagocytose the bubbles. The phagocytosed bubbles became translocalised around the nucleus in approximately 10 min. During the translocalisation, the bubble diameter decreased and the bubbles became distorted in shape.

9.2.3 Diverse Effects of Sonazoid

9.2.3.1 Adverse effects of gas

Microbubbles are composed of gas, and gas embolism has been of concern as an adverse effect of ultrasound contrast agents. However, although next-generation contrast agents, including Sonazoid, have been administered more than 2 million times around the world, there have been no reports of embolic adverse effects. These cases include patients with right-to-left shunts such as patent foramen ovale; however, embolic adverse effects have never been reported. This is because microbubbles easily pass through the capillaries due to the small bubble diameter, the bubbles do not merge together and only a very small amount of gas is administered.

The gas content in the prepared Sonazoid solution is 1 vol%. Therefore, if one vial (2 mL) of Sonazoid solution is injected intravenously, the total amount of gas administered into the blood vessel is 20 μL. It can be said that even if all the gas coalesces to produce embolism, no adverse effects inducing organ dysfunction would be expected.

9.2.3.2 Adverse effects of the shell

As described above, the shell of microbubbles is composed of phosphatidylcholine from egg yolk. Therefore, the package insert states that "the agent should not be given to patients with egg allergy". However, although Sonazoid has been administered to more than 20,000 patients worldwide, including Japan, to date, there have been no reports of serious allergic adverse effects.

9.2.3.3 Physiological effects

If high acoustic power ultrasound is irradiated to the contrast agent incorporated into Kupffer cells (Fig. 9.7), the bubbles collapse, and in such a case, the physiological effects on the cells would be of concern. However, even if the contrast agent (Levovist or Sonazoid) incorporated into the liver is completely eliminated, no events, such as increase in blood LDH levels, have been found that are likely to reflect cell damage. In addition, theoretically, when the recommended dose is administered, one bubble is phagocytosed per 50–100 Kupffer cells. Therefore, even if the bubble collapses, cell necrosis or apoptosis will not be clinically observed as an adverse event despite some effect on the cells.

9.2.4 Administration of Sonazoid

9.2.4.1 Preparation of Sonazoid (Fig. 9.8)

Sonazoid, which is supplied as a lyophilised powder, is dissolved in 2 mL of distilled water to prepare the Sonazoid solution. As described above, the gas contained in the prepared Sonazoid bubbles is perfluorobutane, a fluorocarbon, ($C4F10$; common name, perflubutane). The gas contained in the vials both before and after preparation is perflubutane. If the vial caps are left open, the gas will be replaced by air and gas exchange would occur between the gas in the bubbles and air, leading to a reduction in the contrast effect of Sonazoid due to the gradual replacement of the gas in the bubbles by air. Similarly, if degassed water is not used to dissolve Sonazoid, the perflubutane in the bubbles will be replaced by nitrogen and other gases dissolved in water. Therefore, degassed distilled water for injection must be used to dissolve Sonazoid.

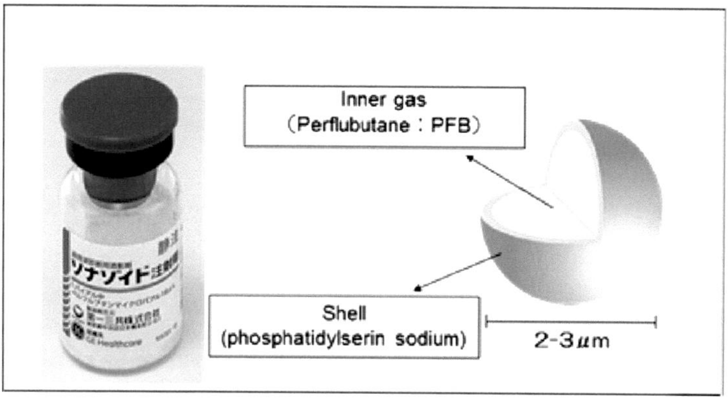

Figure 9.8 Schema for the Sonazoid vial and microbubbles. The lyophilised powder of Sonazoid is dissolved in 2 mL of distilled water. The gas in the vial is perfluorobutane; therefore, conditions should be avoided in which the gas in the upper solution is replaced by air, such as leaving the cap of the vial open, after preparation of the solution. Even after dissolving the Sonazoid and preparing the solution, gas exchange occurs between the gas in the bubbles and air outside the solution.

Sonazoid is usually administered by intravenous bolus injection into a peripheral blood vessel. An elbow vein is selected to increase the peak blood concentration after intravenous injection. The intravenous injection volume of Sonazoid solution is as small as ≤1 mL, and the peak blood concentration can be increased by flushing with 10–20 mL of physiological saline.

Usually, an extension tube filled with the Sonazoid solution is used, and after flushing with 10–20 mL of physiological saline, the intravenous bolus injection is given. At this time, high pressure should not be applied to the Sonazoid bubbles in the extension tube. If pressure above a threshold is applied, the bubbles in the tube will disappear within a moment.

9.2.4.2 Dose of Sonazoid

The package insert states that the recommended dose of Sonazoid is 0.015 mL/kg body weight. Therefore, the recommended dose is 0.9 mL in adults with a body weight of 60 kg. However, when using a sensitive harmonic imaging method, a good contrast effect can be obtained even with half the dose. At our institution, a dose of 0.5 mL is used, irrespective of the body weight. Four injections can be given per vial.

9.2.5 Apparatus

9.2.5.1 Ultrasonic diagnostic apparatus suitable for imaging

An ultrasonic diagnostic apparatus with an imaging mode suitable for the imaging is used. Basically, nonlinear imaging methods (harmonic methods) are used. In the newly developed apparatus, tissue harmonics of the phase modulation method are often used in the usual non-enhanced B-mode, and these are used for imaging by reducing the transmission acoustic power and number of frames.

The conventional fundamental-wave B-mode can also be used for imaging; however, the quality of imaging is worse in this mode than in the harmonic mode. The quality of imaging is improved by increasing the contrast agent dose; however, an overdose would cause shadowing, resulting in a worse contrast effect.

9.2.5.2 Moving and still image recording

Moving images obtained are recorded as moving and still images. Dynamic studies are recorded for 3 min, and moving Kupffer images are recorded for 10–15 s by scanning a single window. It is important to carry our diagnosis on moving images, and it is expected that the moving images obtained can be sent to a reading room and diagnosed on a monitor by radiologists in the near future.

9.2.6 Imaging Modes

9.2.6.1 Principles

9.2.6.1.1 Harmonic imaging method

Ultrasound images are obtained by irradiating ultrasound and visualising the ultrasound reflected by the scatterers in living bodies. Living tissues produce strong linear scattering, while microbubbles produce strong non-linear scattering; namely, the waveform of the irradiated ultrasound is different from that of the ultrasound scattered by the microbubbles. This non-linear scattering shows a frequency distribution with peaks at integral multiples of the frequency of the irradiated ultrasound. For example, if ultrasound with a frequency of 2 MHz is irradiated, ultrasound with peak frequencies of twice (4 MHz) and three times (6 MHz) the irradiated frequency is reflected. Components at twice or three times the fundamental frequency are called the second and third harmonics, respectively. Furthermore, ultrasound with a peak frequency of one-half (1 MHz) is also reflected, which is termed a subharmonic.

The second harmonic is well understood as a signal from the bubbles and used in images. In the frequency distribution, the ultrasound generated by the resonance of bubbles shows high peaks, but when the bubbles are collapsed by a stronger ultrasound, the valleys are filled.

The phase and amplitude modulation methods described below use the second harmonic and harmonic components of the fundamental wave, respectively, for visualisation.

9.2.6.1.2 Phase Modulation (PM) and Amplitude Modulation (AM) methods (Fig. 9.9a,b)

In the phase modulation (PM) method, two pulse waves with a phase difference of 180° are transmitted, and the two scattered waves received are added together for visualisation. There are differences in the degree of distortion of the waveform between ultrasonic waves scattered from tissues and from bubbles. Better contrast sensitivity is obtained with bubbles than with living tissues due to the more non-linear components.

On the other hand, in the amplitude modulation (AM) method, two pulse waves with the same phase and different amplitudes (for example, equal amplitude and one-half amplitude) are transmitted and the wave obtained by subtracting the doubled reflected wave of the one-half amplitude wave from the reflected wave of equal amplitude is visualised.

The PM and AM methods differ in that the spatial resolution is better in the PM method and the sensitivity in deeper areas is better in the AM method. In particular, when relatively high acoustic power is used instead of low acoustic power, like in Sonazoid, non-linear signals from the tissues (tissue harmonics) are strong. Therefore, methods, such as the AM method, that can suppress tissue harmonics are effective for increasing the bubble-tissue ratio in signals.

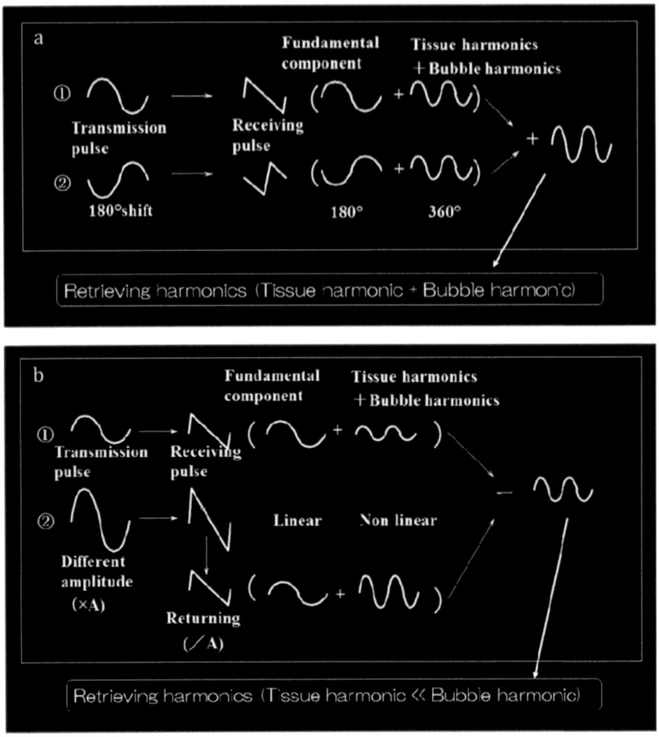

Figure 9.9 a Principle of the phase modulation method. In this method, two pulse waves with a phase difference of 180° are transmitted and received to extract nonlinear components scattered by tissue and bubbles. This method is characterised by the ability to extract the second harmonics, with twice the frequency, from both tissues and bubbles. b Principle of the amplitude modulation method. In this method, two pulse waves with different amplitudes are transmitted and received to extract nonlinear components scattered by tissue and bubbles. This method is characterised by the ability to extract nonlinear components of equal amplitude (fundamental wave band), mainly from bubble harmonics.

9.2.6.2 Imaging condition

9.2.6.2.1 Acoustic power (Fig. 9.10)

Among the imaging conditions, acoustic power is the most important. Next-generation ultrasound contrast agents are generally said to be low-acoustic-power contrast agents; however, the optimum acoustic power varies somewhat among contrast agents.

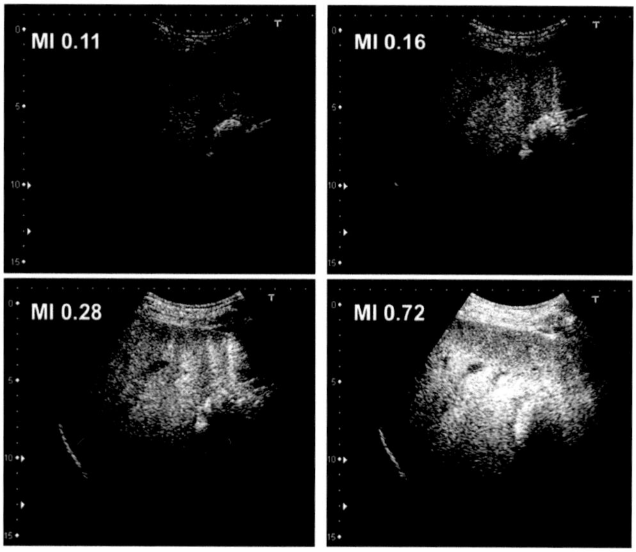

Figure 9.10 The quality of imaging according to the acoustic power setting. When the acoustic power was changed 10 min after injection of 0.5 mL of Sonazoid, the imaged areas expanded with increase in the acoustic power. However, at an MI of 0.72, the bubbles disappeared and signals became attenuated near the surface of the liver. Thus, the resonance and destruction of bubbles strongly depend on the acoustic power.

Vibrating acoustic power and destroying acoustic power: When acoustic power is gradually increased in living bodies containing bubbles, the state in which the bubbles are not vibrated or destroyed is changed to the state in which the bubbles begin to vibrate. When the acoustic power is further increased, the bubbles are destroyed in a short time. The acoustic power required to destroy bubbles varies among contrast agents, and the destroying acoustic power is highest in Optison, followed by SonoVue, Imagent, Definity and Sonazoid, in that order. In the case of Sonazoid, the bubbles begin to collapse at an MI of more than 0.4.

9.2.6.2.2 Particle size distribution of the bubbles and acoustic power

Ultrasound contrast agents composed of microbubbles, such as Sonazoid, have a characteristic distribution of bubble diameters. It is said that Sonazoid has a narrower particle size distribution (higher homogeneity) than other ultrasound contrast agents.

In general, the smaller the bubble diameter, the higher the acoustic power required to produce vibration and destruction of the bubbles. For example, after transmission at an MI of 0.5 to obtain images, when the MI is increased to 1.0, the bubbles that were not destroyed at an MI of 0.5 were destroyed and generated signals.

In addition, the spleen is also a reticuloendothelial organ, and bubbles accumulate in the spleen over 10 min after intravenous injection of Sonazoid. It is known that the size of the particles incorporated into the macrophages in the spleen is smaller than that of the particles incorporated into the Kupffer cells in the liver. As a result, sonazoid accumulated in the spleen is less susceptible to destruction than that accumulated in the liver.

9.2.6.2.3 Bubbles circulating in the blood and bubbles within cells

Sonazoid bubbles remain in the circulating blood after intravenous injection, but are phagocytosed by macrophages throughout the whole body with time, except those that naturally collapse. When looking at the process of phagocytosis of Sonazoid bubbles by cultured Kupffer cells, the bubbles are incorporated into the cytoplasm from the cell surface and move towards the nucleus (Fig. 9.7); this process takes a few to 20 min.

Bubbles in the cells are less susceptible to vibration and destruction than bubbles in the circulating blood. This is because the viscosity of the cytoplasm is higher than that of the blood, and bubbles in the cells are covered by a membrane. In addition, another reason is that after being phagocytosed by the Kupffer cells, the Sonazoid bubbles become smaller than the extracellular bubbles. Therefore, higher acoustic power is required to obtain Kupffer images than to obtain vascular images.

9.2.6.2.4 Number of frames

Sonazoid imaging is characterised by real-time images. The susceptibility to destruction varies according to the number of frames per unit time (frame rate, fps), even under the same acoustic power, and the greater the number of frames (frame rate), the greater the susceptibility to destruction. Particularly in the focus area, the acoustic power is high and beams overlap in each transmission at short distances, causing destruction of bubbles.

Therefore, the quality of the imaging is worse than that in other areas.

It is necessary to obtain images with the minimum number of frames while maintaining real-time imaging, and approximately 15 frames/s (fps) is usually recommended.

9.2.6.2.5 Focus (Fig. 9.11)

In the focus area, acoustic power is high, and bubbles only in this area are vibrated or destroyed, producing heterogeneous liver images. To avoid this, multistage focusing and depth-independent beam forming are used. When a single focus is used, it is necessary to find the appropriate focus setting, such as focusing on the deepest area.

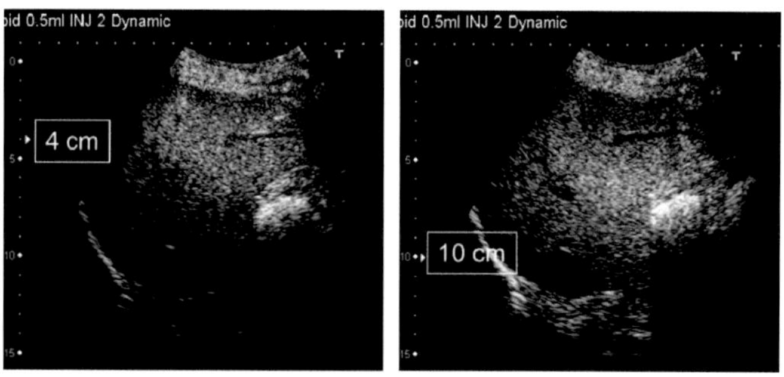

Figure 9.11 Changes in the quality of imaging according to the focus. The focus depth is 4 cm in the left panel and 10 cm in the right panel. As the focus became deeper, bubbles from deeper areas began to vibrate and images could be obtained. Images obtained 10 min after injection of 0.5 mL of Sonazoid are shown.

9.2.7 Actual Image Reading

9.2.7.1 Dynamic study (Figs. 9.12–9.14)

After intravenous injection of Sonazoid, real-time imaging of the liver can be performed and recorded. Usually, the hepatic artery is imaged in 15–20s, followed by the portal vein. The portal vein is predominantly imaged between 20 and 30s, and after 30s, the equilibrium phase is reached, in which the artery and portal vein are similarly imaged. This equilibrium phase is recorded until 3 min after the intravenous injection. Thereafter, Kupffer images become predominant.

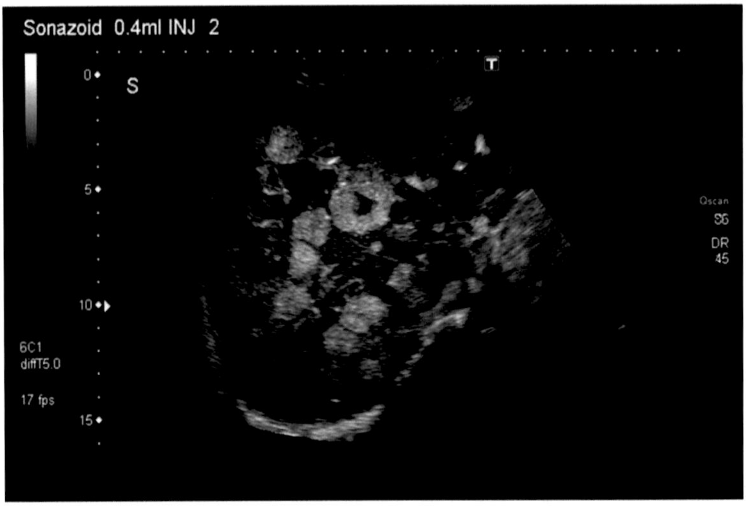

Figure 9.12 Sonazoid images of HCC. The arterial phase after intravenous injection of 0.4 mL of Sonazoid is shown. Many tumours with numerous densely stained arteries are seen in the liver.

9.2.7.2 Flash Replenishment Imaging (FRI) and Micro-Flow Imaging (MFI) (Fig. 9.15, 9.16)

If transmission is made at a high acoustic power (MI $\geq$ 1.0) and 10–30 frames are scanned when the parenchyma is filled with the contrast agent, bubbles in the scan volume are totally destroyed. Thereafter, low acoustic power contrast harmonic imaging can be used to visualise the reperfusion of blood containing new bubbles in the scan volume under low acoustic power. This imaging of reperfusion is called flash replenishment imaging (FRI; Fig. 9.15, 9.16).

Furthermore, when visualising the reperfusion after destruction of bubbles with FRI, fine blood vessels can be continuously visualised by adopting the maximum intensity holding technique, which maintains the maximum intensity in each pixel. This method is called micro-flow imaging (MFI; Fig. 9.15, 9.17–9.19).

MFI, as well as FRI, can be repeatedly performed at any time phase in any part of the liver. In MFI in the arterial phase, tumour vessels that control arterial blood flow are more clearly depicted than non-tumour vessels. In addition, in the portal venous predominant and equilibrium phases, both the arterial and portal venous blood vessels are depicted. In dynamic studies, at the first arrival of the contrast agent after intravenous bolus injection, the hepatic artery is imaged, followed by the portal vein after a delay of 7–10 s. However, in MFI, microbubbles in the blood vessels immediately outside the scan volume flow in when the microbubbles are reperfused into the scan volume. Therefore, if MFI is performed in the portal venous and equilibrium phases, microbubbles flow into the hepatic arterial and portal venous branches almost at the same time. Accordingly, in MFI, performed in this time phase, it is usually difficult to separately depict the hepatic artery and portal vein.

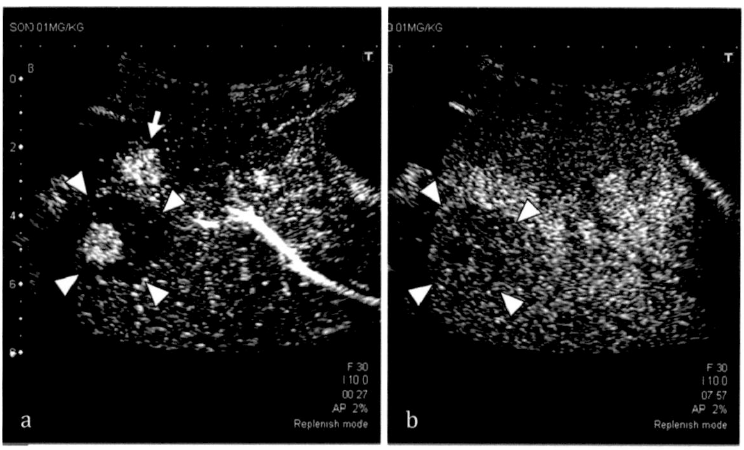

Figure 9.13 Harmonic images obtained 27 s after intravenous injection of 0.01 mL/kg of Sonazoid. In the arterial phase (a), the hepatic artery is imaged and a tumour is visualised near the surface of the liver (arrowhead). Within the tumour, nodules with decreased arterial blood flow contain nodules with increased arterial blood flow. The former were diagnosed as well-differentiated HCC and the latter as moderately differentiated HCC. An AP shunt is seen in the vicinity (arrow). A Kupffer image obtained 8 min after intravenous injection of Sonazoid in the same patient is shown in (b). The uptake of Sonazoid bubbles is slightly reduced in the tumour as compared with that in the surrounding liver parenchyma (arrowhead). Furthermore, uptake of bubbles is absent in the inner nodules with increased arterial blood flow, indicating that these nodules lacked Kupffer cells and represented areas of dedifferentiated carcinoma.

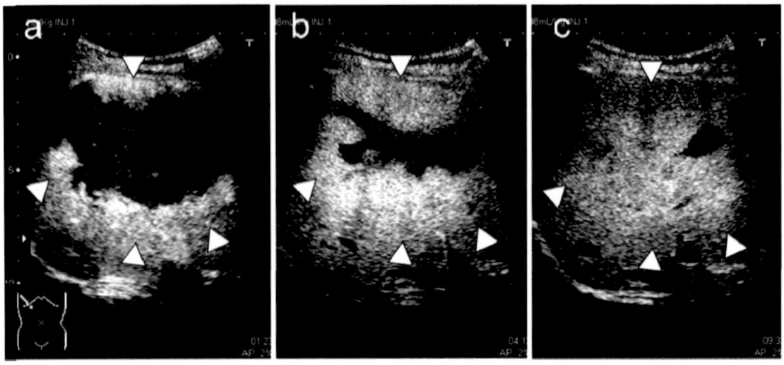

Figure 9.14 Low acoustic power harmonic images obtained after intravenous injection of 0.008 mL/kg of Sonazoid. In the image obtained at 1 min 23 s (a): heterogeneous imaging started from the periphery. At 4 min 12 s, (b) a considerable part of the tumour is stained, and at 9 min 32 s (c) the tumour is almost completely stained. A slow imaging pattern such as this is a characteristic feature of HEM.

9.2.7.3 Kupffer images (Fig. 9.20)

Kupffer images are obtained more than 10 min after intravenous injection of Sonazoid. The timing of Kupffer image acquisition is when bubbles circulating in the hepatic artery and portal vein are reduced so that they do not affect signals from bubbles accumulated in the parenchyma. This varies according to the dose of Sonazoid and the imaging mode used. The lower the dose, the higher the disappearance rate in the blood and the earlier the timing of the image acquisition.

Usually, the scan is initiated 10 min after the intravenous injection. However, if the recommended dose of 0.015 mL/kg is given, the blood level of the contrast agent is still high at 10 min, and purer Kupffer images can be obtained after 20 min. By using the recommended dose, Kupffer images sufficient for diagnosis can usually be obtained for 3 h.

If Sonazoid contrast-enhanced ultrasonography is performed mainly for detecting liver tumours, the throughput of the examination can be increased by moving patients to the examination room more than 30 min after intravenous Sonazoid injection only for obtaining Kupffer images.

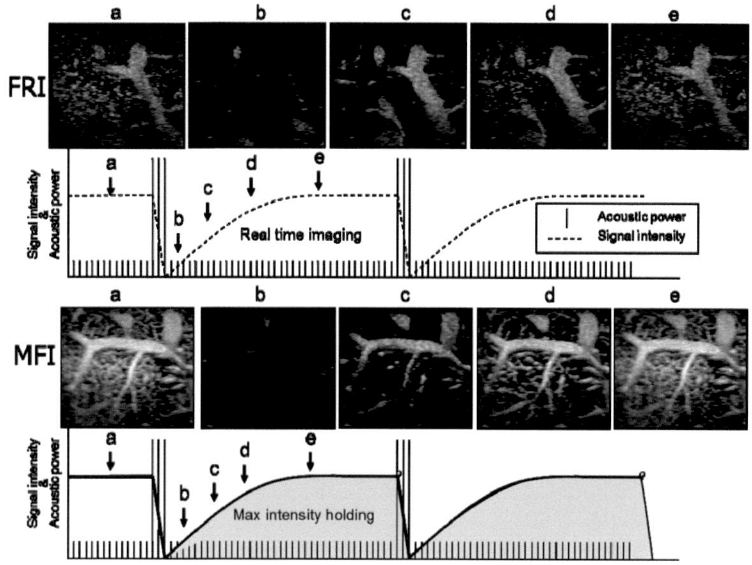

Figure 9.15 Flash replenishment imaging (FRI) and microflow imaging (MFI) methods. The observation is conducted by the low acoustic power harmonic method, and when the scan volume is filled with bubbles, high acoustic power is applied to eliminate the bubbles entirely from the scan volume. Then, reperfusion with new bubbles can be observed by low acoustic power contrast harmonic imaging. This is the FRI method. On the reperfusion images, when the signal intensity of each pixel reaches the maximum, the intensity is maintained to trace the stream of bubbles and visualise fine blood vessels. This is the MFI method.

As described above, somewhat higher acoustic power is required for Kupffer imaging than for vascular imaging. This is because bubbles in the blood are preferentially destroyed as a result of vibration by lower acoustic power than those in the cytoplasm.

The low acoustic power contrast harmonic imaging mode is used for real-time observation of the whole liver. The whole liver can be repeatedly observed if the acoustic power is optimal.

If too much contrast agent is administered, deeper areas cannot be examined due to the strong attenuation. In addition, if the acoustic power is too high, bubbles are destroyed and the quality of imaging is lost in the superficial areas.

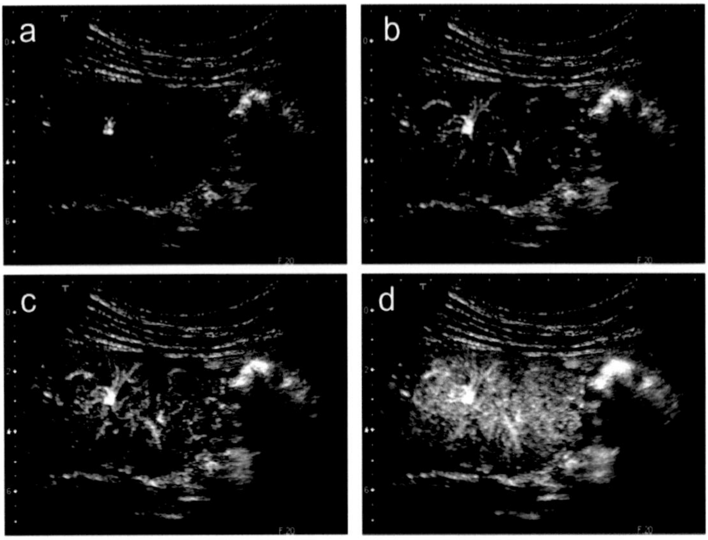

Figure 9.16 MFI images of FNH. Images obtained 1 s (a), 2 s (b), 3 s (c) and 5 s (d) after flushing are shown. Tumour vessels radiating from the centre of the axle are continuously visualised, and at 5 s, they are strongly stained.

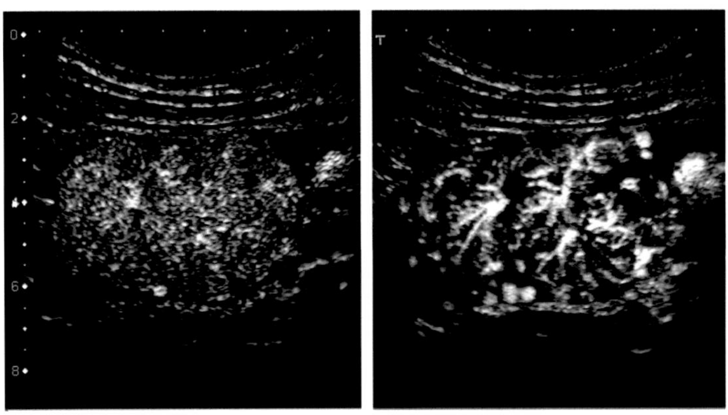

Figure 9.17 Sonazoid imaging of FNH. The left panel shows a normal low-acoustic power harmonic image in the arterial phase obtained 16 s after intravenous injection of 0.5 mL of Sonazoid. The parenchyma is visualised, the tumour vessels are only poorly visualised. The right panel shows the MFI image obtained 25 s after intravenous injection of Sonazoid. Bubbles in the parenchyma have disappeared and axle-like tumour vessels are well depicted. Four central scars are seen.

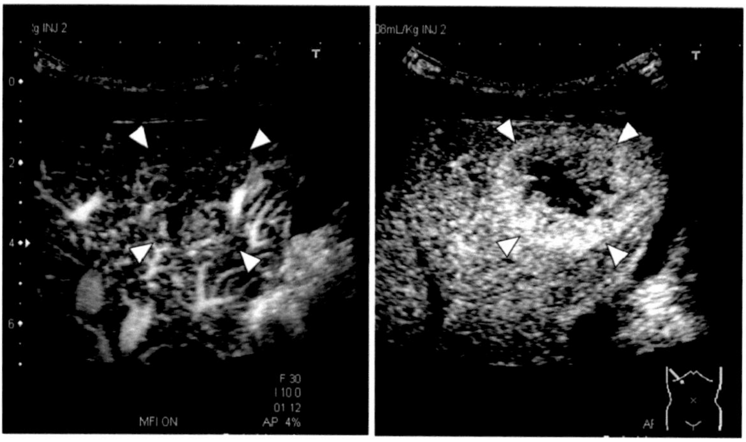

Figure 9.18 MFI images of a metastatic liver cancer obtained using Sonazoid. Tumour vessels irregularly distributed within the tumour periphery are gradually visualised, with the parenchyma also eventually visualised. The vessels within the tumour are distributed in the tutor periphery, and the parenchyma in the central portion is not stained due to necrosis.

9.2.8 Differential Diagnosis of Liver Tumours

9.2.8.1 Hepatocellular Carcinoma (HCC)

Blood vessels from around nodules to tumours are observed in the early phase of serial imaging. Thereafter, outflow blood vessels are sometimes observed. In addition, perfusion images are often observed within tumours in the late vascular phase. Similar patterns are also observed in many hypervascular HCCs measuring ≤20 mm in diameter. In MFI, the structures of the vessels within the tumours are depicted in great detail. The degree of differentiation could be diagnosed based on the blood vessel pattern visualised by MFI, and future studies are awaited. Anyway, in the case of HCC, the degree of tumour differentiation and haemodynamics are varied, but diagnosis can be made based on the disparity, displacement and blocking of blood vessels in many cases (Fig. 9.12, 9.13).

It is difficult to diagnostically image borderline lesions (dysplastic nodules) in a rigorous manner, but the abundance of portal venous and arterial blood flow in the nodules can be revealed by using Sonazoid as a minimally invasive method. In addition, repeated observation of continuity with the portal vein using MFI and intratumoural reperfusion using the above described FRI makes it easier to diagnose intratumoural vessels. Furthermore, nodules and the state of the surrounding Kupffer cells, such as changes in their number and reduction in their function, can be observed simultaneously by Kupffer-phase imaging (Fig. 9.21).

9.2.8.2 Haemangioma (HEM)

In hepatic haemangioma, if the tumour itself has a high echo density, it needs to be distinguished from early HCC with fatty metamorphosis and underlying chronic hepatitis or cirrhosis. We have reported that speckle fluctuation and fluttering signals within tumours in the B-mode are characteristic features of HEM and also that HEM imaging patterns are varied in Levovist contrast-enhanced ultrasonography. Blood pooling in tumours is a characteristic feature of many HEMs, and in the vascular phase, it is observed that blood gradually fills in tumours from the surrounding tissues after injection of Sonazoid (Fig. 9.14, 9.19). Sequential imaging of blood in fine blood vessels, which look like feathers in the tumours, can be observed with Sonazoid, but not with Levovist. If the tumours are large, blood sometimes flows in very slowly, and such an imaging pattern is a characteristic feature of haemangiomas.

9.2.8.3 Focal Nodular Hyperplasia (FNH)

In focal nodular hyperplasia, in the vicinity of the fibrous septa, arterial blood flows into the sinusoids from the muscular blood vessels in the fibrous septa through capillaries, and blood in the sinusoids flows out into the hepatic vein directly or through the sinusoids in the nodules. In any image, spoke-like blood vessels are a characteristic feature, but such a feature is not always observed in all blood vessels. Levovist imaging is characterised by abundant arterial blood flow, because of the lack of the portal vein, the blood flow extending from almost the central portion to the periphery in the vascular phase, being equivalent to that in the surrounding liver in the late vascular phase and being almost equivalent to that in the surrounding liver parenchyma reflecting Kupffer cells in the Kupffer phase. However in Sonazoid imaging, drainage from the tumours to the hepatic veins is often found in addition to the imaging pattern extending from the tumour to the periphery. This is frequently found by careful observation. In addition, normal imaging of FNH with Sonazoid can be performed, because serial imaging of tumours is possible because of the increased frame rate. Radial blood vessels can also be easily recognised. Central scars are clearly depicted in the late vascular phase. Furthermore, the use of MFI increases the diagnostic usefulness of tumours in an axle-like pattern (Fig. 9.16, 9.17). In the Kupffer phase, tumours are imaged as in Levovist imaging.

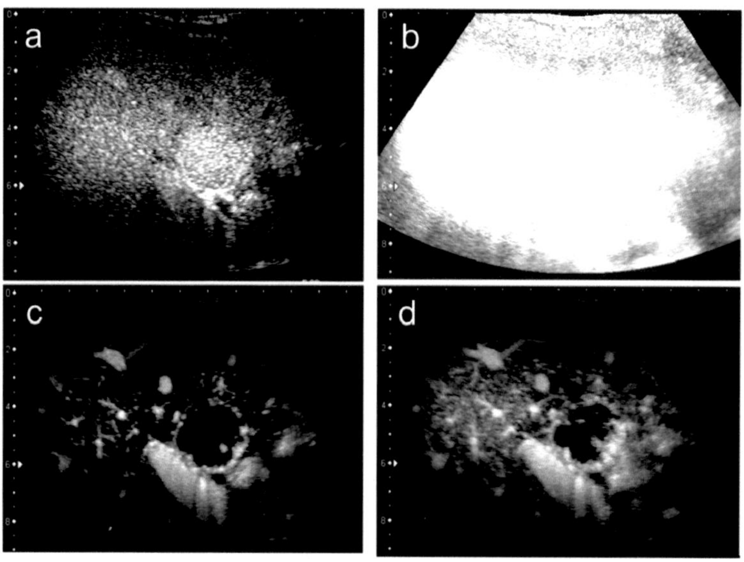

Figure 9.19 MFI images of a haemangioma. a: In the image obtained 1 min 12s after intravenous injection of 0.5 mL of Sonazoid, the entire tumour is visualised; however, discrimination from other tumours such as HCC is difficult. b: High acoustic power scanning eliminated the bubbles in the scan volume. c: In the MFI image obtained 2 s after reperfusion, the vessels within the tumour are sparsely imaged in a dot pattern. d: In the MFI image obtained 5 s after reperfusion, the blood vessels in the liver are visualised up to the periphery, and within the tumour, a loose extension of fine blood vessels is imaged in a cotton-like form.

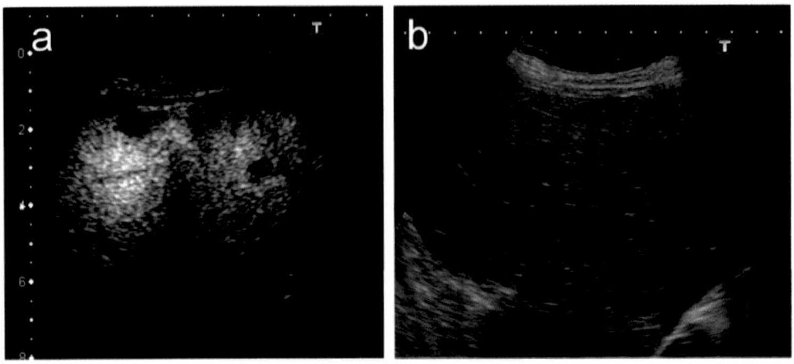

Figure 9.20 A Kupffer image obtained 15 min after intravenous injection of 1 mL of Sonazoid (a). Scanning of the entire liver in the harmonic mode with an acoustic power represented by a MI of 0.2 revealed 8 and 5 mm sized lesions as shadow-less images. The patient had undergone duodenal carcinoid surgery 2 years earlier. The non-contrast-enhanced B mode image shows no tumours in the liver (b).

9.2.8.4 Metastatic liver cancer

Metastatic liver cancer is characterised by poor arterial blood flow in the central tumour area because of the vulnerability of this region to necrosis and hypervascularity in the peripheral areas (Fig. 9.18). Therefore, in the vascular phase, ring-shaped blood flow signals are often seen in the tumour periphery, as in contrast-enhanced CT and other diagnostic imaging methods. The portal vein branches in the liver tissue surrounding the tumours are occluded, which results in a compensatory increase of the arterial blood flow to the tumours. In addition, in the Kupffer phase, metastatic lesions are often completely not visualised. Sonazoid imaging is useful for visualising small metastatic foci ≤5 mm in diameter, which cannot be detected by the usual B-mode ultrasonography (Fig. 9.20).

9.2.9 Summary

The actual procedures of Sonazoid contrast-enhanced ultrasonography for the diagnosis of hepatic mass lesions are outlined. In Japan, only Levovist was available for use for approximately 10 years, and Sonazoid ultrasonography was not

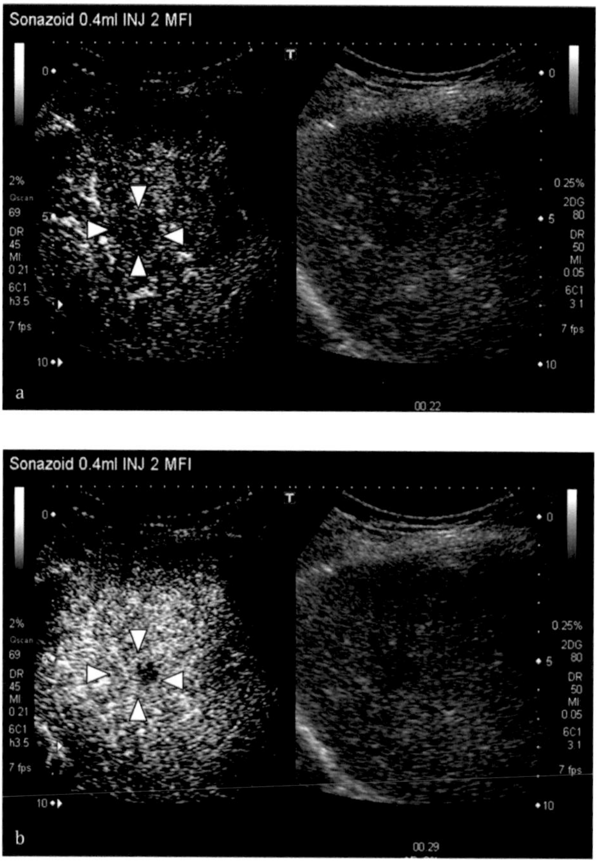

Figure 9.21 a Sonazoid images of a 1.5 cm sized, poorly demarcated low-echoic nodule. The images were obtained by low acoustic power B-mode ultrasonography in the monitor mode (right panel) and low acoustic power B-mode ultrasonography in the harmonic mode (left panel). The number of arteries visualised in the tumour (arrowhead) is reduced in the arterial phase imaged 22s after intravenous injection of 0.4 mL of Sonazoid. b Image in the portal venous phase obtained 29s after intravenous injection of Sonazoid in the same patient. Portal venous blood flow shows iso-signal intensity (equivalent to that of the surrounding liver) in most part of the tumour, and a borderline lesion was suspected; however, the tumour also contains nodules partially lacking in portal venous blood flow; biopsy of this area showed well-differentiated HCC.

widely used because of reasons such as the complicated imaging procedures. However, both vascular-phase and Kupffer imaging can be obtained by a single method, and Sonazoid ultrasonography is now considered to be highly advantageous for patients, also considering the health insurance coverage. Furthermore, it is considered that Sonazoid will contribute significantly to the dissemination of contrast-enhanced ultrasonography because this method is minimally invasive and can be easily performed in outpatient settings by only learning the procedures. It would be of great advantage to patients if invasive tests can be avoided to the extent possible and the QOL of the patients can be improved by the use of this diagnostic method.

References

1. Suzuki S, Iijima H, Moriyasu F, Sasaki S, Yanagisawa K, Miyahara T, Oguma K, Yoshida M, Horibe T, Ito N, Kakizaki D, Abe K, Tsuchiya K (2004) Differential diagnosis of hepatic nodules using delayed parenchymal phase imaging of levovist contrast ultrasound: comparative study with SPIO-MRI, *Hepatol Res*, **29** (2), 122–126.
2. Giorgio A, Ferraioli G, Tarantino L, de Stefano G, Scala V, Scarano F, Coppola C, Del Viscovo L (2004) Contrast-enhanced sonographic appearance of hepatocellular carcinoma in patients with cirrhosis: comparison with contrast-enhanced helical CT appearance, *AJR Am J Roentgenol,* 183(5), 1319–1326.
3. Quaia E, Calliada F, Bertolotto M, Sidhu PS, Leen EL, Basilico R, Pilcher JM, Bushby LH, Hoffmann CW, Harvey CJ, Lynch M, MacQuarrie J, Paul D, Cosgrove DO (2004) Improved characterization of liver lesions with liver-phase uptake of liver-specific microbubbles: prospective multicenter study, *Radiology*, **232**(3), 799–809.
4. Iijima H, Sasaki S, Moriyasu F, Suzuki S, Yoshida M, Horibe T, Tsuchiya K (2007) Dynamic US contrast study of the liver: Vascular and delayed parenchymal phase, *Hepatol Res*, **37**, 27–34.
5. Sugimoto K, Moriyasu F, Kamiyama N, Metoki R, Yamada M, Imai Y, Iijima H (2008) Analysis of morphological vascular changes of hepatocellular carcinoma by microflow imaging using contrast-enhanced sonography, *Hepatol Res*, **38**(8), 790–799.
6. Sugimoto K, Moriyasu F, Saito K, Taira J, Saguchi T, Yoshimura N, Oshiro H, Imai Y, Shiraishi J (2012) Comparison of Kupffer-phase Sonazoid-enhanced sonography and hepatobiliary-phase gadoxetic

acid-enhanced magnetic resonance imaging of hepatocellular carcinoma and correlation with histologic grading, *J Ultrasound Med*, **31**(4), 529–538.

7. Moriyasu F, Itoh K (2009) Efficacy of perflubutane microbubble-enhanced ultrasound in the characterization and detection of focal liver lesions: phase 3 multicenter clinical trial, *Am J Roentgenol*, **193**(1), 86–95.

8. Sugimoto K, Moriyasu F, Shiraishi J, Saito K, Taira J, Saguchi T, Imai Y (2012) Assessment of arterial hypervascularity of hepatocellular carcinoma: comparison of contrast-enhanced US and gadoxetate disodium-enhanced MR imaging, *Eur Radiol*, **22**(6), 1205–1213.

Chapter 10

Fluid Dynamics Field—Drag Reduction of Ships

Yoshiaki Kodama[a] and Munehiko Hinatsu[b]

[a]*CFD Group, Fluids Engineering Department, National Maritime Research Institute, Tokyo 181-0004, Japan*

[b]*Fluid Control Research Group, Fluids Engineering Department, National Maritime Research Institute, Tokyo 181-0004, Japan*

kodama@nmri.go.jp, hinatsu@nmri.go.jp

10.1 Introduction

By injection air through small holes on a surface of a ship advancing in water, the surface is covered with minute air bubbles, and the frictional drag reduces. This phenomenon is called microbubble drag reduction, and research is going on for its application as a clean energy-saving device suited to ships.

Although the word "microbubble drag reduction" gives an impression that micron-order air bubbles are formed, and indeed those generated by electrolysis for laboratory drag reduction experiments are of size 50 μm, those formed through injection on

Micro- and Nanobubbles: Fundamentals and Applications
Edited by Hideki Tsuge

ISBN 978-981-4463-10-2 (Hardcover), 978-981-4463-11-9 (eBook)
www.panstanford.com

ships are of mm-order, mostly ranging from 0.5 mm to 1 mm, and they tend to grow larger by coalescence in downstream. Therefore, it is now more often called air lubrication drag reduction. In this chapter, the word "air bubble drag reduction" is adopted.

Study on air bubble drag reduction was first conducted by McCormick and Bhattacharyya [1]. They generated hydrogen bubbles by electrolysis on copper wires wound on the aft part of an axi-symmetric body of the maximum diameter 0.27 m and obtained drag reduction of a few tens of %. The degree of the drag reduction increased as the electric current increased, but decreased at higher speeds.

In the 1970s, comprehensive air bubble drag reduction study was carried out in the old Soviet Union (Bogdevich et al. [2]). In the 1980s, the study was mainly carried out by the group at the Pennsylvania State University, whose results are compiled in a review paper by Merkle and Deutsch [3]. The experiment by Madavan et al. [4] showed that air bubble drag reduction reaches up to 80%. Figure 10.1a shows a bubble generation plate (102 mm × 78 mm) and a friction measurement plate (102 mm × 254 mm) placed in the channel test section. A sintered stainless steel plate of nominal 1.5 μm pore diameter was used for bubble generation. Figure 10.1b shows the measured frictional drag reduction. In the figure, the horizontal axis shows the amount of injected air in terms of "equivalent air layer thickness" t_a (unit: mm) defined as

$$t_\mathrm{a} = \frac{Q_\mathrm{a}}{U_\infty B_\mathrm{a}} \tag{10.1}$$

where Q_a denotes the volume of injected air per unit time, U_∞ denotes free-stream speed (or ship's advancing speed) and B_a denotes the width of the injection area, resulting that t_a denotes the thickness of the injected air layer assuming that it flows along the solid wall uniformly with the speed U_∞. The parameter serves to understand how much air is injected. The vertical axis of the figure shows the skin friction ratio $C_\mathrm{f}/C_{\mathrm{f}_0}$, where C_f denotes the skin friction with air bubble injection and C_{f_0} without it. The friction reduction effect increases as Q_a increases and reaches 80% at the maximum. Thus, air bubbles have significant skin friction reduction effect. It should be noted that, although the data, taken at immediate downstream of injection point, do not show speed dependence, those data taken at further downstream show smaller reduction effect at higher speeds, as will be shown later.

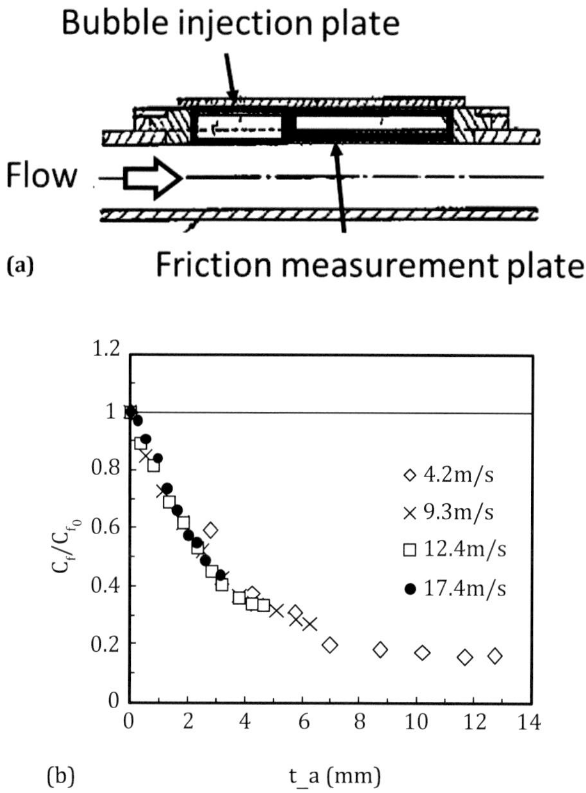

Figure 10.1 Skin friction reduction effect of air bubbles measured at immediate downstream of injection point. (a) Bubble injection plate and friction measurement plate. (b) Skin friction reduction by air bubbles.

Starting 1990s, air bubble drag reduction study has been carried out in Japan. Ships play a dominant role in the transportation of crude oil, iron ore and grain. The largest ship is longer than 400 m and wider than 60 m. The hull shapes are boxy for high loading efficiency, and called full hull forms. These types of ships run relatively slowly (e.g. a typical cruising speed of a large tanker is 14 knots, or 7.2 m/s, resulting that the wave-making drag component is very small and that the frictional drag component is dominant, reaching 80% of the total drag, thus its reduction becomes important. Figure 10.2 shows an image of air bubble drag reduction applied to a full hull form such as tankers. A full hull form has wide and flat bottom and air bubbles injected at the bottom near the bow are expected to stay close to the hull surface by buoyancy as they go downstream, thus covering a wide area and having large skin friction reduction effect.

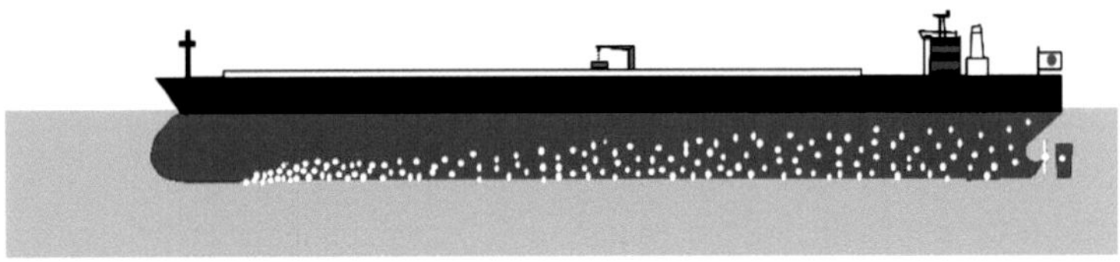

Figure 10.2 An image of air bubble drag reduction applied to a full hull form such as tankers.

10.2 Skin Friction Reduction Mechanism of Air Bubbles

In applying air bubbles to ships, it is important to reduce energy spent to inject them at the bottom of the hull against the hydrostatic pressure; therefore, it is important to elucidate the skin friction reduction mechanism by air bubbles. The research is steadily progressing, and some interesting hydrodynamic phenomena have been clarified. Although they have not yet been used, further progress will lead to significant improvement of air bubbles. In the following, some mechanisms are introduced in order from simple to complicated.

The first mechanism is so-called "density effect". Many experimental evidences show that the higher the local void ratio near the solid wall, that is the more air bubbles near the solid wall, the higher the skin friction reduction effect. This mechanism is easily explained. It may be intuitively explained that the air layer next to the solid wall isolates the wall from surrounding water, and the more detailed explanation is as follows. Suppose that there is flow in the x-direction and there is solid wall at y = constant plane. The shear stress τ acting on a y = constant plane close to the solid wall is expressed as

$$\tau = \mu\frac{\partial \bar{u}}{\partial y} - \rho\overline{u'v'}, \tag{10.2}$$

where the first term of the right-hand side (RHS) consists of μ, the molecular viscosity of water and time-averaged x-component velocity gradient, and the second term is the Reynolds stress, which consists of ρ, the density of the fluid, and the time average of the product of the velocity fluctuations in x- and y-directions. The second term, which is dominant, becomes smaller as the local void ratio becomes higher, resulting in smaller τ. Incidentally this mechanism does not depend on the bubble size. Moriguchi and Kato [5] changed the air bubble diameter from 0.5 mm to 2 mm in their air bubble channel flow experiment and report that the skin friction reduction did not change.

The second mechanism is increase of effective viscosity in particle-laden flow. Einstein [6] has derived a formula that shows that the existence of minute particles (air bubbles in this case) in liquid (water in this case) increases the effective viscosity. Let α be the local void ratio, μ_1 be the molecular viscosity of the liquid and μ_{eff} be the effective viscosity of the liquid with particles. Then,

$$\mu_{\text{eff}} = \mu_1(1 + 2.5\alpha) \tag{10.3}$$

In Eq. (10.2), although this effect increases the first term of RHS by increasing μ, which is actually μ_{eff}, it considerably decreases the 2nd term by decreasing the velocity fluctuations, and the shear stress decreases as a result. According to experimental results collected by Gore and Crowe [7], the presence of solid particles or gas bubbles in turbulent flows increases or decreases the turbulence intensity depending on whether the particles (or bubbles) are large or small. Therefore, small bubbles near a solid wall will decrease turbulence intensity and also Reynolds stresses.

The third mechanism is lifting of near-wall streamwise vortices. Turbulent boundary layers have near-wall streamwise vortices called "streaky structures", which generate Reynolds shear stress, and thus large skin friction. Therefore, if air bubbles are present in a turbulent boundary layer, their size should be compared with the size of the turbulence structure near the solid wall called "wall unit"

$$l \equiv \nu / u_\tau, \tag{10.4}$$

where ν is kinematic viscosity of the fluid, $u_\tau \equiv \sqrt{\tau_w/\rho}$ is called friction velocity and ρ is fluid density. Thus, the parameter for the bubble size should be

$$d_b^+ \equiv \frac{d}{l} = \frac{u_\tau d}{\nu} \tag{10.5}$$

where d is the bubble diameter. Ferrante et al. [8] numerically simulated high Reynolds number turbulent bubbly flows on a flat plate, containing air bubbles of 40 μm diameter ($d_b^+ = 2.4$), and obtained 19%–22% skin friction reduction at α = 1% by moving "streaky structures" away from the solid wall.

Bubble generation using electrolysis can easily generate bubbles of micron orders and is suited to basic studies on the effect of bubble size on the skin friction reduction effect. Ortiz-Villafuerte et al. [9] and Gutiérrez-Torres et al. [10] generated bubbles of 30 μm average diameter in channel flows and investigated the effect of microbubbles on the turbulence structure using PIV. They discuss analogy of their experimental results with the numerical simulation by Ferrante.

The fourth mechanism is turbulence modulation. Kitagawa et al. [11], in a turbulent channel flow with air bubbles of average 0.53 mm diameter, the channel width being 15 mm, the average flow speed being 5 m/s and the average void ratio being 0.5%, measured air bubble motion/deformation and surrounding fluid velocity simultaneously, and elucidated that deformation of air bubbles in wall turbulence does not change the magnitude of turbulence but decreases the Reynolds shear stress. Figure 10.3 shows a measured result. This mechanism suggests that larger bubbles, more easily deformed than smaller bubbles, can reduce Reynolds shear stress, and consequently skin friction, more significantly, which is in contrast to the second mechanism. This mechanism suggests another parameter for the bubble size, i.e. the ratio of wall shear stress and surface tension

$$\frac{\tau_w}{\sigma/d} = \frac{\rho u_\tau^2 d}{\sigma} \equiv We_\tau, \tag{10.6}$$

which is the Weber number based on friction velocity. The experiment by Kitagawa et al. corresponds to $We_\tau = 0.34$. The bubble diameter of 0.53 mm is approximately 130 times the wall unit, and therefore too large to directly interact with the near-wall streaky structure.

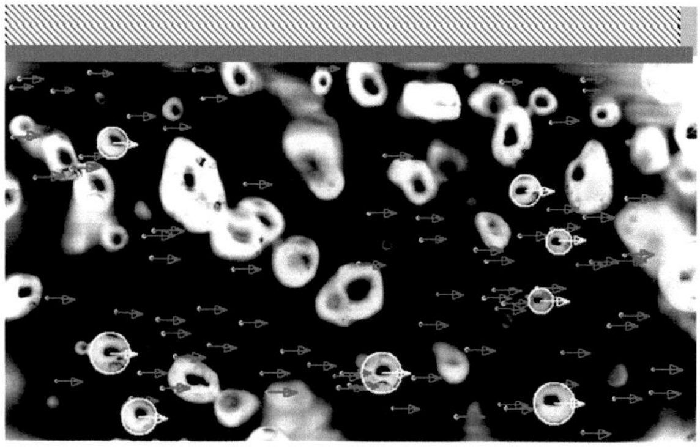

Figure 10.3 Simultaneous measurement of bubble motion/deformation and flow velocity [11].

Serizawa et al. [12] carried out what they called "milky bubbly flow" experiment in a pipe flow. They added a small amount, the void ratio being of the order of a few 0.1%, of air bubbles of 40 μm average diameter in the flow, and obtained significant reduction in skin friction possibly by re-laminarisation. Considering that the phenomenon occurred in the transition regime, the reason for the possible re-laminarisation may be the increase of effective viscosity. It is necessary that bubbles are sufficiently small in this case.

Summarising all the mechanisms listed above, although the research is going on and no conclusions can be drawn, the author thinks that the following comments apply to the skin friction reduction by air bubbles.

1. In case microscopic, i.e. of the same order size with turbulence structures, air bubbles are present in the near-wall region of the boundary layer, the bubbles directly interact with the turbulence structure and there is a chance that significant skin friction reduction is obtained with a small amount of air volume.
2. In case macroscopic, i.e. larger than turbulence structures and large enough to allow deformation against surface tension (1 mm diameter, say), air bubbles are present in the near-wall region, bubble deformation influences the turbulence structure and thus reduces skin friction.
3. In the transitional regime of relatively low Reynolds numbers, increase of effective viscosity caused by microscopic bubbles

relaminarises the flow, and skin friction reduces. This effect cannot be expected in high Reynolds number flows.

4. Density effect applies to bubbles of any size, and if there are a large amount of air bubbles in the near-wall region, significant skin friction reduction always occurs.
5. Effect of co-presence of microscopic and macroscopic air bubbles, as always in real situations, has not been studied yet.

10.3 Sustainability of Skin Friction Reduction Effect in Downstream Direction

Whatever the skin friction reduction effect by air bubbles is, the more bubbles (and thus more air volume) there are, and the closer to solid wall they are, the larger the skin friction reduction effect. Also it is desirable that the skin friction reduction effect sustains in a downstream direction over a long distance. Watanabe et al. [13] towed a flat plate of 40 m length and 0.6 m width in a towing tank at speed of 7 m/s, a typical cruising speed of a large tanker, and carried out bubble experiments. They clarified that the skin friction reduction effect of injected air bubbles sustains over a distance of a few 10 ms. The author and others [14,15] carried out similar experiments in a towing tank of 400 m length and 18 m width by towing a flat plate of 50 m length and 1 m width at speed of 7 m/s. Figure 10.4 shows one of the experiments.

The author and others [16] prevented bubble leakage from the sides of the flat plate by putting a pair of vertical plates called end plates and enhanced two-dimensionality of the flow. Figure 10.5 shows a distribution of the skin friction reduction effect thus obtained. Air bubbles were injected at 3 m from the bow and at the flow rate of $Q = 1.39$ or $1.44\ m^3/min$. The horizontal axis shows the downstream distance from the point of injection, and the vertical axis shows the ratio of the local skin friction with and without injected air bubbles. The skin friction reduction effect reaches a maximum at immediate downstream of the point of injection, and gradually decreases towards downstream. The effect of the end plates is significant, and if applied, the skin friction reduction effect maintains approximately 50% all the way to the downstream end of the measurement. Sanders et al. [17] carried out a similar flat plate experiment at higher speeds (maximum 18 m/s).

Figure 10.4 An air bubble experiment using a flat plate of 50 m length and 1 m width [14,15].

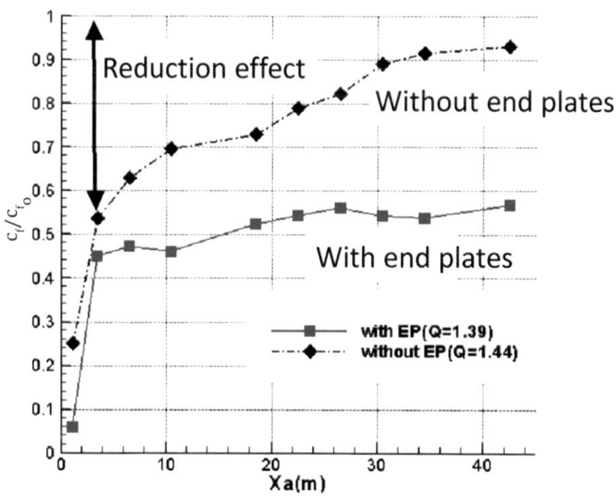

Figure 10.5 Sustainability of skin friction reduction effect of injected air bubbles in downstream direction (towing speed 6.173 m/s) [16].

10.4 Full-Scale Air Bubble Experiment Using Seun-Maru

In September 2001, the SR239 Research Committee of the Shipbuilding Research Association of Japan carried out a full-scale air bubble experiment using a training ship "SEIUN-MARU" of the National Institute for Sea Training. Figure 10.6 shows the bubble injectors attached near the bow. Air bubbles were injected through the holes drilled in the three horizontal parts (two lower ones were aligned) to which air was fed through pipes shrouded in the vertical strut. In the experiment, the drag reduction effect was measured by measuring ship speeds with and without air injection while keeping propeller revolution and course. In case air was injected from all the injectors, the propeller thrust reduced because the injected air was drawn into it, and the ship speed decreased. However, in case air was injected only from the upper horizontal parts, the ship speed increased and the engine power decreased, resulting in 3% power reduction if the ship speed was maintained at 14 knots, which was amazing because the hull had hardly any horizontal flat bottom. In this case, the necessary power to inject air against the hydrostatic pressure at 1 m below water level was 1%, and therefore the net power gain (although nominal because the mechanical loss was not taken into account) was 2%. Thus, it was shown that skin friction reduction by air bubbles can be obtained on a full-scale ship.

Figure 10.6 Bubble injectors (BI-1 through BI-3) attached near the bow of SIRUN-MARU [18].

10.5 Full-Scale Air Bubble Experiment Using a Cement Carrier

Another full-scale air bubble experiments were carried out by the authors using a cement carrier (Fig. 10.7). A cement carrier, which carries material of relatively low density, has a full hull form, and therefore suited to air bubble drag reduction. In the first experiment [20] using the cement carrier in 2006, the drag reduction effect remained only 1% because the bubble injectors were placed on the sides and the injected bubbles hardly reached the hull bottom. In the second experiment [21,22] carried out in 2008, when the bubble injectors were located at the flat bottom, the net energy-saving rate of 5.3%, as an average of the full load and the ballast (light load) conditions, the nominal power for air injection against hydrostatic pressure (adiabatic compression) being taken into account, was obtained. Table 10.1 shows the relation between the number of blowers used for bubble injection and the net energy-saving rate (adiabatic compression). Based on the measured changes in the propeller thrust, the maximum drag reduction rate by bubble injection was estimated as 11% in the ballast condition. Thus, it has been demonstrated that air bubble injection has a potential sufficient as a drag reduction device for ships. But, at the same time, there were cases in which stern vibration increased by bubble injection. Murakami et al. [23] carried out numerical simulation of the bubbly flow corresponding to the second experiment.

Figure 10.7 The cement carrier "PACIFIC SEAGULL" used in the full-scale air bubble experiments [21] (L_{pp} = 120 m, width = 21.4 m, gross tonnage = 7809 tons, draft at full load = 7.215 m).

Table 10.1 Net energy-saving rate of a cement carrier by bubble injection [22].

Number of active blowers	2	3
Ballast condition	5.2%	6.4%
Full-load condition	3.6%	4.1%
Average	4.4%	5.3%

10.6 Towards Practical Application to Ships

Although it was a significant progress that net energy saving rate of 5.3% was obtained by air bubble injection in full-scale experiments, further progress is needed to apply the method to large ships with depth larger than 20 m, which are the majority of ships in world transportation. Also, the stern vibration problem met with the cement carrier needs special care, because injected air bubbles tend to coalesce in high wake zone, and because ship hull form is designed such that a propeller operates in such high wake zone. Kawashima et al. [24,25] carried out propeller test in bubbly flows, where a model propeller of the cement carrier was used. They obtained a result that the propulsive efficiency of a propeller near the operation point is hardly deteriorated by air bubbles.

Ship hull forms are curved in many places, and understandings on the effects of surface curvature and pressure gradients on the skin friction reduction effect are necessary. Kawakita and Takano [26] realised those conditions in their experiment in a cavitation tunnel, and showed that the skin friction reduction effect in favourable pressure gradient is the same as that at zero pressure gradient, and that it increases in adverse pressure gradient, especially in the aft part of convex surface.

Currently, several projects to install bubble injection devices for commercial use in large ships are under way in Japan. The related recent publications are Hinatsu M [27], Mizokami S et al. [28] and Mizojiri T et al. [29].

References

1. McCormick ME et al. (1973) Drag reduction of a submersible hull by electroysis, *Naval Engineers Journal*, **85**(2), 11–16.

2. Bogdevich VG (1977) Gas-Saturation Effect on Near-Wall Turbulence Characteristics, in *2nd International Conference on Drag Reduction*, BHRA Fluid Engineering, Paper D2, Cambridge, UK, 25–37.
3. Merkle CL, Deutsch S (1990) Drag reduction in liquid boundary layers by gas injection, *Progress in Astronautics and Aeronautics,* **123**, AIAA, 351–412.
4. Madavan NK, Deutsch S, Merkle CL (1984) Reduction of turbulent skin friction by microbubbles, *Physics of Fluids*, **27**(2), 356–363.
5. Moriguchi Y, Kato H (2002) Influence of microbubble diameter and distribution on frictional resistance reduction, *J Mar Sci Technol,* **7**(2), 79–85.
6. Batchelor GK (2000) *An Introduction to Fluid Dynamics*, Cambridge Mathematical Library.
7. Gore RA, Crowe CT (1989) Effect of bubble size on modulating turbulent intensity, *Int J Multiphase Flow,* **15**(2), 279–285.
8. Ferrante A, Elghobashi S (2004) On the physical mechanisms of drag reduction in a spatially developing turbulent boundary layer laden with microbubbles, *J Fluid Mech*, **503**, 345–355.
9. Ortiz-Villafuerte J, Hassan YA (2006) Investigation of microbubble boundary layer using particle tracking velocimetry, *J Fluids Eng*, **128**, 507–519.
10. Gutiérrez-Torres CC, Hassan YC, Jimenez-Bernal JA (2006) Turbulence structure modification and drag reduction by microbubble injections in a boundary layer channel flow, *J Fluids Eng*, **130**, 507–519.
11. Kitagawa A, Hishida K, Kodama Y (2005) Flow structure of microbubble-laden turbulent channel flow measured by PIV combined with the shadow image technique, *Experiments in Fluids* **38**, 466–475.
12. Serizawa A et al. (2003) Laminarization of Microbubble Containing Milky Bubbly Flow in a Pipe, paper presented at 3rd European-Japanese Two-Phase Flow Group Meeting, Certosa di Pontignano.
13. Watanabe O, Masuko A, Shirose Y (1998) Measurements of drag reduction by microbubbles using very long ship models, *J of the Society of Naval Architects of Japan*, **183**, 53–63 (in Japanese).
14. Takahashi T, Kakugawa A, Kodama Y et al. (2001) Experimental Study on Drag Reduction by Microbubbles Using a 50 m-long Flat Plate Ship, paper presented at Proceedings of TSFP-2, Second International Symposium on Turbulence and Shear Flow Phenomena.
15. Takahashi T, Kodama Y, Makino M et al. (2003) Experimental study on drag reduction by microbubbles using three kinds of air injection

plates, in *Conference Proceedings of the Society of the Naval Architects of Japan,* **2**, 121–122 (in Japanese).

16. Kawashima H, Kawashima H, Hori T et al. (2006) An experiment of the bubble injection method for frictional drag reduction using 50 m flat plate with end plates, in *Conference Proceedings of the Japan Society of Naval Architects and Ocean Engineers,* **3**, 285–288 (in Japanese).
17. Sanders WC, Winkel ES, Dowling DR et al. (2006) Bubble friction drag reduction in a high-Reynolds-number flat-plate turbulent boundary layer, *J Fluid Mech*, **552**, 353–380.
18. Kodama Y, Kakugawa A, Takahashi T et al. (2002), A full-scale experiment on microbubbles for skin friction reduction using "Seiun-maru" Part 1: the preparatory study, *J of the Society of Naval Architects of Japan*, **192**, 1–13 (in Japanese).
19. Nagamatsu T, Kodama Y, Kakugawa A et al. (2002) A full-scale experiment on microbubbles for skin friction reduction using "Seiun-maru" Part 2: the full-scale experiment, *J of the Society of Naval Architects of Japan*, **192**, 14–27 (in Japanese).
20. Kodama Y, Hori T, Kawashima H et al. (2005) A full scale microbubble experiment using a cement carrier, in *Proceedings of the 5th General Meeting of the National Maritime Research Institute*, 165–168 (in Japanese).
21. Hinatsu M, Kodama Y, Hor, T et al. (2008) A full-scale air lubrication experiment using a large cement carrier for energy saving (preparation), in *Conference Proceedings of the Japan Society of Naval Architects and Ocean Engineers*, **6**, 161–162 (in Japanese).
22. Kodama Y, Hinatsu M, Hori T et al. (2008) A full-scale air lubrication experiment using a large cement carrier for energy saving (result and analysis), in *Conference Proceedings of the Japan Society of Naval Architects and Ocean Engineers*, **6**, 163–166 (in Japanese).
23. Murakami A, Kawamura T, Hinatsu M (2008) Numerical simulation of flow around a full scale ship equipped with bubble generators, in *Conference Proceedings of the Japan Society of Naval Architects and Ocean Engineers,* **6**, 153–156 (in Japanese).
24. Kawashima H, Hinatsu M, Kodama Y et al. (2009) An experimental study on propeller characteristics in bubbly flows Part 1: measurement of local void ratio and velocity field at propeller with bubbly flow, *J of the Japan Society of Naval Architects and Ocean Engineers,* **9**, 63–70 (in Japanese).

25. Hinatsu M, Kawashima H, Kodama Y et al. (2009) An experimental study on propeller characteristics in bubbly flows Part 2: result of propeller open water test in bubbly flow, *J of the Japan Society of Naval Architects and Ocean Engineers*, **9**, 71–78 (in Japanese).

26. Kawakita C, Takano S (2000) Microbubble skin friction reduction under the influence of pressure gradients and curved surface, *J of the Society of Naval Architects of Japan*, **188**, 11–21 (in Japanese).

27. Hinatsu M (2011) Energy saving in ships through air lubrication, *J of Japan Institute of Marine Engineering*, **46**(6), 68–74 (in Japanese).

28. Mizokami S, Kawakita C, Kodan Y et al. (2010) Development of Air lubrication system snd verification by the full scale ship test, *J of the Japan Soceity of Naval Architects and Ocean Engineers*, **12**, 69–77 (in Japanese).

29. Mizojiri T, Fujita H, Koyama H et al. (2011) An application of air lubrication method to 28,000DWT bulk carrier as energy saving system, in *Conference Proceedings of The Japan Society of Naval Architects and Ocean Engineers*, **12**, 425–428 (in Japanese).

Part 3

Application of Nanobubbles

Chapter 11

Nanobubbles: An Introduction

Masayoshi Takahashi
National Institute of Advanced Industrial Science and Technology, Tsukuba, Ibaraki Japan
m.taka@aist.go.jp

11.1 Fundamental Properties of Nanobubble

Bubbles are gas bodies surrounded by water, and ordinary bubbles rise to the water surface and burst. These bubbles are common and familiar to us, but it has recently become clear that smaller bubbles have a potential for new applications that will have a great impact on our daily life as well as on a wide variety of industries.

Microbubbles are tiny bubbles with diameter of less than 50 μm, and the bubbles decrease in size and eventually disappear underwater because of the rapid dissolution of the interior gas. Smaller bubbles shrink faster, and so those with nanoscale dimensions disappear almost instantaneously. However, our experimental studies have suggested that stabilised nanobubbles are created when microbubbles collapse in aqueous electrolyte solutions. Also, we are now realising several astonishing properties of the aqueous solutions

Micro- and Nanobubbles: Fundamentals and Applications
Edited by Hideki Tsuge

ISBN 978-981-4463-10-2 (Hardcover), 978-981-4463-11-9 (eBook)
www.panstanford.com

after the dispersion of microbubbles in the water. Microbubbles has been started to be used in the field of agriculture in around 2000, and the good results have been announced in soil cultivation of tomatoes and hydroponic cultivation of root crops by farmers. However, the mechanism for the improvement of productivity was unknown. We had been attempting to interpret the mechanism based on the dispersed microbubbles, but it was difficult to expect the direct effect of microbubbles on the plans. That is because most of microbubbles tend to be disappeared without reaching to the roots of the plants. In addition, through laboratory experiments with a liquid particle counter, we have been recognising the sign of the existence of stabilised tiny objects smaller than the ordinary size distribution of microbubbles. As shown in Fig. 11.1, while the microbubble generator was operating, the observed distribution showed a broad peak around 10 μm. But after the stoppage of the generator, we recognised about the appearance of a sharp peak near the lower limit of the measurement of 2 μm. This peak grew up in accordance with the disappearance of the peak of microbubbles, and thereafter the peak maintained for a long period of time. The particles indicated as

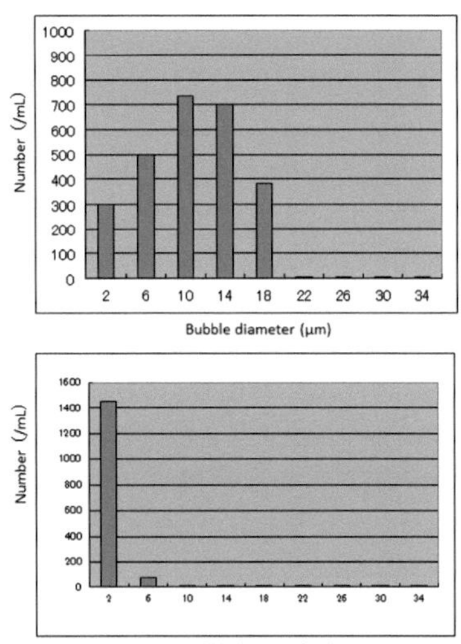

Figure 11.1 Size distribution of microbubble. The upper diagram shows the distribution during the operation of microbubble generator, and the lower diagram shows the distribution after the stoppage of the generator.

the peak were stabilised with half-life of from several hours to several days depending on the water conditions, but they tended to disappear rapidly by the operation of the microbubble generator. Considering these results, we have been proposing the ideas of the generation of stabilised tiny bubbles as residues of collapsing microbubbles. Also, since the tiny bubble should be unstable because of the rapid gas dissolution due to the pressurised interior gas, we needed more specific information about the mechanism of stabilisation. Through the zeta potential observation during the collapsing microbubbles, it has been demonstrated that the ionic cloud around the shrinking gas–water interface of the bubble might contribute to stabilise the residues by suppressing the gas dissolution to the surrounding water.

In various fields of science, the term of nanobubble is not very novel. The formation of nanobubbles at solid–liquid interfaces has been studied using the atomic force microscopy imaging technique. These bubbles are tentatively stabilised, and many researchers are trying to clarify the mechanism of stability. In addition, it is conventionally used for air bubbles less than 1 μm which may be generated by ultrasound radiation or formed under the stabilised effect of surfactant. On the other hand, the bubbles presented here are the tiny bubbles generated by microbubble dispersion in an aqueous solution. These bubbles are tentatively stabilised as bulk nanobubbles, but the patented inventions have enabled to stabilise them more strongly. REO Institute Co., Ltd. and National Institute of Advanced Industrial Science and Technology (AIST) have also succeeded in the long stability by introducing the collapsing technology of microbubbles. They have a long experience in waste water treatment by the microbubble technology regarding the generation of hydroxyl radicals through collapsing microbubbles. They found the method of condensing the ionic cloud around the gas–water interface of collapsing microbubble. The patented work use a physical stimulus to increase the collapsing speed of microbubble and consequently to form a rigid ionic cell around the bubble for the suppression of gas dissolution from the bubble. Two types of nanobubble are now commercially available; oxygen-nanobubble and ozone-nanobubble. They have different functions, and the application research is now underway in a variety of fields such as medical and food industry. Nanobubble is invisible, and its smallness leads the difficulty of measurement of nanobubbles. Considering the further applications of nanobubble the evaluation method will be one of the most important targets of the research work; so, AIST is working to establish a measurement technique depending on the understanding of the basic properties of nanobubbles.

11.2 Stabilisation Mechanism of Nanobubbles

Because bubble is a gas body surrounded by water, the interior gas is pressurised because of the surface tension of the gas–water interface. The interior gas of smaller bubble is more strongly compressed, and tiny bubbles shrink under water owing to the dissolution of the gas. So it is obvious, from a purely physical point of view, that a tiny bubble with nanoscale dimension disappears almost instantaneously. But now we are realising the stabilised nanobubbles in aqueous solutions after the dispersion of microbubbles.

We are now considering the effect of surrounding ions around the bubble as the mechanism of stabilisation. Surface electrical charge is an important factor in understanding the properties of microbubbles and nanobubbles. In an electrophoresis cell, microbubbles move towards the electrode with an opposite electrical charge. The surface charge of each microbubble can be determined from the speed and direction of its movement in the presence of an electrical potential and can be evaluated by the value of the ζ potential. In distilled water, microbubbles are electrically charged to a ζ potential of approximately −35 mV. This value changes according to the precise conditions of the water. Adsorbed OH^- and H^+ are crucial factors influencing the gas–water interface charge; electrolyte ions are attracted to the interface by the electrostatic force and generate an electrical double layer. The ζ potentials of microbubbles under specific water conditions are similar, regardless of their size. Therefore, the amount of electrical charge around the gas–water interface is the same per unit area. However, observations of collapsing microbubbles over time have shown that the ζ potential increases according to the rate of shrinkage, which is itself inversely proportional to the bubble size (Fig. 11.2). These observations suggest that the rate of movement of electrolyte ions in water is not sufficiently high to counteract the increasing rate of shrinkage of the microbubbles. Consequently, it is likely that some excess ions become trapped at the gas–water interface, thereby increasing its ζ potential during the process of collapse. At present, there are insufficient data to conduct a detailed analysis. However, the accelerative increase of the ζ potential of a microbubble suggests the extreme accumulation of ions during the final stage of the collapse process. Salting out is a well-known effect in which the solubility of a substance in a certain solvent is reduced by the presence of a second solute dissolved in the solvent. Also, in the case of the accumulated ionic cloud of collapsing microbubble, the electrolyte ions squeeze out the gas molecule from the surrounding region of the bubble and consequently the gas dissolution from the bubble is strongly suppressed. This is a present explanation of the stability of nanobubble in aqueous solutions.

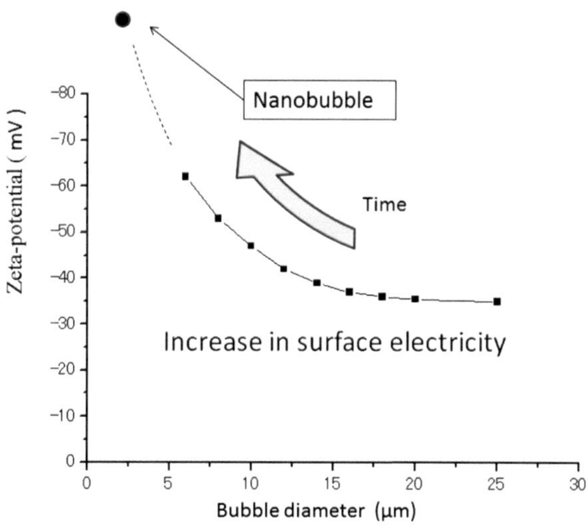

Figure 11.2 Nanobubble generation as a residue of *collapsing microbubble*. The increase of zeta-potential means the accumulation of ions at the shrinking gas–water interface.

11.3 Measurement of Nanobubbles

Considering the application of nanobubble in a wide variety of fields, it is very important to establish the measuring method of nanobubble. The nanobubbles generated by collapsing microbubble might be analysis by laser diffraction depending on the generating condition. The peak may be confirmed in 300 nm to 1 μm level. On the other hand, for the bubbles stabilised at smaller level, we need other measuring devices such as Dynamic Light Scattering (DLS). But unfortunately these measuring devices for nanoparticles principally cannot distinguish between solid particles and gas bubbles, so we must make a decision from a comprehensive perspective, taking account of not only these optical methods but also the measurement of another physical phenomenon related to nanobubbles.

11.3.1 Dynamic Light Scattering

Suspended particles and emulsions in aqueous solutions undergo Brownian motion induced by the bombardment by water molecules, which themselves are moving, depending on their thermal energy. If the particles are illuminated with a laser, the intensity of the scattered light fluctuates at a rate that is dependent upon the size of the particles and thc smaller particles are moved more rapidly. By analysing the intensity of fluctuations, we can evaluate the size distribution of the particles.

Ozone-nanobubble was used as a sample. As a pretreatment we filtered a sample by a membrane filter of aperture diameter 100 nm to exclude the influence of the suspended solids or floating impurities. Ozone-nanobubble is pinkish colour indicating its oxidation ability, and the filtration did not change the colour; without the filtration as the pretreatment, the convergence of the DLS data of the sample was not good because of the influence of the solid particle. As a result of the DLS measurement of the pretreated sample, a clear peak was confirmed to 100–200 nm as shown in Fig. 11.3. Because we could not distinguish between the signal derived from solid particles and that from gas bubbles, the DLS data are not self-sufficient. But these data have important meaning to evaluate the invisible target, a nanobubble. It is strange that the peak was recognised at the position bigger than 100 nm although the sample had passed through a filter of 100 nm. We are considering the two possible reasons about that.

One could be the flexibility of gas bubble. The bubble might go through the smaller hole than the bubble size by changing its shape. But it is hard to think about this phenomenon because tiny bubble is pressurised by the surface tension. Since the interior gas pressure of the bubbles is inversely proportional to the bubble diameter due to the Young–Laplace equation, the bubble must not be deformed at the filtration. The bubbles of 100 nm in diameter are pressurised to approximately 3 MPa. In contrast, the driving pressure at the filtration is less than 0.1 MPa to suck the water through the filter. It may not happen that the strongly pressurised bubble change its form by the weak suction pressure. At all events, it is hard to assume that the sample could contain the bubble larger than 100 nm in diameter after the filtration.

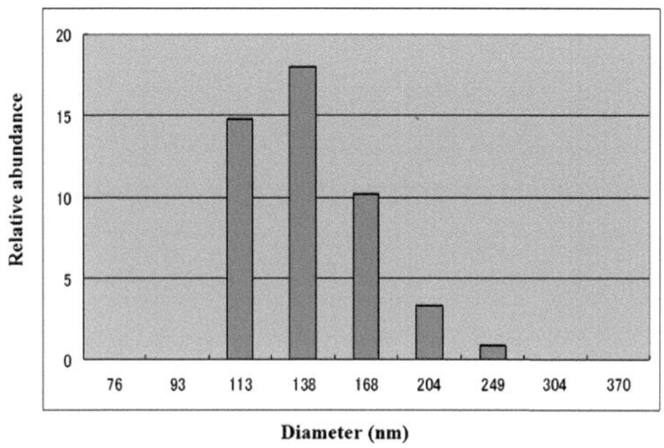

Figure 11.3 Size distribution of ozone-nanobubble measured by DLS after filtration through a 100 nm porous membrane.

Another possible reason for the appearance of peak at larger than 100 nm may be related to the analytical method of DLS. The DLS measurements are depending on the Brownian motion of the particles, and the analysis software must not intend for the observation of gas bubbles. If the Brownian motion of a gas bubble was different from that of a particle having the same size with the bubble, we could not get the correct size distribution of nanobubbles in aqueous solutions.

If we permit the data including some errors, the DLS measuring system will be a useful tool for the evaluation of nanobubbles. But since it is difficult to distinguish gas bubbles depending only on other particles precisely, we cannot conclude the presence of nanobubble form a provided data of DLS. Since other measuring systems for particle distribution have the same problem, we are now trying to establish the evaluation method of nanobubble by taking account of another physical phenomenon, the bubble as well as these measuring data such as DLS.

11.3.2 Measurement by Electron Spin Resonance

Since nanobubble is a residue of collapsing microbubble, nanobubble occasionally shows the same properties as that of microbubble. One of the most significant phenomena of microbubble is a radical generation at the extinction of the collapsing bubble under the specific conditions. So, we are trying to evaluate quantitatively the nanobubbles depending on the electron spin resonance (ESR) data.

As described above, the electricity of the gas–water interface of microbubble is increasing during the collapsing process because of the accumulation of adsorbed ions. This is also relating to the main reasons of the stability of nanobubble, but in the case of collapsing microbubble the extinction of the gas–water interface lead the dispersion of surface energy as the radical generation. The details of the phenomenon are as follows. The accelerative increase of the zeta potential of a microbubble suggests the extreme accumulation of ions during the final stage of the collapse process; the extinction of the gas–water interface might cause a drastic environmental change for the ions originally concentrated at the bubble surface. The extinction of the gas–water interface might trigger radical generation via dispersion of the elevated chemical potential that has accumulated around the interface.

Figure 11.4 Observation of Brownian movement of ozone-nanobubbles visualised by NanoSight.

At the experiment of radical generation from nanobubble, 5,5-dimethyl-1-pyrroline N-oxide (DMPO) was used as the spin-trap reagent because it is appropriate for observation of oxygen-centred free radicals such as hydroxyl radicals. The samples of oxygen-nanobubble and ozone-nanobubble were used after having mixed with DMPO for the radical observation with an ESR spectroscopy. As the results of the tests, the signal of DMPO-R was observed from oxygen-nanobubble and DMPO-OH from ozone-nanobubble. The signal of DMPO-OH was also obtained from oxygen-nanobubble by the addition of hydrochloric acid at the mixing with DMPO. Since the signal of DMPO-R and DMPO-OH indicated the generation of alkyl radicals and hydroxyl radicals, respectively, at the extinction of gas–water interface, mixing with the reagent and the acid might cause the destabilisation of nanobubble and consequently the radical generations.

If the generation of free radicals is a particular phenomenon of tiny bubbles, we might confirm the presence of nanobubble by measuring the ESR signals. There are other sources of radical generation such as hypochlorous acid in tap water, so a study is needed to delete the error factor at the radical measurement for the quantitative evaluation of nanobubbles.

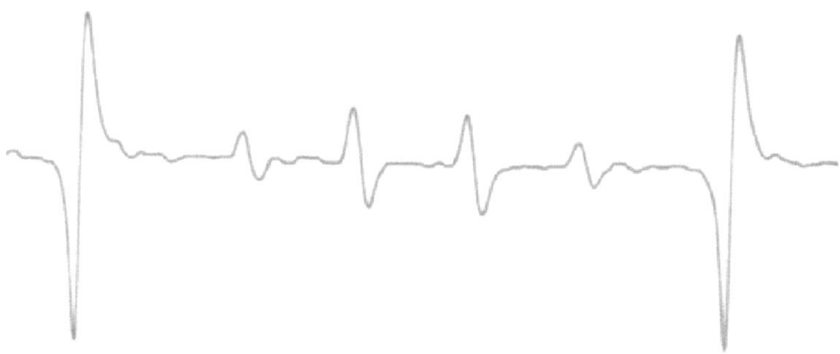

Figure 11.5 Radical generation at the extinction of *oxygen-nanobubble* by a chemical stimulus.

11.4 Conclusion

The nanobubble is the object that we should consider as one of the tools like a microbubble. Therefore, it is the most important subject of the future research work to establish the methods how to use the nanobubble in the fields of science and technology. The technical accumulations have been accomplished about the generation methods and the evaluation techniques of the nanobubbles. On the contrary, it is not easy to realise the applications because we need deeper insight for nanobubble to use it in the practical usages. So, it was a big pleasure for me that an opportunity to ask the superior experts in the field of nanobubble to introduce us to their knowledge, and I hope the information in this book will help you to develop the nanobubble technology further more.

Chapter 12

Medical Field

12.1 Medical Application of Nanobubble Water

Yoshihiro Mano
Hyperbaric Medical Center, University Hospital of Medicine,
Tokyo Medical and Dental University, Tokyo, Japan

12.1.1 Introduction

Nanobubble water is known to cause mysterious phenomena. For example, it enables not only saltwater fish and fresh water fish, but also tropical and deep-water fish to live in the same environment water. In nanobubble water, in addition to aquatic plants in general, a wide variety of plants from the Phalaenopsis orchid to cactus can live and grow under the conditions beyond the threshold for their survival. What is the origin of this world of nanobubble water—the world that makes it possible for all the organisms on earth to live in the same environment?

A somewhat radical shifting of perspective, or even reversal of perspective, makes it easier to understand the following explanation. Tracing back the Darwinian theory of evolution, when living organisms emerged on the earth, that is, when monocellular organisms appeared three billion and several hundred million years ago, these organisms are thought to have lived in water. It is speculated that these organisms were able to survive because the habitat environment was extremely protective and enabled these

Micro- and Nanobubbles: Fundamentals and Applications
Edited by Hideki Tsuge

ISBN 978-981-4463-10-2 (Hardcover), 978-981-4463-11-9 (eBook)
www.panstanford.com

organisms to have unimaginably potent immune capacities. I believe that these first organisms gradually differentiated into different species adapting to one new environment after another in the evolutionary process as described in Darwinian theory of evolution. The evolutionary process is thought to involve renewal of immune capacity and resistance in accordance with new environment. When it happens, early immune capacities cease to function and are updated to work best for new species. Humans have undergone such transformation most extensively among all living organisms, but the potent immune function that existed when living organisms first emerged has become unavailable. Then, did we lose the early potent immune capacity and resistance in exchange for various new capacities we obtained in the evolutionary process? I do not think it is the case. Instead, I visualise these immune capacities stored in a locked drawer. I speculate that this drawer is unlocked and original functions come alive in the environment consisting of nanobubble water, that is, they are maintained in the environment I described above as one that is extremely protective and enables the organisms to exert unimaginably potent immune capacities. Thus, I believe that the environment consisting of nanobubble water is one of the distinctive environments that made the emergence of living organisms on the earth possible. An example of such an environment would be a volcanic crater on a deep seabed, which is also called "hydrothermal deposit (hydrothermal vent)". It develops from under the seabed, forming a structure called "chimney", and continues even today to shoot up black and white smoke (Fig. 12.1).

Figure 12.1 Active chimney by heated water from a deposit presented by JaMSTeC.

In the environment around hydrothermal deposits, high-temperature gas heated by geothermal heat (4500°C) shoots up from the seabed several thousand meters under water. This environment is considered similar to the flow field that causes crush necessary for the creation of nano nuclei (Fig. 12.2). While the production of nanobubble water requires the flow field that resembles the flow field created when extremely high pressure and extremely high temperature were applied instantly, such an environment is always created around the chimneys of hydrothermal deposits. Studies on deep seabed revealed that the areas around these chimneys were forming living spheres for rudimentary small organisms. This is the environment that triggered the emergence of living organisms on the earth. These organisms gradually evolved and then further evolved into the organisms suited to various environments as explained in the Darwinian theory of evolution. Thus, I speculate that the environment consisting of nanobubble water brings back the extremely potent immune capacities that are considered to exist in the living organisms when they came into being. If this is the case, the environment consisting of nanobubble water, which can revive the strong homeostasis of organisms, would also function to restore the part of physical function becoming inactive, which is called disease. Therefore, I believe that nanobubble water can be applied in medicine in the form of the liquid that has various capacities including preservation (Fig. 12.3), repair (Figs. 12.4 and 12.5) and regeneration (Figs. 12.5 and 12.6) of tissue and bactericidal action (Figs. 12.7 and 12.8).

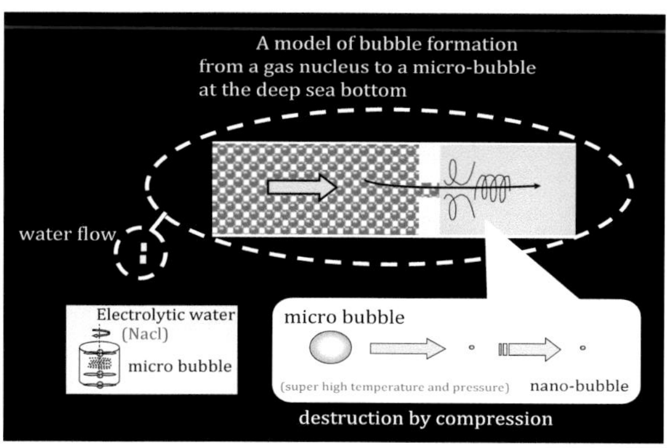

Figure 12.2 Manufacture method of nanobubble.

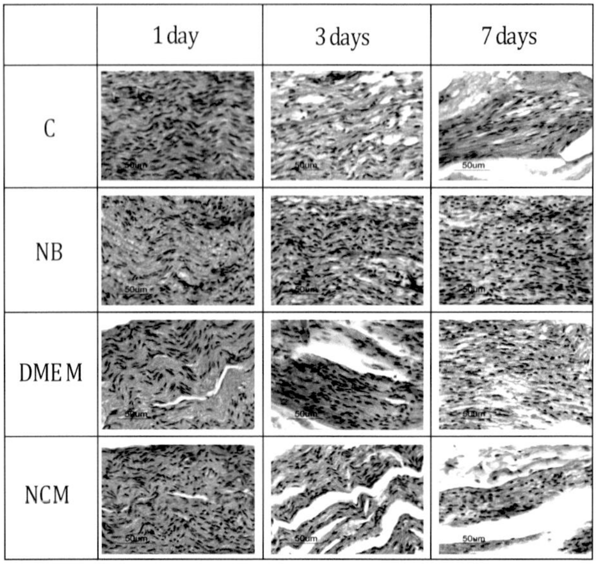

Figure 12.3 Tissue Preservation by NB compared with others N. Vagus of a rat (H.E. Dyeing : × 400). C Solution of Salt; NB Oxy-Nano-Water; DMEM Dulbecco's Modified Eagle Medium; NCM Tissue preservation at 4°C in each liquids for 1, 3, and 7 days and then fixed and preserved again for a week at 4°C, then sectioned.

	Pre	1 week	2 weeks
Whole	2.32 ± 0.30	2.06±0.26	2.02 ± 0.25
Maxilla	2.30 ± 0.38	0.08 ± 0.32	2.00 ± 0.29
Mandible	2.34 ± 0.47	2.05 ± 0.41	2.03 ± 0.41
Left jaws	2.36 ± 0.46	2.07 ± 0.40	2.04 ± 0.39
Right jaws	2.29 ± 0.06	2.06 ± 0.35	1.99 ± 0.33

	Improvement rate	improved sites/total sites
Whole	61.8%	139/225
Maxilla	66.7%	74/111
Mandible	58.8%	67/114
Left jaws	64.1%	75/117
Right jaws	59.3%	64/108

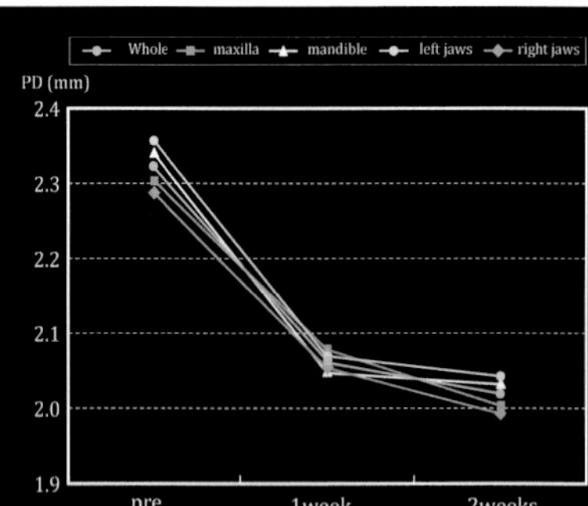

Figure 12.4 Improvement of Probing Depth (mm) (Two 20 s rinses per day using Ozone-Nano water).

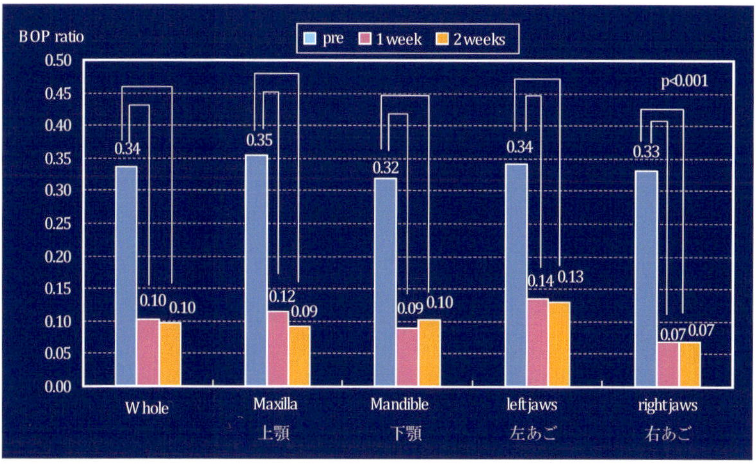

Figure 12.5 Changes in Bleeding on probing rate (Two 20 s rinses per day using Ozone-Nano water).

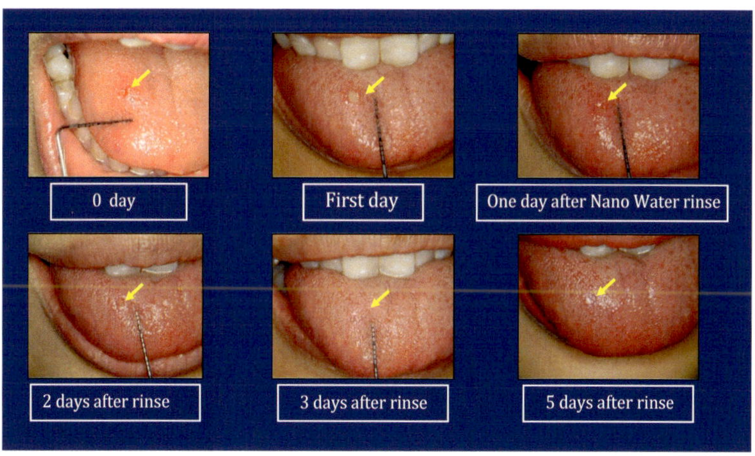

Figure 12.6 Recurrent Aphtha stomatitis. Sense of incompatability occurs (0 day), Aphtha stomatoplastic pain (first day). Pain disappears with a 20 s rinse using Oxy-Nano water a few times per day (one day after), Aphtha almost disappeared (3 days after). Recovery (5 days after).

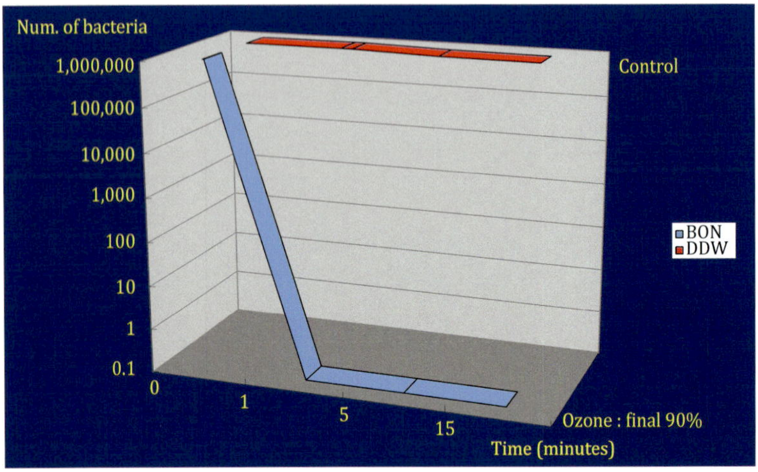

Figure 12.7 Staphylococcus aureus N315.

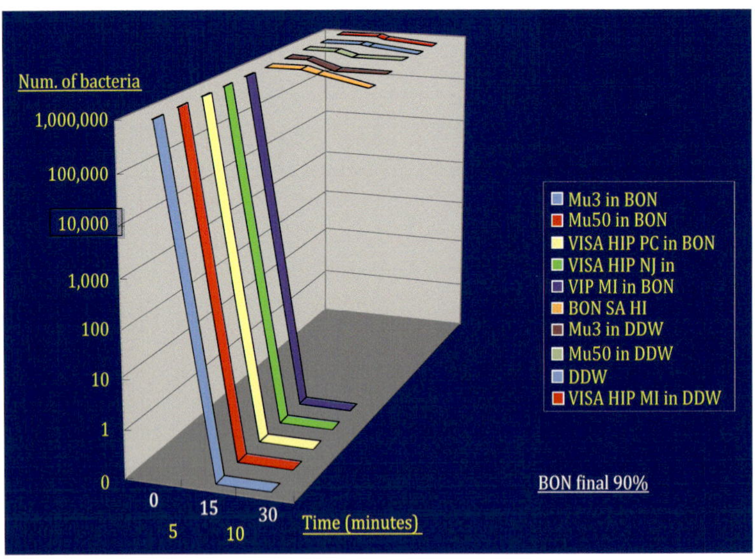

Figure 12.8 Multidrug resistant Staphylococcus aureus.

12.1.2 Strong Capacity of Nanobubble Water to Preserve Tissue

An example of the application of the capacity for tissue preservation is use of nanobubble water in the production of Sasakamaboko (fish cake made of fish paste which is a local specialty of Sendai, Japan). By just adding a small amount of oxygen nanobubble water in fish paste in the manufacturing process of Sasakamaboko, the myofibril of fish paste is preserved without being broken even though no food additives are used. This helps maintain the taste when it is manufactured. Moreover, no growth of bacteria including thermoduric spore-forming bacteria was noted in a two-week study on shelf life. Thus, by using nanobubble water, safe and fresh unperishable Sasakamaboko is shipped out from manufacturers (Fig. 12.9).

Boiled fish paste 10°C day keeping Test result (Oxygen nano bubble Addition products/Preservative additive-free)

	Coliform group	Standard bacteria	Lactic acid bacteria	Heat-resistant bacterial spore bacillus	Bacillus cereus
Start	–	–	–	–	–
Day 4	–	–	–	–	–
Day 6	–	–	–	–	–
Day 9	–	–	–	–	–
Day 14	–	–	–	–	–

(Unit ; CFU/g – ; It has not been detected (<10) I n : outside inspection body)

Figure 12.9 Results by bacillus test of Kamaboko under 10°C. Daily preservation test using only nanobubble water but not using preservation drugs. Every bacterium included sporo-bacterium was killed by nanobubble water; so, it is possible to make preserve Kamaboko without preservation drugs.

This capacity for tissue preservation is not limited to muscular tissue (myofibril). In a study on rats, the vagus nerve was dissected and cultured in four different environments, in physiological saline which was used as control, oxygen nanobubble water (NB), a common tissue preservation solution or a tissue preservation solution for nerve tissue for 24 h, three days and seven days as shown in Fig. 12.3. In the nerve cultured in NB, cell nuclei were shown to best maintain their original conditions and tissue as a whole was also firmly maintained, indicating the strong preserving capacity of NB [1]. This result suggested much stronger capacity of NB in preserving body tissue compared with common tissue preservation agents (Fig. 12.3). In another example, when oysters are frozen at the temperature of −20°C or lower, kept completely frozen for a whole day and then defrosted, they resume filtration (Fig. 12.10). This indicates that oysters are still alive but in so-called hibernating state when they are frozen and show regular vital reactions when they are defrosted.

Figure 12.10 Preservative oyster under −20°C for 48 h (upper) and the result after that (there are arrive after that).

From this result, we can speculate that tissue destruction is unlikely to happen in NB water and the potency of this water in tissue preservation is maintained at a level sufficient for the activation of vital reactions. Therefore, NB water might very well be able to preserve various organs for an extended period of time and can become a powerful tool for future medical transplantation.

12.1.3 Capacity for Sterilisation, Repair and Regeneration

In addition to tissue preservation capacity, NB water has various capacities including repair (Figs. 12.4 and 12.5) and regeneration (Figs. 12.5 and 12.6) of tissue and bactericidal action (Figs. 12.7–12.9).

A study on its effects on periodontal disease conducted to explore its tissue repair capacity demonstrated that rinsing the mouth with ozone NB water twice per day in the morning and evening not only made periodontal pockets significantly and evenly shallower regardless of the position of affected teeth in a week (Fig. 12.4), but also markedly improved bleeding. Practicing dentists are already using this water for measures against periodontal disease (Fig. 12.5) [2–4]. When the patients were asked to rinse their mouths with oxygen NB water twice a day after a bout of stomatitis, the pain disappeared after rinsing once or twice and ulcerative lesions were healed in two or three days, showing the acceleration of tissue repair (Fig. 12.6) [5,6].

Moreover, when the bactericidal action of ozone NB (90%) solution was examined, both *Staphylococcus* and its multidrug resistant strains were almost completely killed (Figs. 12.7 and 12.8) and comparable sterilisation against *Escherichia coli*, *Enterococcus*, *Pseudomonas* and other bacteria including their multidrug resistant strains was observed [6]. Therefore, this capacity would be useful when applied in cleaning and disinfection of medical devices or sterilisation apparatus. In addition, ozone NB water is highly effective in eliminating odour by breaking down odour-causing organic substances. Thus, when ozone NB water is clinically applied for wound healing, specifically disinfection of wound, it would also be able to repair and regenerate the affected tissue simultaneously [3,4].

I believe that the therapeutic technique using this water can be effective in the treatment of infectious skin diseases including bedsores and refractory trichophytosis [4,6]. Furthermore, by spraying ozone NB water in the form of fumes, operating rooms and patients' rooms can be disinfected. It is also suitable for hygiene management of bathrooms and care facilities such as nursing homes as it can be sprayed for the purpose of disinfection and odour elimination. Another possible application of ozone NB water based on its effects in the form of fumes would be its use in the measures to prevent the infection with avian influenza virus. Various applications of its strong odour eliminating action would also be possible in healthcare. Because ozone in ozone NB water would not be dispersed in the air in the form of gas, but is degraded and disappears in the water, it causes almost no oxygen toxicity in the body as active oxygen, which makes it suitable for medical applications. The same is true for oxygen NB water. There is almost no room for adverse reactions to these NB solutions. They can be ingested safely without any problems. More than a few thousand consumer testers have tried NB water without any complaint. Moreover, their remarkable efficacy has also been reported. Patients with virus poisoning or infection with *Helicobacter pylori* who drank NB water have experienced immediate improvement in gastrointestinal disorder.

12.1.4 Homeostatic Function of Nanobubble Water

In the application of NB water in medicine, different effects are intended or expected, depending on its type, its composition represented by, for example, mixing ratio and the methods of ingestion or addition. However, it can be assumed that NB water basically produces effects by enhancing the homeostatic function of the living organisms and facilitating the maintenance of normal biological equilibrium and that it not only lowers the elevated blood pressure, blood sugar and cholesterol to the levels within the normal range, but is also effective for the treatment of lung cancer and prostate cancer. Therefore, in combination with appropriate exercise regimen, NB water is expected to be used for the improvement or prevention of conditions including metabolic syndrome, which is also an ageing phenomenon or to prevent lifestyle diseases [3,4]. For these purposes, NB water should be ingested every day for at least three weeks. Continuation is important and once the users become aware of its effects, it is desirable to continue the ingestion for at least two to three months. This is because it takes three months or longer for most of the components of cells in our bodies to be replaced by the components of NB water through metabolism. Even if conditions of users are improved temporarily, the effects will inevitably disappear when the ingestion is discontinued. I speculate that, after three weeks of ingestion, NB water remains in our bodies and comes to comprise a part of body fluid. Therefore, the effects of oxygen NB water in particular are considered to persist for at least six months. However, if ingestion of NB water is discontinued, its effects are naturally reduced. Most of the consumer testers who have continued to drink NB water for the period from one year to more than five years for the purpose of preventing lifestyle diseases made positive comments about the efficacy of the water, including, "I'm in good shape", "I don't catch a cold", "I haven't had gastrointestinal disorders", "I don't get tired", "I have a good appetite", "I gained strength" and "I can sleep well" [4]. These comments indicate that the immunity of individual testers has improved and they are in control of their lives. They seem to be able to perform their daily activities smoothly in a manner suitable for their age and physical strength. Necessity to enhance immunity and resistance of individuals has important meaning. NB water is also highly effective in the treatment of atopic dermatitis and psoriasis vulgaris, which can be called representative autoimmune diseases of skin. In many cases of patients who used NB water, refractory dermatitis which had not been alleviated for years completely resolved in three weeks to three months. Figure 12.11 shows a case of psoriasis vulgaris. This case had not responded to various therapies, but the condition resolved by ingestion of NB water, and the disease has not recurred since then. I speculate that restoration of autoimmune function contributed to the recovery.

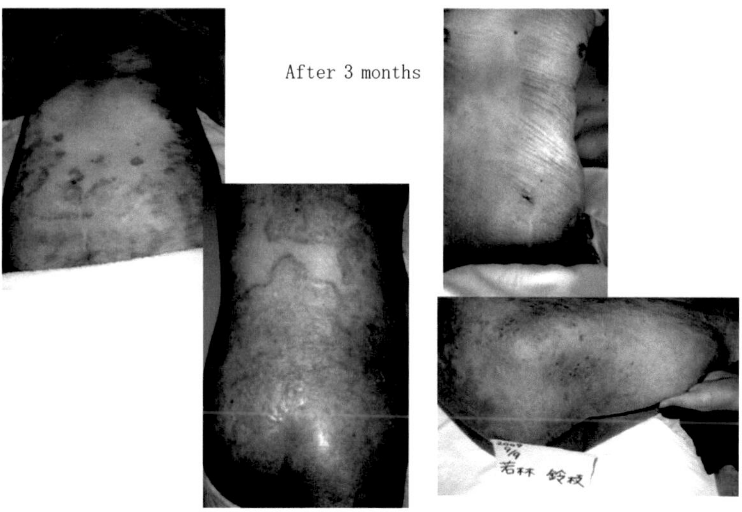

Figure 12.11 Eighty-six-year-old female patient: Senile and lack of cognition.

Japan is a rare country in the world with a high longevity rate, having a life expectancy of well over 80 years. It used to be a virtue to live long, but people think differently today. It is not too much to say that to live long no longer means to live happily. The time people spend being bed-ridden or living with dementia cannot be included in healthy life expectancy. To live a long, healthy, active and fulfilling life is not only important for individuals but also has an essential social implication as it reduces healthcare costs for the elderly. To achieve such long healthy life expectancy, lifestyle diseases should be reduced as much as possible.

As in the last example, the results of an experiment using Hela cells, which is a model of squamous cell carcinoma, are presented in Figs. 12.12 and 12.13. Figure 12.12 shows that, treatment with physiological saline reduced cancer cells after a 24 h incubation by several percentages compared with those cultured only with culture medium, whereas treatment with oxygen nanobubble water with 0% salt resulted in an increase of cancer cells by several percentages. However, with 1% oxygen NB water, cancer cells were reduced by 70%, and most of the remaining cells became apoptotic (Fig. 12.13) [3,4]. Using ozone NB water with 1% salt showed a 30% reduction probably because the mixture ratio was 25% (Fig. 12.12). It is too far-fetched to associate the results of in vitro experiments directly with the results in the human body. Studies on animals and further research on topics including mechanism of action are necessary to consider how to use NB water and which cancer can be treated in the future. However, the efficacy of NB water in restoring the normal environment in human body by attacking abnormal cells certainly shows its function to maintain homeostatic action in the human body.

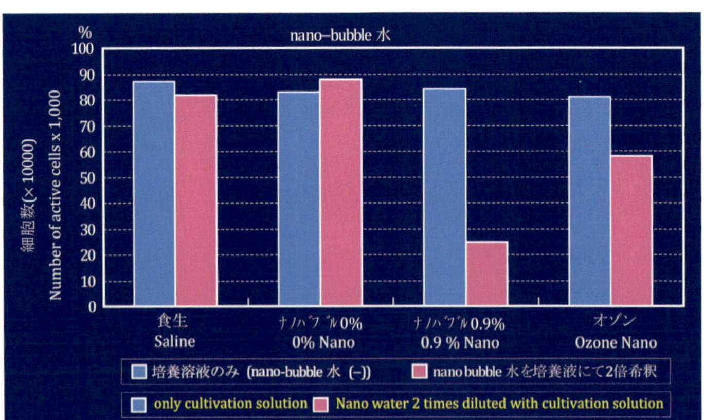

Figure 12.12 Cultivation Test of Squamocellular Cancer cells for 24 h (Effect by Nanobubble Water).

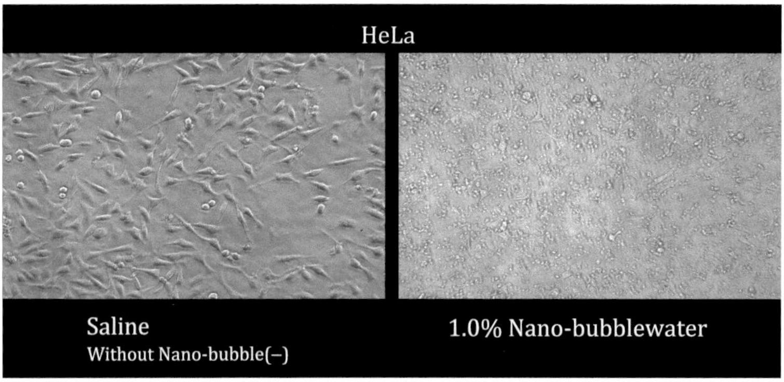

Figure 12.13 Effect by Nanobubble Water. Cultivation test of Squamocellular Cancer cells for 24 h.

Thus, I showed the results of only a small portion of the studies available today on the effects of NB water on living organisms. I believe that these studies will lead to the discovery of the potential impact of NB water on people's health as well as prevention, improvement and treatment of pathological conditions.

NB water has been shown to be possibly effective in the treatment of conditions ranging from intractable diseases including autoimmune diseases and cancer to the ageing phenomena often called lifestyle diseases. When we think about health issues of the most important people in our lives, it is meaningful that we now know that NB water can enable the achievement of age-appropriate quality of life. I predict that in the immediate future, NB techniques will be key to the efforts to narrow the gap between life expectancy and healthy life expectancy.

References

1. Mano Y, Chiba K, *et al.* (2006) Preservation solutions of tissue (Patented).
2. Mano Y, Chiba K, *et al.* (2006) Repair and regeneration solutions of tissue (Patented).
3. Mano Y, Chiba K, *et al.* (2006) Sterilization and disinfection solutions of tissue (Patented).
4. Kamiyama K, Mano Y, *et al.* (2008) Effects of nanobubble water on periodontal disease, *Jpn. J. Hyperbaric and undersea medicine*, **43**, 53–60.

5. Mano Y (2009) An example of clinic application developed from fundamental researches of hyperbalic medicine, *Jpn. J. Hyperbaric and undersea medicine*, **44**, 2–7.
6. Mano Y, *et al.* (2009) Application of nanobubbles to medical field, *Material integration*, **22**, 30–35.

12.2 Anti-Atherogenic Action of Oxygen Nanobubbles in Vascular Endothelial and Smooth Muscle Cells

Yukihiro Hojo
Department of Cardiology, Jichi Medical University, Shimotsuke, Tochigi, 329-0398, Japan
yhojo@jichi.ac.jp

12.2.1 Introduction

The number of elderly people is increasing in the 21st century, and most elderly are affected by diseases, resulting in a growing problem in advanced countries. Healthy ageing is becoming important from social, medical and economical points of view. Atherosclerotic diseases such as myocardial infarction, angina pectoris, stroke and renal failure are major diseases that lower quality of life in elderly people. In western countries, these atherosclerotic diseases are the most common cause, of death and becoming a big problem in other countries [1].

Endothelial cells are composed of a layer of internal lumen of blood vessels that have important roles in maintaining normal vascular function. For example, endothelial cells prevent thrombus formation by secreting various substances, that maintain cellular permeability, and produce vasoactive substances to regulate vascular function. Thus, endothelial cells play important roles to prevent atherosclerotic diseases. The formation of atherosclerosis is initiated by interaction between activated white blood cells and endothelial cells [2]. Activated white blood cells attach to the surface of the endothelial cells, migrating into the vessel wall and transform into foam cells that is a major composition of atherosclerotic plaque. In addition to activation of white blood cells, inflammation of endothelial cells caused by various extrinsic stimulations is a major cause of atherosclerotic plaque formation. In humans, hypertension, dyslipidemia, diabetes mellitus, smoking and genetic predisposition are major cause of atherosclerosis. Overlap of these risk factors synergistically increases the morbidity of atherosclerotic diseases. In addition to classical risk factors, obesity (visceral fat), mental stress, chronic kidney disease and sleep apnoea syndrome have been shown to be risk factors for development of atherosclerosis [3–5]. These factors are involved in the formation of atherosclerosis by activating white blood cells and vascular cells.

It has been clarified that inflammation plays a central role in the formation and progression of atherosclerosis. Cytokines are bioactive substances released from various cells and one of the important causes of inflammation. To date, a number of cytokines are identified. Cytokines are composing positive and negative feedback networks. Among these cytokines, tumour necrosis factor α (TNFα) interleukin-1β and interferon γ are major inflammatory cytokines. Thus, it is very important to reduce inflammation in vascular cells to prevent atherosclerosis. The treatment of hypertension, dyslipidemia and diabetes mellitus has been developing drastically in the last several decades. However, it is still not enough. In particular, we do not have effective tools to regulate chronic vascular inflammation.

Nanobubbles are supposed to have various biological activities including the facilitation of plant growth, sterilising bacteria and acceleration of wound healing process of sea fish. We hypothesised that nanobubbles have anti-inflammatory properties and can be applied in the prevention of atherosclerosis. In this chapter, we demonstrate the results of in vitro experiments using vascular cells to investigate the biological action of nanobubbles. To date, two types of nanobubbles are available: oxygen-nanobubbles (OXNBs) and ozone-nanobubbles (OZNBs). Because OXNB has been used in preliminary experiments in plants and fish, we used OXNB in our experiments.

12.2.2 Anti-Inflammatory Action of Oxygen-Nanobubbles in Cultured Endothelial Cells

Endothelial cell maintains vascular homeostasis by releasing various biological active substances such as nitric oxide (NO), tissue plasminogen activator, adrenomedullin, endothelin and C-type natriuretic peptide. NO, adrenomedullin and natriuretic peptide have anti-inflammatory activity; thus, endothelial cells protect themselves from atherosclerotic stimuli. In addition, appropriate fluid shear stress is essential to exert anti-inflammatory action of endothelial cells via a NO production [6] and a modulation of cellular redox state [7]. In contrast, various stimuli such as ageing, inflammatory cytokines, mechanical stress and oxidative stress induce atherogenic responses of endothelial cells. For example, adhesion molecules are upregulated on the surface of endothelial cells by inflammatory cytokines. The expression of adhesion molecules such as intercellular adhesion molecule-1 (ICAM-1) and vascular adhesion molecule (VCAM-1) are established markers to represent inflammation of endothelial cells [8].

First, we investigate the effect of OXNB on the inflammatory reaction of endothelial cells induced by cytokine stimulation. Rat aortic endothelial cells (RAOECs) were cultured and stimulated by TNFα. As shown in the Western blot in Fig. 12.14, TNFα induced the expression of ICAM-1 after 24 h (lane 2 from the left). When we incubated RAOECs in the medium containing OXNB before TNFα stimulation, the expression of ICAM-1 was markedly decreased (lane 4 from the left), suggesting the anti-inflammatory action of OXNB. The expression of endothelial NO synthase (eNOS) and α-tubulin were not changed by TNFα stimulation. eNOS is constitutively expressed in the endothelial cells and plays an important role in maintaining vaso-relaxation. α-Tubulin is a cytoskeletal protein and is also constitutively expressed in endothelial cells, composing cellular basic construction. The expression of constitutive molecules was not changed by the stimulations, suggesting that OXNB selectively inhibited the inflammatory responses of endothelial cells.

Next, we investigated the dose-dependent effect of the anti-inflammatory action of OXNB. Because accurate measurement of OXNB is still not established, we used crude OXNB solution (100% concentration) and diluted it with distilled water for the experiment. As shown in Fig. 12.15, OXNB attenuated the TNFα–induced ICAM-1 expression in RAOECs in a dose-dependent manner. Similar results were obtained in the expression of VCAM-1. The expression of eNOS and α-tubulin were not changed in this experiment. Although we do not know the accurate concentration of OXNB, we observed a dose-dependent trend of anti-inflammatory action of OXNB.

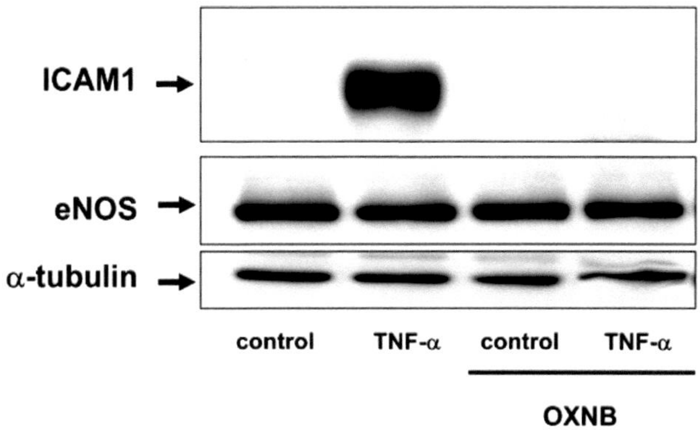

Figure 12.14 Effect of oxygen nanobubbles on the expression of intercellular adhension molecule-1 induced by tumor necrosis factor α in rat aortic endothelial cells. Tumor necrosis factor α(TNFα) induced the expression of intercellular adhesion molecule-1 (ICAM-1) in the cultured rat aortic endothelial cells (RAOECs). Preincubation of RAOECs in the medium containing oxygen-nanobubble (OXNB) for 24 h attenuated the TNFα-induced ICAM-1 upregulation. Endothelial nitric oxide synthase (eNOS) and α-tubulin, constitutive marker for endothelial cells were not changed by TNFα or OXNB. TNFα: 20 ng/mL, incubated with 24 h.

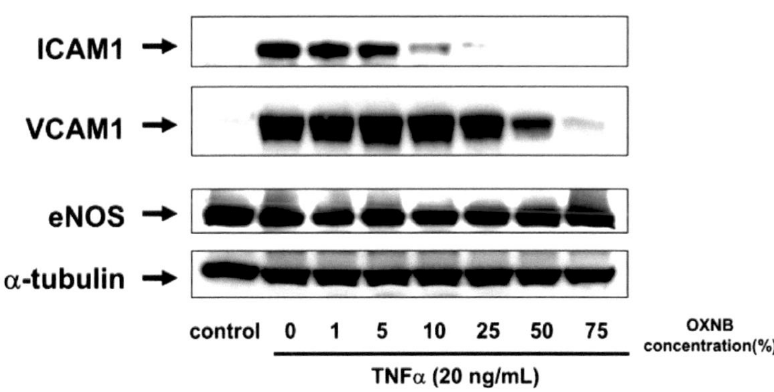

Figure 12.15 Attenuation of the expression of adhesion molecules induced by TNFα in the various concentration of OXNB. Expression of ICAM-1 and vascular adhension molecule-1 (VCAM-1) induced by TNFα were decreasing, increasing the concentration of OXNB. The expression of eNOS and α-tubulin were not changed. TNFα: 20 ng/mL, incubated with 24 h.

To explore the mechanism of how OXNB inhibits the inflammatory response in endothelial cells, we examined the effect of OXNB on mRNA expression of ICAM-1. mRNA is an upstream regulator of protein synthesis, that is, DNA information can be copied into mRNA (transcription), and proteins can be synthesised using the information in mRNA as a template (translation). As shown in Fig. 12.16, preincubation of RAOECs with the culture medium containing OXNB markedly downregulated the mRNA expression of ICAM-1 induced by TNFα. Similar results were obtained regarding the VCAM-1 expression in RAOECs. When we used interleukin-1β instead of TNFα, mRNA expression of adhesion molecules were decreased by OXNB. These results suggest that OXNB inhibits de novo synthesis of adhesion molecules induced by inflammatory cytokines.

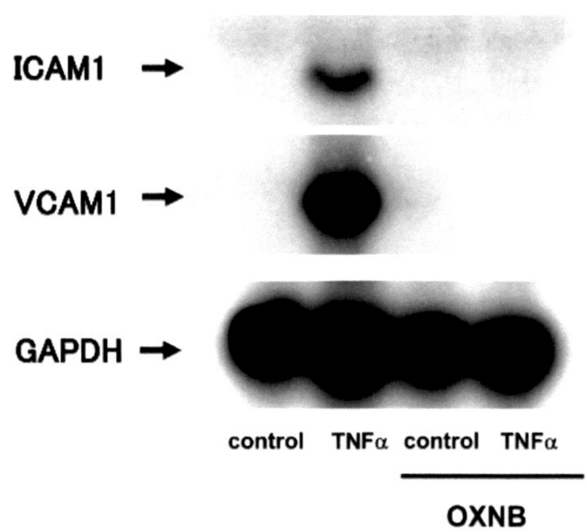

Figure 12.16 Effect of OXNB on the messenger RNA expression of ICAM-1. Stimulation of RAOECS with TNFα for 6 h induced the messenger ribonucleic acid (mRNA) expression of ICAM-1 in RAOECs. Preincubation of RAOECs with the medium containing OXNB inhibited the ICAM-1 mRNA expression. In contrast, expression of GAPDH, a house-keeping gene was not changed by TNFα or OXNB.

We confirmed that the mRNA and protein expression of adhesion molecules were attenuated by OXNB. Next, we investigated the functional effect of OXNB on RAOECs. As shown in Fig. 12.17, rat macrophages attached on the surface of TNFα-stimulated RAOECs. Preincubation of RAOECs inhibited the adhesion of activated macrophages to TNFα-stimulated RAOECs. We observed that OXNB not only attenuated the protein and mRNA expression of adhesion molecules of the endothelial cells but also inhibited the attachment of macrophages to the endothelial cells. These results suggest that OXNB could inhibit the first step of the atherosclerosis.

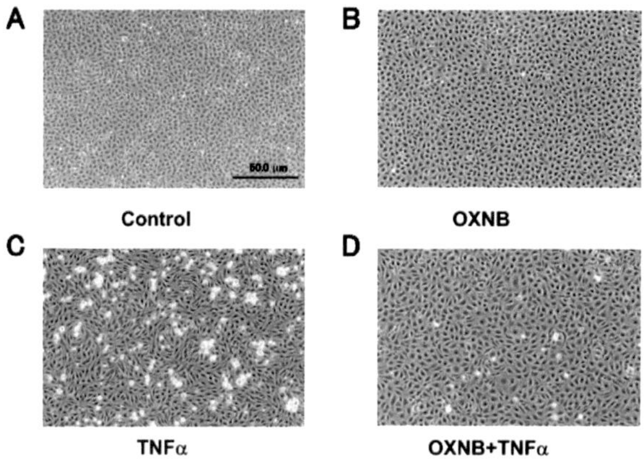

Figure 12.17 Effect of OXNB on the adhesion of rat macrophages to endothelial cells. Panel A: control RAOECs. Panel B: RAOECs incubated in the medium containing OXNB for 24 h. Panel C: Coculture of rat macrophages and TNFα-stimulated RAOECs. A number of macrophages are attached on the surface of RAOECs. Panel D: Coculture of rat macrophages and TNFα-stimulated RAOECs pretreated by OXNB for 24 h. Compared to panel C, adhesion of macrophages onto RAOECs was markedly decreased.

To clarify how OXNB attenuates inflammatory action of endothelial cells, we investigated the effect of OXNB on the downstream signal transduction of TNFα in RAOECs. We checked the effect of OXNB on the phosphorylation of extracellular signal-regulated kinase (ERK) and c-Jun N-Terminal kinase (JNK) by Western blotting; however, no significant change was observed in the activation of these kinases (data not shown). Furthermore, we checked the phosphorylation of nuclear factor-kB (NFkB), one of the important transcription factors that regulate inflammation of endothelial cells. We did not find the effect of OXNB on the TNFα-induced activation of NFkB. We also checked whether OXNB inhibits inflammation by increasing cellular NO production or modulating cellular redox status. Because we could not measure cellular NO levels directly, we used NO synthase inhibitor, l-NAME and found that l-NAME did not change TNFα-induced upregulation of adhesion molecules in RAOECs. The results suggest that the anti-inflammatory action of OXNB was not mediated by NO. Similarly, OXNB did not affect the cellular redox state of glutathione, a major intracellular anti-oxidant substance. A number of signal transduction molecules are NO and redox-sensitive. Our results suggest that anti-inflammatory OXNB is not mediated by the modulation of intracellular signal transduction mediated by NO or cellular redox state.

12.2.3 Anti-Proliferative Action of Oxygen-Nanobubbles in Cultured Vascular Smooth Muscle Cells

Vascular smooth muscle cells (VSMCs) exist in the mid-layer of blood vessels and play an important role regulating blood pressure by its contraction. Transformation, migration and inappropriate proliferation of VSMCs lead to vascular remodelling, progression of atherosclerosis and restenosis of blood vessels after vascular intervention [9]. Not only mechanical factors but also humoral factors play critical role in proliferation of VSMCs. To date, angiotensin, endothelin, insulin, thrombin and platelet-derived growth factor (PDGF) are known to induce proliferation of VSMCs. PDGF is one of the strong growth factors for VSMCs. We examined the effect of OXNB on the PDGF-induced proliferation of VSMCs in vitro.

As shown in Fig. 12.18, PDGF induced an approximately two-fold increase in the protein concentration of rat VSMC lysates. Preincubation of VSMCs in the medium containing OXNB for 24 h significantly decreased PDGF-induced protein synthesis of VSMCs. To clarify how OXNB attenuates PDGF-induced protein synthesis in rat VSMCs, we examined the PDGF signal transduction in VSMCs. As shown in Fig. 12.19a and b, OXNB did not change phosphorylation of PDGF receptor or phosphorylation of ERK1/2.

These results suggest that OXNB dose not affect PDGF signal transduction steps upstream of the ERK1/2 activation. Therefore, we investigated the signal transduction downstream of ERK.

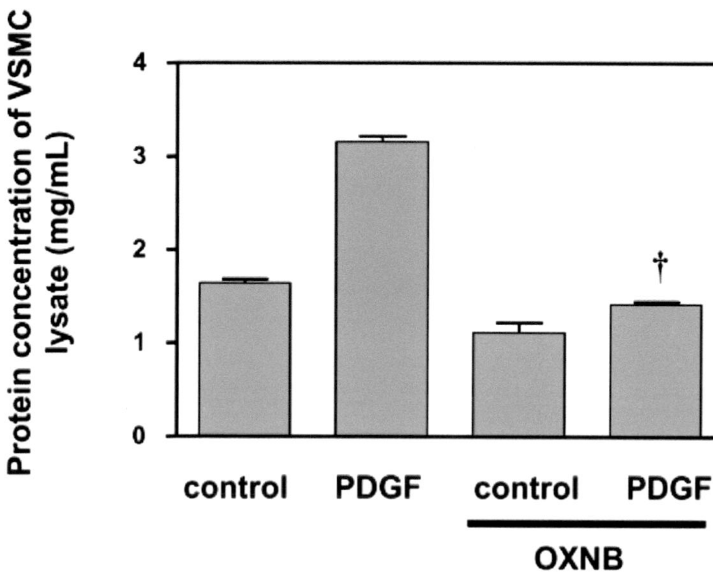

Figure 12.18 Effect of OXNB on the protein synthesis induced by platelet-derived growth factor in rat smooth muscle cells. Platelet-derived growth factor (PDGF, 50 ng/mL 48 h) enhanced protein synthesis in rat vascular smooth muscle cells (VSMCs). Preincubation of rat VSMCs in the medium containing OXNB for 24 h significantly decreased PDGF-induced protein synthesis of rat VSMCs.

Histone acetyltransferase (HAT) is an enzyme of acetylate lysine residue on histone proteins by transferring an acetyl group from acetyl CoA to form ε-N-acetyl lysine [10]. In general, histone acetylation is linked to transcriptional activation and associated with euchromatin. Acetylation of lysine neutralises the positive charge normally present, thus reducing affinity between histone and negatively-charged DNA, which renders DNA more accessible to transcription factors. We hypothesised that OXNB affects HAT activity in VSMCs. As shown in Fig. 12.20, incubation of rat VSMCs in the medium containing OXNB for 24 h significantly reduced the acetylation of Histone H3 and H4 induced by PDGF. The results suggest that OXNB inhibited the HAT activity in VSMCs. The decreased histone acetylation leads to the inhibition of DNA transcription, resulting in the decrease of protein synthesis. So far, the precise mechanisms of OXNB inhibiting HAT is still not clear. It is possible that inhibition of HAT by OXNB results in decreased inflammation in endothelial cells.

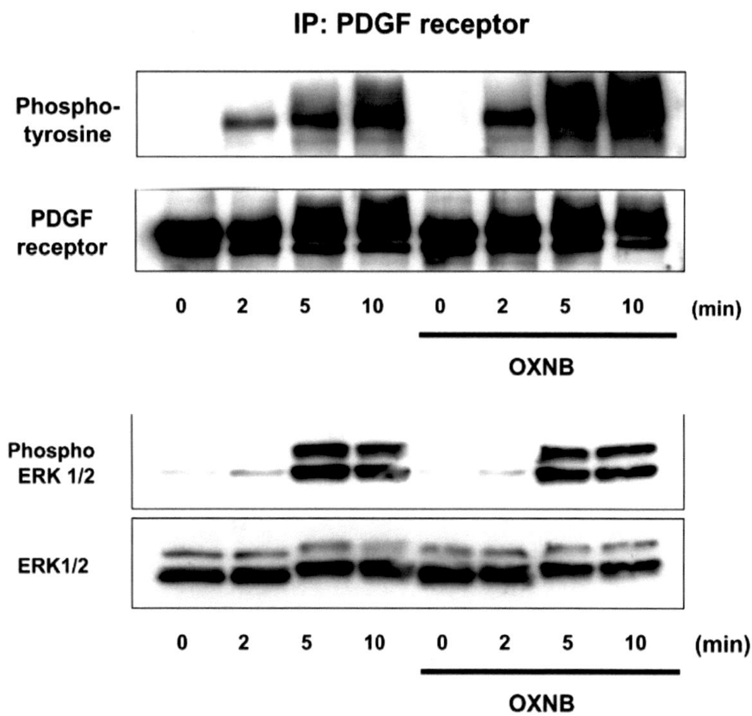

Figure 12.19 (a) Effect of OXNB on the PDGF signal transduction in rat VSMCs. Time course of receptor activation by 50 ng/mL of PDGF demonstrated by western blotting in rat VSMCs. Preincubation of rat VSMCs in the medium containing OXNB did not change PDGF receptor activation after the stimulation. (b) Effect of OXNB on the PDGF signal transduction in rat VSMCs. Time course of activation of extracellular signal-regulated kinase 1/2 (ERK1/2). OXNB did not affect activation of ERK1/2 by PDGF in rat VSMCs.

12.2.4 Future Scope of Nanobubbles in the Medicine

Our results suggest that OXNB could protect vascular cells from atherogenic stimuli in vitro. We still do not know whether OXNB works in animals and humans (in vivo). There are a number of problems to be solved. First, an assay system of nanobubbles is still not established. We do not know the accurate concentration of nanobubbles in the culture medium. Because culture medium contains various kinds of proteins, amino acids, glucose and minerals, nanobubbles might be unstable in such conditions. Besides, dilution of nanobubble solution by distilled water might disrupt nanobubbles. Thus, we should be very careful until the assay system is established. Second, the concentration of OXNB is presumably quite high in vitro. We do not know how to apply nanobubbles in living organisms. Third, the side effect of nanobubbles should be strictly checked before clinical application. Despite these problems, nanobubble technologies are unique and attractive in the medical field. Proper application of nanobubble technologies should be established for medicine.

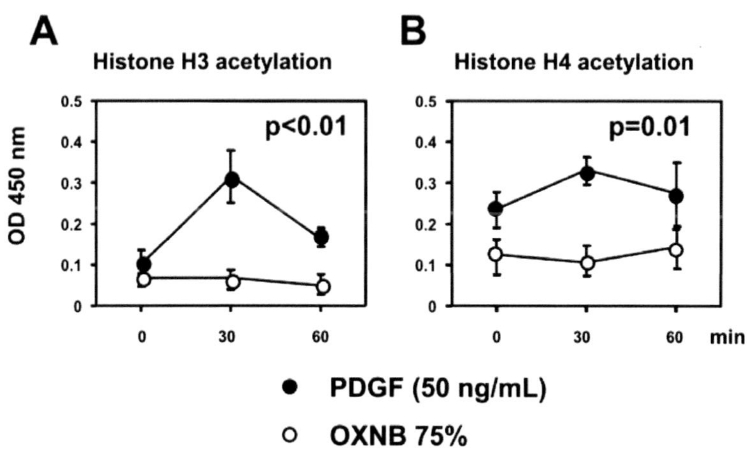

Figure 12.20 Effect of OXNB on histone H3 and H4 acetylation in rat VSMCs. Preincubation of rat VSMCs in the medium containing OXNB for 24 h significantly inhibited the acetylation of histone H3 (panel A) and H4 (panel B).

References

1. Roger VL, Go AS, Lloyd-Jones DM, Adams RJ, Berry JD, Brown TM, Carnethon MR, Dai S, de Simone G, Ford ES, Fox CS, Fullerton HJ, Gillespie C, Greenlund KJ, Hailpern SM, Heit JA, Ho PM, Howard VJ, Kissela BM, Kittner SJ, Lackland DT, Lichtman JH, Lisabeth LD, Makuc DM, Marcus GM, Marelli A, Matchar DB, McDermott MM, Meigs JB, Moy CS, Mozaffarian D, Mussolino ME, Nichol G, Paynter NP, Rosamond WD, Sorlie PD, Stafford RS, Turan TN, Turner MB, Wong ND, Wylie-Rosett J; American Heart Association Statistics Committee and Stroke Statistics Subcommittee (2011) Heart disease and stroke statistics—2011 Update a report from the American Heart Association, *Circulation*, **123**, e18–e209.
2. Ross R (1999) Atherosclerosis–an inflammatory disease, *N Engl J Med*, **340**, 115–126.
3. Arad Y, Newstein D, Cadet F, Roth M, Guerci A (2001) Association of multiple risk factors and insulin resistance with increased prevalence of asymptomatic coronary artery disease by an electron-beam computed tomographic study, *Arterioscler Thromb Vasc Biol*, **21**, 2051–2058.
4. Sarnak M (2003) Cardiovascular complications in chronic kidney disease, *Am J Kidney Dis*, **41**, 11–17.
5. Ninomiya T, Kiyohara Y, Kubo M, Tanizaki Y, Doi Y, Okubo K, Wakugawa Y, Hata J, Oishi Y, Shikata K, Yonemoto K, Hirakata H, Iida M (2005) Chronic kidney disease and cardiovascular disease in a general Japanese population: the Hisayama Study, *Kidney Int*, **68**, 228–236.
6. Korenaga R, Ando J, Tsuboi H, Yang W, Sakuma I, Toyo-oka T, Kamiya A (1994) Laminar flow stimulates ATP- and shear stress-dependent nitric oxide production in cultured bovine endothelial cells, *Biochem Biophys Res Commun*, **198**, 213–219.
7. Hojo Y, Saito Y, Tanimoto T, Hoefen R, Baines C, Yamamoto K, Haendeler J, Asmis R, Berk B (2002) Fluid shear stress attenuates hydrogen peroxide-induced c-Jun NH2-terminal kinase activation via a glutathione reductase-mediated mechanism, *Circ Res*, **91**, 712–718.
8. Nakashima Y, Raines E, Plump A, Breslow J, Ross R (1998) Upregulation of VCAM-1 and ICAM-1 at atherosclerosis-prone sites on the endothelium in the ApoE-deficient mouse, *Arterioscler Thromb Vasc Biol*, **18**, 842–851.
9. Raines E, Ross R (1993) Smooth muscle cells and the pathogenesis of the lesions of atherosclerosis, *Br Heart J*, **69**, S30–S37.
10. Laurenson P, Rine J (1992) Silencers, silencing, and heritable transcriptional states, *Microbiol Rev*, **56**, 543–560.

12.3 Clinical Application of Ozone Nanobubble Water to Periodontal Treatment

Shinichi Arakawa,[a,b] **Sae Hayakumo,**[b] **Yoshihiro Mano,**[c] **and Yuichi Izumi**[b,d]

[a]*Section of Lifetime Oral Health Care Sciences, Graduate School of Medical and Dental Sciences, Tokyo Medical and Dental University, Tokyo 113-8549, Japan*
[b]*Section of Periodontology, Graduate School of Medical and Dental Sciences, Tokyo Medical and Dental University, Tokyo 113-8549, Japan*
[c]*Hyperbaric Medical Center, Hospital of Medicine, Tokyo Medical and Dental University, Tokyo 113-8549, Japan*
[d]*Global Center of Excellence (GCOE) Program, Tokyo Medical and Dental University, Tokyo 113-8549, Japan*
shinichi.peri@tmd.ac.jp, s-arakawa.ltoh@tmd.ac.jp

12.3.1 Introduction

Elimination of pathogen-containing biofilms remains the primary goal of periodontal treatment. Supra- and subgingival mechanical debridement is traditionally used as the initial phase of treatment to achieve this goal even though it is rarely capable of complete removal of putative periodontal pathogens. A review of studies evaluating the effectiveness of various subgingival debridement procedures showed that 5%–80% of treated roots harbour residual plaque or calculus, and as the pockets and furcation involvements get deeper, more deposits are left behind [1].

To augment mechanical debridement, adjunctive antimicrobial agents in the form of topical or systemic antibiotics or topical antiseptics have been employed [2,3]. Considering the potential problems with selectivity of antimicrobial action, possible development of resistant bacteria and adverse host reactions, systemic antibiotic therapies seem to constitute an inferior choice to the topical use of low-cost, broad-spectrum antiseptic agents with low potential for adverse reactions. Subgingival irrigation with various antiseptic agents such as povidone-iodine, chlorhexidine or hydrogen peroxide has been performed in conjunction with scaling and root planing.

Currently, ozone is attracting attention as a possible alternative antiseptic in the dental field as well as food industries. Ozone has strong antimicrobial effects against bacteria, fungi, protozoa and viruses [4] and does not induce microbial resistance, characteristics that were originally noted in the fields of water purification and food preservation [5–7].

This part described the evaluation of the clinical application of ozone nanobubble water (NBW3) as an adjunct to mechanical subgingival debridement for the treatment of periodontitis and analyses of the micro-morphological changes of NBW3-treated bacterial cells.

12.3.2 Periodontitis

Periodontal diseases is a multifactorial infection elicited by a complex of bacterial species, which interact with host tissues and cells causing the destruction of periodontal structures. On the other hand, it has been reported that there was an association between the severity of periodontal disease and the systemic diseases such as cardiovascular diseases, pulmonary disease, adverse pregnancy outcomes (preterm low birth weight) and diabetes [8–11]. The paradigm shift invents the term periodontal medicine which refers that periodontal disease is interrelated with systemic health in important ways.

The initiation of disease is the presence of microbial biofilm that colonises the sulcular regions between the tooth surface and the gingival margin. It has been thought that nearly 700 bacterial taxa, phylotypes and species can colonise the oral cavity of humans [12]. The different microbial complexes have been associated with the sequence of colonisation on the tooth surface and gingiva as well as disease severity [13]. The "red complex" comprises species that are considered periodontal pathogens, such as *Porphyromonasgingivalis* (*P. gingivalis*) and *Tannerella forsythia* (*T. forsythia*) and *Treponema denticola.* The members of red complex are the specific pathogens most frequently associated with this disease, being present in high numbers within deep periodontal pockets in severe periodontal conditions [14–17]. Elimination of pathogen-containing biofilms remains the primary goal of periodontal treatment. Supra- and subgingival mechanical debridement is traditionally used as the initial phase of treatment to achieve this goal even though it is rarely capable of complete removal of putative periodontal pathogens.

Biofilms cause a wide range of human infectious diseases including dental caries, osteomyelitis, bacterial prostatitis and cystic fibrosis. Characteristic of these infections is the persistence and chronicity of the infections as well as the difficulty in their eradication. Antimicrobial agents are unlikely to be effective for the treatment of these infections. The minimum inhibitory concentration for vancomycin for an organism in the planktonic state might be 1 μg/mL, whereas in biofilm mode it will be 1000 μg/mL. Furthermore, it is important to understand the factors leading to antimicrobial resistance in biofilms such as dental plaque [18]. Given the remarkable resistance of organisms in biofilms to host defence mechanisms and antimicrobial agents, the logical first step in the control of these organisms would be their removal by physical means. Fortunately, biofilms in the oral cavity, unlike many other biofilms, are readily accessible allowing their physical removal. Indeed, the most common form of periodontal therapy is the removal of supra- and subgingival plaque by procedures such as self-performed oral hygiene, scaling and root planing. Other mechanical debridement procedures, such as periodontal surgery, have similar or even greater effects on the red complex [19]. The mechanical procedures undoubtedly removed most organisms that colonised the tooth surface. Perhaps 90% or even somewhat higher proportions were removed. However, given the rapid multiplication rates of bacteria, it is not surprising that the majority of taxa examined returned to almost baseline levels at 3 months. Data in the literature suggest that the return to baseline total counts might occur within 4–8 days [20].

12.3.3 Clinical Application of Ozone Nanobubble Water (NBW3) to the Periodontal Treatment

Elimination of pathogen-containing biofilms remains the primary goal of periodontal treatment. To augment mechanical debridement, adjunctive antimicrobial agents in the form of topical or systemic antibiotics or topical antiseptics have been employed [2,3]. Considering the potential problems with selectivity of antimicrobial action and possible development of resistant bacteria and adverse host reactions, systemic antibiotic therapies seem to constitute an inferior choice to the topical use of low-cost, broad-spectrum antiseptic agents with low potential for adverse reactions. Subgingival irrigation with various antiseptic agents such as povidone-iodine, chlorhexidine (CHX) or hydrogen peroxide has been performed in conjunction with scaling and root planing and produced significant clinical benefits relative to conventional mechanical root debridement alone [21]. However, these existing antiseptic agents have serious side effects: e.g. the prolonged use of CHX may cause mucosal desquamation, tooth staining and altered taste sensation. In addition, an increasing number of immediate-type allergies to this agent such as anaphylactic shock have been reported [22]. Povidone-iodine should not be administered to individuals who are allergic to iodine, suffering from thyroid dysfunction or are pregnant or nursing [23–25]. Moreover, a huge amount of povidone-iodine may be dispersed into the air, linen and other nearby surfaces when used as an adjunct to ultrasonic debridement and the surroundings can be stained brown by the iodine. Therefore, an alternative adjunctive antiseptic with high antimicrobial potential, a good safety profile, fewer contraindications and a higher degree of usability would be beneficial for periodontal therapy.

Currently, ozone is attracting attention as a possible alternative antiseptic in the dental field as well as food industries. Ozone has strong antimicrobial effects against bacteria, fungi, protozoa and viruses [4] and does not induce microbial resistance. Recent clinical investigations have reported antimicrobial effects of both gaseous and aqueous forms of ozone on oral pathogens associated with caries and endodontic infections [26–31]. Furthermore, ozone therapy has been reported to considerably reduce the growth of *A. actinomycetemcomitans*, *P. gingivalis* and *T. forsythia* [32,33]. In comparison to ozone gas, it has been reported that aqueous ozone possesses a high level of biocompatibility to fibroblasts, cementoblasts and epithelial cells [34–37], suggesting the properness of its use against oral infectious diseases such as periodontal disease and peri-implantitis. So far, positive clinical effects of ozonated water in reducing signs of gingivitis and periodontitis have been described in two clinical studies [38,39]. However, ozonated water has a half-life of about only 20 min and will degrade back into oxygen very quickly, so should be used within the first 5–10 min to assure its potency.

To overcome this disadvantage of ozonated water, NBW3 was developed using nanobubble generating technology by Chiba and Sugai [40]. The procedure of generating nanobubble water has been protected by patent [40]. The ozone concentration of NBW3 is 1.5 mg/L that is equivalent to the oxidation titre determined by electron spin resonance (ESR). Nanobubble is a form of gas nucleus which is less than 100 nm in diameter and is produced by collapsing micro-bubbles (less or equal 50 μm in diameter) in the electrolyte under ultra-high temperature and pressure. During the collapse of micro-bubbles by applying physical stimulus in the electrolyte, the ions in aqueous solution become concentrated around the gas nucleus and prevent the gas from dispersion (that is called the salting-out phenomenon). Due to this nature of the ions, nanobubbles are stabilised over a long period in aqueous solution. NBW3 retains ozone gas in the form of a gas nucleus and exerts antimicrobial effect for more than 6 months if it is protected against ultraviolet rays [41].

Although the entire mechanisms of NBW3 to inactivate bacteria are as yet not well known, it might be basically similar to those of the existing ozonated water. Ozone is a potent oxidizing agent [42] and ozone in ozonated water has been reported to react with various organic substances and decompose them by free radical-mediated oxidation reactions [43]. Ozone per se also changes to oxygen when it reacts with organic substances. In this process, hydroxyl radicals (•OH) are generated, which are among the most reactive oxidizing species. These free radicals might play a role in the destruction of bacteria by NBW3.

We evaluated the potential of NBW3 in periodontal treatment. A total of 22 healthy, non-smoking subjects, aged 26–72 years old (6 females and 16 males, mean age 45.9 ± 14.8 years), were recruited from new referrals to the periodontics clinic of the Tokyo Medical and Dental University, Dental Hospital and participated in this single-centre, placebo-controlled, randomised trial with blind allocation to two parallel groups. No statistically significant differences in the demographic and clinical variables were found between the two treatment groups at baseline ($p > 0.05$). The research schedule is shown in Fig. 12.21. All recordings were made without knowledge of the previous measurements.

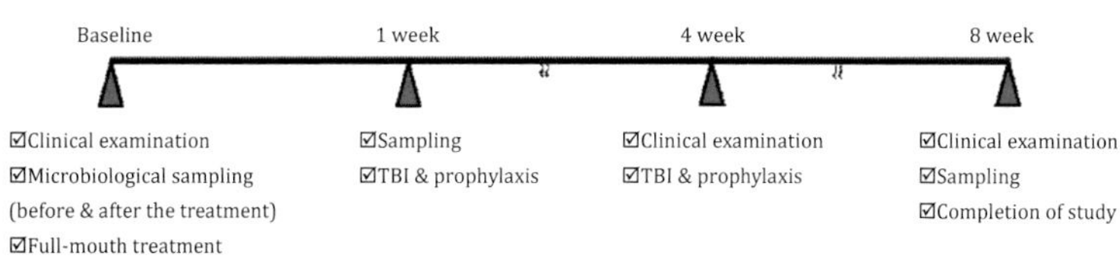

Figure 12.21 Research schedule.

After screening, all subjects received thorough oral hygiene instructions. They were then randomly assigned to one of the two treatment groups based on the treatment protocol and the examiner was blinded to the allocation. After random allocation, the patients in this study were subjected to one of the two treatment protocols: full-mouth mechanical debridement with tap water in a single visit (WATER) or full-mouth mechanical debridement with NBW3 in a single visit (NBW3). Mechanical debridement included supra- and subgingival ultrasonic instrumentation, which was performed by one experienced and trained periodontist with an ultrasonic scaler (Varios 750; Nakanishi Inc., Tochigi, Japan) equipped with a P10 tip. During the experimental period, no additional pocket instrumentation was allowed.

Clinical examination was performed at baseline (before treatment) and 4 and 8 weeks after treatment by a single blinded examiner. Full-mouth clinical measurements of probing pocket depth (PPD), clinical attachment level (CAL) and the percentage of bleeding on probing (BOP (%)) were recorded using a manual constant load periodontal probe (TUCL probe, Williams type; Shioda Dental Co., Tochigi, Japan) at all six sites per tooth (disto-buccal, buccal, mesio-buccal, disto-lingual, lingual, mesio-lingual). The reduction in the full-mouth mean PPD from baseline to 4 and 8 weeks is shown in Table 12.1. Mean PPD was reduced by 0.34 mm and 0.17 mm at 4 weeks and 0.29 mm and 0.14 mm at 8 weeks in NBW3 and WATER groups, respectively. When the treatment groups were compared with each other, there were significant differences in the reduction of the full-mouth mean PPD at both 4 and 8 weeks ($p = 0.013$ and 0.037, respectively). The CAL gain recorded at 4 and 8 weeks is shown in Table 12.1. The CAL gain was 0.31 mm and 0.1 mm at 4 weeks and 0.27 mm and 0.09 mm at 8 weeks in the NBW3 and WATER groups, respectively. When the treatment groups were compared with each other, there were significant differences in the amount of CAL gain at both 4 and 8 weeks ($p = 0.002$ and 0.021, respectively).

Microbiological examination was performed just before and after the treatment and at 1 and 8 weeks post-treatment. Subgingival plaque samples were collected from one selected pocket site with a probing depth of 4 mm or more around a maxillary single-rooted tooth in each patient. The bacterial species identified were *P. gingivalis* and *T. forsythia*. The NBW3 group showed statistically significant reductions in the mean total number of bacteria over an 8 week re-examination period ($p = 0.005$ just after the treatment; $p = 0.006$ at 1 week; $p = 0.009$ at 8 weeks). However, the control group did not show significant reductions in the mean total number of bacteria at any re-examination period. When the treatment groups were compared with each other, no statistical differences were found in the total number of bacteria at any time point during the study.

Table 12.1 Improvements in clinical parameters at 4 and 8 weeks re-examination

		ΔBOP(%)*	ΔPPD*	Attachment gain*
4 weeks	NBW3 ($n = 10$)	15.69 ± 12.5	0.34 ± 0.2#	0.31 ± 0.1#
	WATER ($n = 11$)	8.98 ± 9.2	0.17 ± 0.1	0.10 ± 0.2
8 weeks	NBW3 ($n = 10$)	13.47 ± 9.2	0.29 ± 0.2#	0.27 ± 0.2#
	WATER ($n = 11$)	6.97 ± 10.8	0.14 ± 0.2	0.09 ± 0.2

Note: NBW3, full-mouth debridement with ozone nano bubble water; WATER, full-mouth debridement with water; BOP (%), bleeding on probing (%); PPD, probing pocket depth.

*Mean ± standard deviation.

#Significantly different from FMD + water.

Percentage reductions in the total number of bacteria in subgingival plaque over the study period are tabulated in Table 12.2. When the treatment groups were compared with each other, differences in the mean percentage reductions in the total number of bacteria were not statistically significant at any re-examination period (data not shown). However, the NBW3 group showed a reduction of 95% or more in the total number of bacteria in 40% and 20% of subgingival plaque samples at 1 and 8 weeks, respectively, whereas the WATER group revealed similar reductions in only 9.10% and 0%, respectively, of subgingival plaque samples.

These results suggest that subgingival irrigation with NBW3 may be a valuable antimicrobial adjunct to mechanical instrumentation in the management of periodontal infections. The observed beneficial effects of NBW3 were, however, minor and of unknown clinical significance. Further studies with more severe cases of periodontitis, longer follow-up periods, and the comparison of NBW3 with the established agents would be needed to fully delineate the utility of NBW3 in the treatment of periodontitis.

Table 12.2 Percentage reductions in total number of bacteria in subgingival plaque

Treatment		% of subgingival plaque samples showing >95% reduction	% of subgingival plaque samples showing 95–33% reduction	% of subgingival plaque samples showing <33% reduction
NBW3 ($N = 10$)	Just after treatment	40	60	0
	1 week	40	40	20
	8 weeks	20	40	40
WATER ($N = 11$)	Just after treatment	45.50	45.50	9.10
	1 week	9.10	45.50	45.50
	8 weeks	0	45.50	54.50

Note: NBW3: full-mouth debridement with ozone nano bubble water; WATER: full-mouth debridement with water.

12.3.4 Transmission Electron Microscopic Observation

Changes in the microstructure of bacterial cells exposed to NBW3 were assessed using a transmission electron micrograph (TEM). A TEM of *Escherichia coli* cells exposed to 0.9% NaCl is shown in Fig. 12.22a and indicates that the cells remain intact following exposure to 0.9% NaCl. The cytoplasm of the control cells contains electron-dense material enveloped by a typical cytoplasmic membrane and cell wall structure.

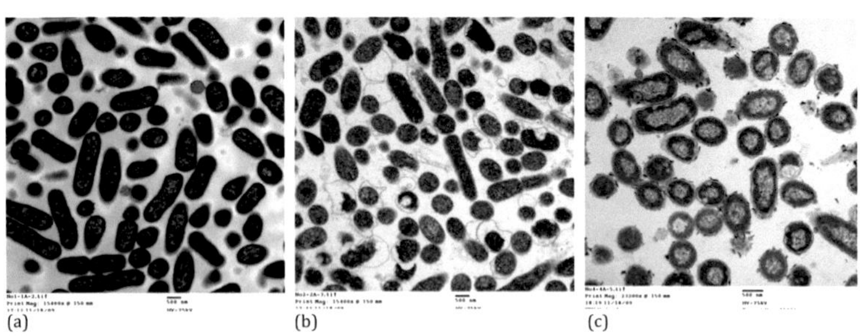

Figure 12.22 Transmission Electron Microscopic Observation. (a) 0.9% NaCl; (b) Concool®; (c) NBW3.

In *E. coli* cells exposed to Concool®, TEM indicated severe cellular damage, such as disruption of both the cytoplasmic membrane and the cell wall (Fig. 12.22b). Cytoplasmic contents leaked through the damaged areas and as a result, most of the cells underwent extensive lysis. On the other hand, in *E. coli* cells exposed to NBW3, disruption of the cell wall and the presence of numerous membrane vesicles projecting from the cell wall were observed (Fig. 12.22c). Moreover, coagulated materials were seen inside the treated cells, in particular, close to the cell membrane and the apical ends. However, there was no apparent lysis or gross leakage of cellular cytoplasmic contents from cells exposed to NBW3. The microbicial mechanism of NBW3 might be different from that of Concool®.

Our TEM analysis also revealed the presence of intracytoplasmic coagulated materials close to the cell wall and the apical ends of *E. coli* cells treated with NBW3. Similar results have been reported for *E. coli* cells exposed to oregano oil [44] or tea tree oil [45]. The coagulated material observed in *E. coli* cells was thought to be a precipitate of abnormal proteins which might have resulted from a quick influx of these substances into the cytoplasm [44,45]. In order to obtain a fuller understanding of the antibacterial mechanism of NBW3, further research is needed to determine how NBW3 damages the cellular structure and how this impacts cellular function.

References

1. Petersilka GJ, Ehmke B, Flemmig TF (2002) Antimicrobial effects of mechanical debridement, *Periodontol 2000*, **28**, 56–71.
2. Drisko CH (1996) Non-surgical pocket therapy: pharmacotherapeutics, *Ann Periodontol*, **1**, 491–566.
3. Greenstein G, Tonetti M (2000) The role of controlled drug delivery for periodontitis. The Research, Science and Therapy Committee of the American Academy of Periodontology, *J Periodontol*, **71**, 125–140.
4. Kim JG, Yousef AE, Dave S (1999) Application of ozone for enhancing the microbiological safety and quality of foods: a review, *J Food Prot*, **62**, 1071–1087.
5. Restaino L, Frampton EW, Hemphill JB, Palnikar P (1995) Efficacy of ozonated water against various food-related microorganisms, *Appl Environ Microbiol*, **61**, 3471–3475.
6. Unal R, Kim JG, Yousef AE (2001) Inactivation of *Escherichia coli* O157:H7, Listeria monocytogenes, and Lactobacillus leichmannii

by combinations of ozone and pulsed electric field, *J Food Prot*, **64**, 777–782.

7. Paraskeva P, Graham NJ (2002) Ozonation of municipal wastewater effluents, *Water Environ Res*, **74**, 569–581.
8. Beck JD, Offenbacher S (1998) Oral health and systemic disease:periodontitis and cardiovascular disease, *J Dent Educ*, **62**, 859–870.
9. Limeback H (1998) Implications of oral infections on systemic dieases in the institutionalized elderly with a special focus on pneumonia, *Ann Periodontal*, **3**, 262–275.
10. Offenbacher S, Katz V, Fertik G, Collins J, Boyd D, Maynor G, McKaig R, Beck J (1996) Periodontal infection as a possible risk factor for preterm low birth weight, *J Periodontol*, **67**, 1103–1113.
11. Gurav A, Jadhav V (2011) Periodontitis and risk of diabetes mellitus, *J Diabetes*, **3**(1), 21–28.
12. Paster BJ, Boches SK, Galvin JL, Ericson RE, Lau CN, Levanos VA, Sahasrabudhe A, Dewhirst FE (2001) Bacterial diversity in human subgingival plaque, *J Bacteriol*, **183**, 3770–3783.
13. Socransky SS, Haffajee AD (2005) Periodontal microbial ecology, *Periodontol 2000*, **38**, 135–187.
14. Dzink JL, Socransky SS, Haffajee AD (1988) The predominant cultivable microbiota of active and inactive lesions of destructive periodontal diseases, *J Clin Periodontol*, **15**, 316–323.
15. Ali RW, Velcescu C, Jivanescu MC, Lofthus B, Skaug N (1996) Prevalence of 6 putative periodontal pathogens in subgingival plaque samples from Romanian adult periodontitis patients, *J ClinPeriodontol*, **23**, 133–139.
16. Ashimoto A, Chen C, Bakker I, Slots J (1996) Polymerase chain reaction detection of 8 putative periodontal pathogens in subgingival plaque of gingivitis and advanced periodontitis lesions, *Oral Microbiol Immunol*, **11**, 266–273.
17. Haffajee AD, Socransky SS, Patel MR, Song X (2008) Microbial complexes in supragingival plaque, *Oral Microbiol Immunol*, **23**, 196–205.
18. Levy SB (1998) The challenge of antibiotic resistance, *Sci Am*, **278**, 46–53.
19. Levy RM, Giannobile WV, Feres M, Haffajee AD, Smith C, Socransky SS, (1999) The short-term effect of apically repositioned flap surgery on the composition of the subgingival microbiota, *Int J Periodontics Restororative Dent*, **19**, 555–567.

20. Furuichi Y, Lindhe J, Ramberg P, Volpe AR (1992) Patterns of de novo plaque formation in the human dentition, *J Clin Periodontol*, **19**, 423–433.

21. Rams TE, Slots J (1996) Local delivery of antimicrobial agents in the periodontal pocket, *Periodontol 2000*, **10**, 139–159.

22. Krautheim AB, Jermann TH, Bircher AJ (2004) Chlorhexidine anaphylaxis: case report and review of the literature, *Contact Dermatitis*, **50**, 113–116.

23. Fleischer W, Reimer K (1997) Povidone-iodine in antisepsis—state of the art, *Dermatology*, **195**(2), 3–9.

24. NobukuniK HN, Namba R, Hayakawa N, Ihara Y, Sato K, Takada H, Hayabara T, Kawahara S (1997) The influence of long-term treatment with povidone-iodine on thyroid function, *Dermatology*, **195**(2), 69–72.

25. Linder N, Davidovitch N, Reichman B, Kuint J, Lubin D, Meyerovitch J, Sela BA, Dolfin Z, Sack J (1997) Topical iodine-containing antiseptics and subclinical hypothyroidism in preterm infants, *J Pediatr*, **131**, 434–439.

26. Baysan A, Whiley RA, Lynch E (2000) Antimicrobial effect of a novel ozone-generating device on micro-organisms associated with primary root carious lesions in vitro, *Caries Res*, **34**, 498–501.

27. Baysan A, Lynch E (2004) Effect of ozone on the oral microbiota and clinical severity of primary root caries, *Am J Dent*, **17**, 56–60.

28. Nagayoshi M, Kitamura C, Fukuizumi T, Nishihara T, Terashita M (2004) Antimicrobial effect of ozonated water on bacteria invading dentinal tubules, *J Endod*, **30**, 778–781.

29. Arita M, Nagayoshi M, Fukuizumi T, Okinaga T, Masumi S, Morikawa M, Kakinoki Y, Nishihara T (2005) Microbicidal efficacy of ozonated water against Candida albicans adhering to acrylic denture plates, *Oral Microbiol Immunol*, **20**, 206–210.

30. Hems RS, Gulabivala K, Ng YL, Ready D, Spratt DA (2005) An in vitro evaluation of the ability of ozone to kill a strain of *Enterococcus faecalis*, *Int Endod J*, **38**, 22–29.

31. Huth KC, Quirling M, Maier S, Kamereck K, Alkhayer M, Paschos E, Welsch U, Miethke T, Brand K, Hickel R (2009) Effectiveness of ozone against endodontopathogenic microorganisms in a root canal biofilm model, *Int Endod J*, **42**, 3–13.

32. Bezrukova IV, Petrukhina NB, Voinov PA (2005) Experience in medical ozone use for root canal treatment, *Stomatologiia (Mosk)*, **84**, 20–22.

33. Huth KC, Quirling M, Lenzke S, Paschos E, Kamereck K, Brand K, et al. (2011) Effectiveness of ozone against periodontal pathogenic microorganisms, *Eur J Oral Sci,* **119,** 204–210.

34. Huth KC, Jakob FM, Saugel B, Cappello C, Paschos E, Hollweck R, Hickel R, Brand K (2006) Effect of ozone on oral cells compared with established antimicrobials, *Eur J Oral Sci*, **114**, 435–440.

35. Filippi A (2001) The effects of ozonized water on epithelial wound healing (in German), *Deutsche Zahnärztliche Zeitschrift*, **56**, 104–108.

36. Ebensberger U, Pohl Y, Filippi A (2002) PCNA-expression of cementoblasts and fibroblasts on the root surface after extraoral rinsing for decontamination, *Dent Traumatol*, **18**, 262–266.

37. Nagayoshi M, Fukuizumi T, Kitamura C, Yano J, Terashita M, Nishihara T (2004) Efficacy of ozone on survival and permeability of oral microorganisms, *Oral Microbiol Immunol*, **19**, 240–246.

38. Brauner A (1991) Clinical studies of therapeutic results from ozonized water for gingivitis and periodontitis, *Zahnarztl Prax*, **42**, 48–50.

39. Kshitish D, Laxman VK (2010) The use of ozonated water and 0.2% chlorhexidine in the treatment of periodontitis patients: a clinical and microbiologic study, *Indian J Dent Res*, **21**, 341–348.

40. Chiba K, Sugai T (2011) Functional water production method, JP, 2010–167365, A. 2010–08–05.

41. Grebenshchikov SY, Qu ZW, Zhu H, Schinke R (2007) New theoretical investigations of the photodissociation of ozone in the Hartley, Huggins, Chappuis, and Wulf bands, *Phys Chem Chem Phys*, **9**, 2044–2064.

42. Broadwater WT, Hoehn RC, King PH (1973) Sensitivity of three selected bacterial species to ozone, *Appl Microbiol*, **26**, 391–393.

43. Staehelin J, Hoigne J (1985) Decomposition of ozone in water in the presence of organic solutes acting as promoters and inhibitors of radical chain reactions, *Environ Sci Technol*, **19**, 1206–1213.

44. Becerril R, Gómez-Lus R, Goñi P, López P, Nerín C (2007) Combination of analytical and microbiological techniques to study the antimicrobial activity of a new active food packaging containing cinnamon or oregano against *E. coli* and *S. aureus*, *Anal Bioanal Chem*, **388**(5–6), 1003–1011.

45. Gustafson JE, Liew YC, Chew S, Markham J, Bell HC, Wyllie SG, Warmington JR (1998) Effects of tea tree oil on *Escherichia coli*, *Lett Appl Microbiol*, **26**(3), 194–198.

Index